Springer Geophysics

More information about this series at http://www.springer.com/series/10173

Shuanggen Jin • Nader Haghighipour
Wing-Huen Ip
Editors

Planetary Exploration and Science: Recent Results and Advances

Editors
Shuanggen Jin
Shanghai Astronomical Observatory
Chinese Academy of Sciences
Shanghai, China

Bulent Ecevit University
Zonguldak, Turkey

Wing-Huen Ip
National Central University
Taoyuan, Taiwan

Nader Haghighipour
University of Hawaii-Manoa
NASA Astrobiology Institute
Honolulu, HI, USA

ISBN 978-3-662-45051-2   ISBN 978-3-662-45052-9 (eBook)
DOI 10.1007/978-3-662-45052-9
Springer Heidelberg New York Dordrecht London

Library of Congress Control Number: 2014957128

© Springer-Verlag Berlin Heidelberg 2015
This work is subject to copyright. All rights are reserved by the Publisher, whether the whole or part of the material is concerned, specifically the rights of translation, reprinting, reuse of illustrations, recitation, broadcasting, reproduction on microfilms or in any other physical way, and transmission or information storage and retrieval, electronic adaptation, computer software, or by similar or dissimilar methodology now known or hereafter developed. Exempted from this legal reservation are brief excerpts in connection with reviews or scholarly analysis or material supplied specifically for the purpose of being entered and executed on a computer system, for exclusive use by the purchaser of the work. Duplication of this publication or parts thereof is permitted only under the provisions of the Copyright Law of the Publisher's location, in its current version, and permission for use must always be obtained from Springer. Permissions for use may be obtained through RightsLink at the Copyright Clearance Center. Violations are liable to prosecution under the respective Copyright Law.
The use of general descriptive names, registered names, trademarks, service marks, etc. in this publication does not imply, even in the absence of a specific statement, that such names are exempt from the relevant protective laws and regulations and therefore free for general use.
While the advice and information in this book are believed to be true and accurate at the date of publication, neither the authors nor the editors nor the publisher can accept any legal responsibility for any errors or omissions that may be made. The publisher makes no warranty, express or implied, with respect to the material contained herein.

Printed on acid-free paper

Springer is part of Springer Science+Business Media (www.springer.com)

Preface

With the development of space techniques, more and more curious solar system bodies are being explored by humans. For example, several countries have launched orbiters and landers to the moon recently, focusing on unprecedented resources, origins and evolutions of the moon, including Japan's SELenological and ENgineering Explorer (SELENE), China's Chang'E-1/2/3 and India's Chandrayaan-1 and US's Lunar Reconnaissance Orbiter (LRO) and Gravity Recovery and Interior Laboratory (GRAIL). These missions provided direct observations on space environments, surface processes, rocks and minerals, water ice, interior structure and the origin of the moon. Furthermore, a number of upcoming lunar missions programmes have been planned, e.g., India's Chandrayaan-2, (2014), Russia's Lunar Glob 1 and 2 (2014/2015), China's Chang' E-4 (2017), and International Lunar Network (2018), which will enable us to answer more unknown questions on lunar exploration and sciences. In addition, with recent Mars Global Surveyor (MGS), Mars Express, Mars Odyssey, Mars Reconnaissance Orbiter (MRO), Venus Express, Phoenix, and so on, the atmosphere, surface processes and interior structure of the Mars, Venus and other planets were well explored and understood. However, the origin, formation and evolution on planets and exoplanets are still unclear, as well as seeking life beyond Earth.

This book will present the recent developments of planetary exploration techniques and the latest results on planetary science as well as future objectives of planetary exploration and science, e.g., lunar surface iron content and Mare Orientale basalts, Earth's gravity field, Martian radar exploration, crater recognition, ionosphere and astrobiology, exoplanetary atmospheres and planet formation in binaries. It will help readers to quickly familiarize themselves with the field of planetary exploration and science. In addition, it is also useful for planetary probe designers, engineers and other users' community, e.g., planetary geologists and geophysicists. This work was supported by the National Basic Research Program of China (973 Program) (Grant No. 2012CB720000) and Main Direction Project of Chinese Academy of Sciences (Grant No. KJCX2-EW-T03). Meanwhile, we would like to gratefully thank Springer Publisher for their processes and cordial cooperation to publish this book.

Shanghai, China
Honolulu, HI, USA
Chung-Li, Taiwan
May 2015

Shuanggen Jin
Nader Haghighipour
Wing-Huen Ip

Contents

1 Partial Least Squares Modeling of Lunar Surface FeO Content with Clementine Ultraviolet-Visible Images 1
Lingzhi Sun and Zongcheng Ling

2 Quantitative Characterization of Lunar Mare Orientale Basalts Detected by Moon Mineralogical Mapper on Chandrayaan-1 21
S. Arivazhagan

3 Gravity Changes over Russian River Basins from GRACE 45
Leonid V. Zotov, C.K. Shum, and Natalya L. Frolova

4 Gravimetric Forward and Inverse Modeling Methods of the Crustal Density Structures and the Crust-Mantle Interface ... 61
Robert Tenzer and Wenjin Chen

5 Radar Exploration of Mars: Recent Results and Progresses 77
Stefano Giuppi

6 Automatic Recognition of Impact Craters on the Martian Surface from DEM and Images 101
Tengyu Zhang and Shuanggen Jin

7 Upper Ionosphere of Mars During Solar Quiet and Disturbed Conditions 119
S.A. Haider

8 Mars Astrobiology: Recent Status and Progress 147
Antonio de Morais M. Teles

9 Classical Physics to Calculate Rotation Periods of Planets and the Sun 247
Sahnggi Park

10 Estimates of the Size of the Ionosphere of Comet 67P/Churyumov–Gerasimenko During Its Perihelion Passage in 2014/2015 271
Wing-Huen Ip

11 Photometric and Spectroscopic Observations of Exoplanet Transit Events 279
Liyun Zhang and Qingfeng Pi

12 Photochemistry of Terrestrial Exoplanet Atmospheres 291
Renyu Hu

13 Planet Formation in Binaries 309
P. Thebault and N. Haghighipour

About the Editor

Shuanggen Jin is a Professor at the Shanghai Astronomical Observatory, CAS and Bulent Ecevit University, Turkey. He received the B.Sc. degree in Geodesy/Geomatics from Wuhan University in 1999 and the Ph.D. degree in GNSS/Geodesy from University of Chinese Academy of Sciences in 2003. His main research areas include satellite navigation, remote sensing, satellite gravimetry and space/planetary exploration as well as their applications. He has published over 200 papers in JGR, IEEE, EPSL, Icarus, GJI, JG, Proceedings etc., seven books/monographs and seven patents/software copyrights. He has been President of the International Association of Planetary Sciences (IAPS) (2013–2017), Chair of the IAG Sub-Commission 4.6 (2011–2015), Editor-in-Chief of *International Journal of Geosciences*, Associate Editor of *IEEE Transactions on Geoscience and Remote Sensing* (2014–), Associate Editor of *Journal of Navigation* (2014–), Associate Editor of *Advances in Space Research* (2013–), Editorial Board member of *Journal of Geodynamics* (2014–), *Planetary and Space Science* (2014–) and other seven international journals. He has received Special Prize of Korea Astronomy and Space Science Institute (2006), 100-Talent Program of Chinese Academy of Sciences (2010), Fellow of International Association of Geodesy (IAG) (2011), Pujiang Talent Program of Shanghai (2011), Fu Chengyi Award of Chinese Geophysical Society (2012), Second Prize of Hubei Natural Science Award (2012), Second Prize of National Geomatics Science and Technology Progress Award (2013/2014), Outstanding Young Scientist Award of Scientific Chinese (2013), Liu Guangding Geophysical Youth Science and Technology Award (2013), and Second Prize of Shanghai Science and Technology Progress Award (2014).

Chapter 1
Partial Least Squares Modeling of Lunar Surface FeO Content with Clementine Ultraviolet-Visible Images

Lingzhi Sun and Zongcheng Ling

Abstract To accurately predict the iron abundance of the Moon has long been the goal for lunar remote sensing studies. In this paper, we present a new iron model based on partial least squares regression (PLS) method and apply this model to map the global lunar iron distribution using Clementine ultraviolet-visible (UVVIS) dataset. Our iron model has taken into account of more calibration sites other than Apollo and Luna sample-return sites and stations (i.e., the six additional highland or immature sites) in combination with more spectral bands (5 bands and 2 band ratios), in order to derive reliable FeO content and improve the robustness of the PLS model. By comparing the PLS-derived iron map with Lucey's band-ratio FeO map and Lawrence's Lunar Prospector (LP) FeO map, the differences are mostly within 1 wt% in FeO content. Moreover, PLS-derived FeO is more consistent with LP's result which was derived by direct measurement of Fe gamma-ray line (7.6 MeV) rather than the Lucey's experiential algorithm applying only two bands (750, 950 nm) of Clementine UVVIS dataset. With a global mode of 5.1 wt%, PLS-derived iron map is also validated by FeO abundances of lunar feldspathic meteorites and in support of the lunar magma ocean hypothesis.

Keywords Lunar iron content • Partial least squares regression (PLS) • Spectroscopy • Clementine UVVIS

1.1 Introduction

As one of the major rock-forming elements, iron is closely related to lunar mafic mineral assemblages and rock types; thus the accurate estimation of iron abundance would provide important information of lunar geochemistry, petrogenesis, as well

---

L. Sun • Z. Ling (✉)
School of Space Science and Physics, Shandong Provincial Key Laboratory of Optical Astronomy and Solar-Terrestrial Environment, Institute of Space Sciences, Shandong University, Weihai 264209, China
e-mail: zcling@sdu.edu.cn

© Springer-Verlag Berlin Heidelberg 2015
S. Jin et al. (eds.), *Planetary Exploration and Science: Recent Results and Advances*,
Springer Geophysics, DOI 10.1007/978-3-662-45052-9_1

as the crustal evolution. Iron is often expressed as FeO in astrochemistry. The absorption properties in the ultraviolet-visible (UVVIS) and near-infrared (NIR) spectral regions of iron-bearing minerals (e.g., pyroxene, olivine, and ilmenite) are dominated by Fe^{2+} or Ti^{4+}/Ti^{3+} (Lucey et al. 1998). The absorption features from lunar sample or remotely sensed spectra would mix up influences from the exposures of lunar soils to the space environment, i.e., the Moon has been suffering from bombardments by micrometeorites, solar wind ions, cosmic rays, and solar flare particles (Fischer and Pieters 1994, 1996). Sustained bombardments will cause the lunar surface material change in petrography and chemistry. These changes include reduction of mean grain size, the production of nanophase iron (npFe^0) and complex glass-welded aggregates of lithic and mineral fragments (agglutinates), and so on (Fischer and Pieters 1994, 1996; Mckay et al. 1974). This so-called process "space weathering" will bring about the maturation of lunar regolith, i.e., the mature regolith usually has suffered from a longer time of space weathering compared to immature regolith. Space weathering will cause an overall reduction in the reflectance, and reduce the absorption band strengths, creating and steepening a red-sloped continuum (Fischer and Pieters 1994, 1996).

Many authors have obtained the empirical relationships between spectral properties and iron abundance of lunar soils with intent to get a more accurate lunar iron model (Lucey et al. 1995, 2000; Blewett et al. 1997; Gillis et al. 2004; Wilcox et al. 2005). Lucey et al. (1995) firstly provided a method for the derivation of iron from Clementine multispectral images, utilizing the laboratory spectra and iron abundance of lunar soils. A Fe parameter was defined based on compositional and maturity-related trends on a plot of 950 nm/750 nm versus 750 nm reflectance, which was found to have a strong linear relationship with iron abundance (Fig. 1.1) (Lucey et al. 1995). It can be seen from Fig. 1.1 that iron content has an orthogonal effect, where low iron abundance has high reflectance and high ratio whereas high iron abundance has low reflectance and low ratio. This trend has a hypothetical

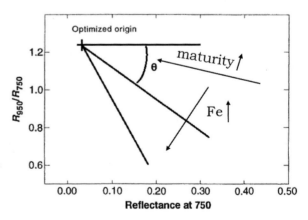

Fig. 1.1 A schematic diagram of NIR/VIS ratio versus VIS for lunar samples. Samples with high iron abundance toward *lower left* on the plot. Samples with same iron abundance but different maturities locate along a line radial to a dark red mature end-member at the *upper left*. The Fe parameter (θ) could decouple iron content from maturity

end-member ("Optimized origin" in Fig. 1.1), which is dark and "red." However, Blewett et al. (1997) pointed it out that the directional-hemispherical laboratory spectra that Lucey used in his algorithm may induce error when applying to the bidirectional spacecraft measurements. He improved Lucey's method by collecting image of lunar landing sites and applying them to iron mapping algorithms. Lucey et al. (1998) examined and quantified important aspects, e.g., maturity, grain size and mineralogy, and topographic shading, in his new iron modeling. He then obtained an improved iron abundance model by using their final processing of Clementine UVVIS datasets (Lucey et al. 2000). Later on, Gillis et al. (2004) noted that TiO_2 abundance has an effect on the relationship between Fe content and Fe parameter, and they optimized this method by adding TiO_2-sensitive regression parameters into the regression of iron content. Wilcox et al. (2005) developed a new algorithm to determine the iron content in lunar mare regions based on the findings that the maturity trends in lunar mare area are more parallel than radial. They collected more than 9,000 craters from mare regions and make a 950/750 nm vs. 750 nm reflectance plot with these data and found the radial trends were disobeyed. While iron abundance was still orthogonal to maturity trends, the maturity trends were parallel to each other, suggesting new trends of iron distribution in lunar mare. Their new iron model has absolute uncertainty similar to Lucey 2000's model (1.5 wt%), while it allows better compensation for the maturity-induced iron uncertainties (<0.5 wt%).

Except for NIR/VIS ratio methods mentioned above, many other approaches like utilizing infrared continuum slope of the spectrum in order to suppress the effect of topography (Le Mouelic et al. 2002) and iron absorption band depth (Fischer and Pieters 1994) have been proposed in the iron modeling. These methods are limited by the data calibration and quality of Clementine NIR dataset. Statistical relationships between spectral and chemical abundance of lunar soils have also been evaluated by Pieters et al. 2002 for their applications of remotely compositional analysis. She firstly applied principle component analysis (PCA) regression method with lunar mare soil spectra produced by Lunar Soil Characterization Consortium (LSCC) to define and evaluate the correlations between chemical abundance and spectral parameters (Pieters et al. 2002). Then she also derived three statistical relations between spectral and mineral parameters using LSCC data and applied them to Clementine UVVIS data (Pieters et al. 2006).

Although many iron models have been put forward as discussed above, a quantitatively accurate iron model is still in need, especially for the exploration of the potentials of multispectral imaging data like Clementine UVVIS and other lunar hyperspectral datasets (e.g., data from Moon Mineralogy Mapper (M^3), Interference Imaging Spectrometer (IIM), etc.). In this paper, we choose to build iron abundance models with partial least squares (PLS) regression method. PLS is known as the second generation of regression method, which performs well in multivariable regression especially when multiple correlations exist among variables. Li (2006) made a comparison between PLS and PCA in deriving chemical and mineral abundances using data from LSCC. He found PLS models use less components and perform better than PCA in the estimation of lunar chemical and mineral abundances. However, Li didn't apply his result to lunar remotely sensed images. Our PLS-derived iron model is developed with intent to explore the potential of the UVVIS imaging dataset. During the modeling, we find it is easy to reach a good regression relationship (high correlation efficient (R^2) value) between spectra parameters and iron contents, while maturity suppressing is more difficult to attain. We have tested many different variables in PLS modeling to find the most applicable one for Clementine UVVIS images and compare our results with previous studies to evaluate the robustness of the PLS model.

1.2 Data

The lunar remote sensing images used in this study are from Clementine UVVIS Digital Image Model (DIM) published by US Geological Survey (USGS) Astrogeology Team at Flagstaff, Arizona (NASA PDS Geosciences Node). The DIM has five bands with a nominal ground resolution at 100 m/pixel, and the center wavelengths (spectra resolutions) of the five filters are A, 415 nm (40 nm); B, 750 nm (10 nm); C, 900 nm (20 nm); D, 950 nm (30 nm); and E, 1,000 nm (30 nm) (Eliason et al. 1999). This dataset is archived in the NASA Planetary Data System, and each image has undergone radiometric and geometric correction, spectral registration, and photometric normalization by Integrated Software for Imagers and Spectrometers (ISIS) processing system.

The PLS modeling data points include those extracted from Apollo and Luna sampling stations (from Wilcox et al. (2005), Table 2) and supplementary data from farside highlands and optically fresh (immature) areas (shown in Table 1.1). From our experiments, the statistical prediction of chemical abundance rely significantly on the input variables, i.e., an obvious deviation of iron abundance would appear when the modeling didn't include data points from farside highland and immature areas. Note that the iron abundances of supplementary data in Table 1.1 were calculated with Lucey's (2000) parameters.

Table 1.1 Iron abundance and reflectance values of supplementary data from Clementine UVVIS

Sites	Clementine spectra					FeO (wt%)
	415 nm	750 nm	900 nm	950 nm	1,000 nm	
Farside-1	0.1196	0.2022	0.2178	0.2250	0.2302	2.3
Farside-2	0.1195	0.2032	0.2183	0.2271	0.2336	1.7
Farside-3	0.1330	0.2271	0.2372	0.2447	0.2546	3.8
Fresh-1	0.0794	0.1265	0.1140	0.1108	0.1111	17.2
Fresh-2	0.0767	0.1229	0.11155	0.1088	0.1103	17.4
Fresh-3	0.0811	0.1353	0.1202	0.1174	0.1203	16.8

1.3 Partial Least Squares Regression Method and Data Processing

PLS is a new kind of multivariate statistics regression method, which was developed by Herman Wold in 1966 (Li 2006). Comparing to other regression methods (like PCA regression), PLS has many advantages, especially in resolving mutual influence problems among variables. PLS has already been utilized for analyzing material compositions from laboratory and remote sensing spectra datasets. Li (2006, 2008) resampled LSCC bidirectional reflectance data into the airborne visible/infrared imaging spectrometer (AVIRIS) spectral resolution and derived several composition derivation models such as iron and TiO_2 with PLS regression method (Li 2006, 2008, 2011). Li's model was based on laboratory data and was not applied to remotely sensed data, making it difficult to evaluate the ability of the model in maturity suppressing.

As an advanced statistical method, the principle of PLS analyzing can be expressed as: PLS = PCA + CCA + MLR (CCA, classical component analysis; MLR, multiple linear regression). The key to PLS modeling is to determine the number of latent variables (LVs), which are also called the components. Covariance between each corresponding component of independent variable and dependent variable should be kept maximum; this can be considered as a combination of LVs searching conditions of PCA and CCA.

Assuming the independent variance is an $n \times m$ matrix X, and dependent variance is an $n \times p$ matrix Y, we first standardize matrixes X and Y before modeling in order to reach a more stable result. Following PLS rules while regressing X and Y, finally, we can get the relations listed below (Eqs. 1.1 and 1.2). Both X and Y are decomposed into two parts: a matrix product term and a residual term. The matrix product term consists of a score matrix and a loading matrix, score matrixes are T for X and U for Y, and they are both $n \times a$ matrixes; loading matrixes are P for X and Q for Y, and they are both $m \times a$ matrixes. E and F are residual matrixes. The goal of regression is to find the correlative relation between X and Y (Eq. 1.3) while keeping residual matrixes E and F minimum:

$$X = TP^T + E = \sum_a t_a p_a^T \tag{1.1}$$

$$Y = UQ^T + F = \sum_a u_a q_a^T \tag{1.2}$$

$$Y = XB + F \tag{1.3}$$

In order to find better relations between spectral data and iron content, we transfer reflectance spectra into effective absorbance spectra first. With zero transmittance given, absorbance can be roughly expressed as log reflectance based on Beer's Law (Eq. 1.4), where R denotes reflectance and α is absorbance. The derived absorbance

α is assumed to have a linear relationship with the abundance of composition (Li 2006; Yen et al. 1998; Whiting et al. 2004; Milliken and Mustard 2005):

$$-\ln R = -\ln (1 - \alpha) \tag{1.4}$$

1.4 PLS Modeling

1.4.1 Iron Modeling

In our PLS model, modeling data points including 47 lunar sampling sites from Wilcox et al. (2005) and 6 added data from lunar farside highlands and fresh areas, so there are 53 modeling sites in total, and X is a 53×5 matrix, and Y is a 53×1 matrix. After transforming reflectance into absorbance, we standardized both X and Y in order to get a more stable model. While modeling, the most important thing is to derive reasonable iron content as well as suppress the space weathering effect at the same time. All of the five bands are included in the dependent variables to keep the maximum potential, and they are expressed by $A_1 - A_5$, respectively. Band ratios are helpful especially when extracting chemical abundances, and they are also indications of maturity degree. Our model takes account of the typical NIR/VIS ratio (950 nm/750 nm), which is used in Lucey's algorithm. Pieters et al. (2002) have tested the correlations between composition and spectral ratios, and experiments showed that the highest correlation for iron is 1,000/400 nm. Hence, we also bring it into our model, expressed by 1,000/415 nm. Finally, all the variables chosen to build model are listed in Eq. 1.5, $c_0 - c_7$ are regression coefficients, and $A_1 - A_5$ represent five absorption bands of Clementine data:

$$\text{Iron} = c_0 + \sum_{i=1}^{5} A_i c_i + \left(\frac{A_4}{A_2}\right) c_6 + \left(\frac{A_5}{A_1}\right) c_7 \tag{1.5}$$

After inputting all the data into PLS toolbox, leave-one-out cross-validation is executed during modeling. The cross-validation means modeling with one variable left out until all the variables have been left out once; thus we would derive a regression model in each cross-validation and compute the root mean square error of cross-validation (RMSECV) for every leave-one-out model by Eq. 1.6; k is the number of variable that is left out. Usually, the one with minimum RMSECV will be chosen as the best LV number. After the number of components is determined, the total root mean square (RMSE) can be calculated by Eq. 1.7.

Figure 1.2 is the plot of RMSE and RMSECV values, the minimum RMSECV is 1.51 wt%, and the corresponding LV number is 1. Measured abundances of iron and those derived from the PLS model are plotted in Fig. 1.3. Correlation coefficient

1 Partial Least Squares Modeling of Lunar Surface FeO Content... 7

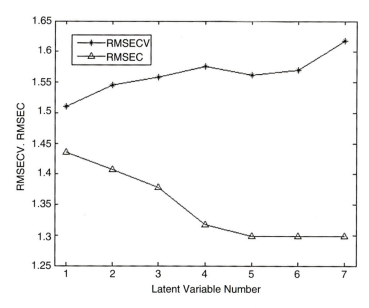

Fig. 1.2 RMSE and RMSECV distribution plot. The minimum RMSECV is 1.51, indicating the optimal number of LVs is 1

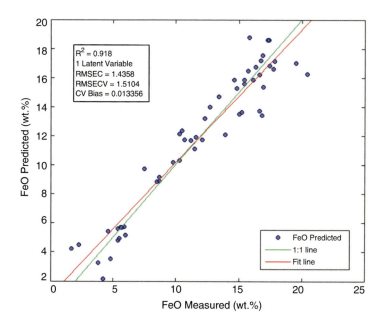

Fig. 1.3 Regression result of PLS modeling. The correlation coefficient is 0.918, and RMSE is 1.44, and RMSECV is 1.51

(R^2) of this model is 0.918, and RMSE is 1.44 wt%, indicating a good regression of iron abundance has been achieved:

$$RMSECV_k = \sqrt{\frac{\sum_{i=1}^{8} (\widehat{y}_i - y_i)_k^2}{8}} \qquad (1.6)$$

$$RMSE = \sqrt{\frac{\sum_{i=1}^{8} (\widehat{y}_i - y_i)^2}{8}} \qquad (1.7)$$

1.4.2 Results and Analysis

We apply the PLS model mentioned in Sect. 1.4.1 to a small area near the southern rim of Mare Crisium (Fig. 1.4a) for preliminary check and validation. This region is chosen for two reasons. First, its location is near the boundary of mare and highland, so bimodal distribution of FeO is expected to exist; moreover, there are many small spectrally fresh craters (low maturity degree) in the mare area, which could be used as an indicator for the maturity-suppressing ability of our model (Ling et al. 2011).

PLS-derived iron map is shown in Fig. 1.4c in comparison with the result of Lucey's work (Fig. 1.4b). Considering the maturity-suppressing ability can be indicated by the small fresh craters in the mare region, the difference between our model and Lucey's is subtle, i.e., most of small fresh craters (bright spots in the 750 nm reflectance image (Fig. 1.4a)) are invisible in the PLS-derived FeO map in Fig. 1.4b, c. The distribution of FeO in the mare area is relatively homogenous, which indicates that the maturity-suppressing ability of our model is comparable to Lucey's algorithm.

It can be seen from Figs. 1.4 and 1.5 that FeO abundance is high in mare regions and low in highland regions. Generally, FeO abundance of mare regions is higher than 10 wt%, which is due to the large concentration of iron-bearing silicates in mare basalts such as pyroxene, olivine, ilmenite, etc. Highland region is deficient in iron, as its rock type is dominated by anorthosite (Lucey et al. 1995, 1998). From the histogram of iron abundance (Fig. 1.5), we can distinguish a bimodal distribution of FeO in this region; the peak on the left represents FeO concentration in highland region and right peak represents that in mare region. Comparing the PLS model to Lucey et al.'s (2000) algorithm, it can be recognized that the two models have similar peak positions for the mare region (right peak), and the peak FeO abundances are about $\sim$16.5 wt%; while the FeO abundances for the highland (left peak) have a little difference, our result is around 9 wt%, about 1 wt% higher than Lucey's. This discrepancy may result from various causes such as model input parameters (e.g., different bands or sampling sites). Further discussions will be given in Sect. 1.4.3.

1 Partial Least Squares Modeling of Lunar Surface FeO Content... 9

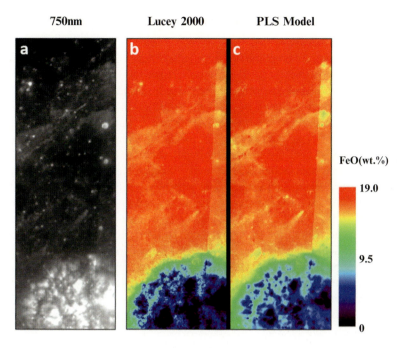

Fig. 1.4 (**a**) 750 nm reflectance of a small area from southern Mare Crisium, (**b**) iron map derived from Lucey's algorithm, and (**c**) iron map derived from PLS model. It can be seen that most of the bright fresh craters are invisible in our iron map

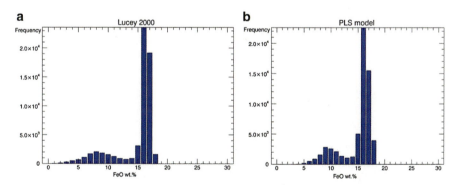

Fig. 1.5 Statistical result of Fig. 1.4 (**a**) Lucey's result, corresponding to Fig. 1.4b; (**b**) our result, corresponding to Fig. 1.4c. Both of them have obvious bimodal structure, the left peak represents iron concentration in highland area, and the right peak represents iron concentration in mare area. Iron peak in highland area is 9 wt% in our model and 8 wt% in Lucey's, while iron peak in mare area is about 16.5 wt%, which is the same with Lucey's

1.4.3 PLS Modeling of Highland Areas

As discussed above, although the two iron maps behave similarly in iron abundance and maturity suppressing, they still exhibit discrepancy in FeO modeling of highland regions. The iron abundance of our model for highland region is a little higher than Lucey's, which is shown both in iron map (Fig. 1.4b, c) and the statistical results of iron map (Fig. 1.5). As is known to all, statistical methods strongly depend on the sampling data points, i.e., when the sampling data points lack of a specific range of FeO abundance, the result may tend to behave deviate from that range. During the modeling, although six supplementary data are added for lunar sampling stations, the highland data sources are only composed of Apollo 16 sampling stations and 3 added lunar farside sites. The limited proportion of highland spectra to the total modeling data may lead to the overestimation of iron abundance in highland areas during the PLS modeling. To testify this hypothesis, we derive another iron model using only Apollo 16 and Apollo 17 sampling sites, in order to increase the proportion of highland sampling sites. Data processing pipeline follows the first PLS model (Eq. 1.5).

Applying this highland model to test area, we derive a new iron map. Iron abundance derived by highland model (Fig. 1.6c) is obviously less than that

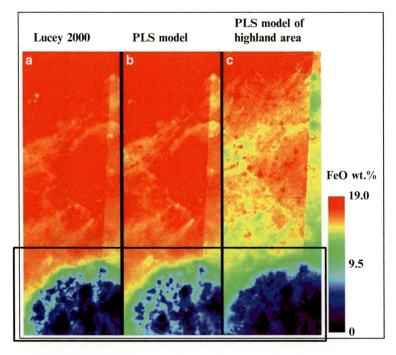

Fig. 1.6 Iron map comparison, highland regions are indicated by a black frame. (**a**) Lucey's result; (**b**) the first PLS model result; (**c**) highland modeling result. The iron abundance of **c** in the highland regions is obviously less than **b** and **a**

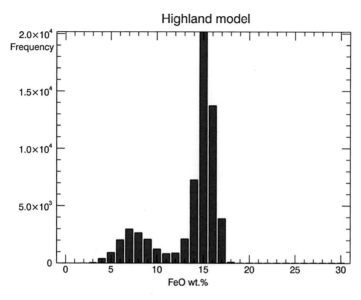

Fig. 1.7 Statistical result of iron abundance of the model build with highland spectra. The FeO abundance peak of highland region is 7 wt%, which is lower than the first model (Fig. 1.6b, 9 wt%) as well as Lucey's result (Fig. 1.6a, 8 wt%)

modeling with all the data presented in Table 1.1 (Fig. 1.6b). From the statistical results, we can find that the iron concentration peak of highland regions reduces to 7 wt% (Fig. 1.7), as compared to the former 9 wt% (Fig. 1.5b). Given the proportion of highland data increases and PLS regression procedures remain the same, we can conclude that the PLS model relies on the input modeling data, i.e., the spectral types and iron abundances range of the modeling data could affect PLS model behavior significantly. Although it looks like one can improve the PLS modeling behavior by adding supplementary data to the original lunar sampling sites, the number of added data should be in caution. As stated above, elemental abundance of added data is usually calculated by empirical methods, which may induce uncertainty or even correct conclusions. We have done tens of experiments with the number of data points varying from dozens to hundreds; the PLS model presented in Sect. 1.4.1 is the best one when all the available Apollo and Luna ground truth data are considered. As supplementary data for highlands and fresh areas, the added data only accounts for a very small proportion in the modeling data compared to lunar sampling stations. For future work, we will focus on trying to find more effective variables or anticipating more typical sampling sites in the future missions.

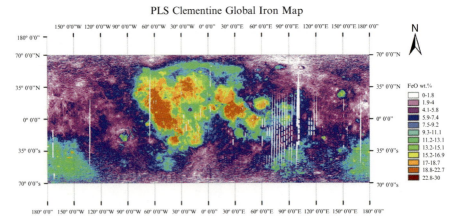

Fig. 1.8 Global iron map of PLS model. Data greater than 30.0 wt% or less than 0.0 wt% are set to NAN (not a value) in this map

1.5 Global Iron Mapping and Analysis

1.5.1 Global Iron Mapping

Applying the PLS model from Sect. 1.4.1 to Clementine DIM global mosaic, we can obtain a lunar global iron map (Fig. 1.8). As shown in Fig. 1.8, FeO is obviously rich in the large basins that spread in lunar nearside such as Mare Imbrium, Oceanus Procellarum, Mare Serenitatis, Mare Tranquillitatis, etc., all of which are known to have been flooded by large amounts of iron-rich basalt lava flow. The histogram of global iron distribution is shown in Fig. 1.9c. Global mean FeO abundance is 7.6 wt% by PLS model. The bimodal structure represents decoupled iron distribution in mare and highland region. The global mode of FeO is 5.1 wt%, corresponding to the left peak in Fig. 1.9c, and FeO abundance peak in mare is 16.9 wt%, corresponding to the right peak.

1.5.2 Comparison with Former Works

To understand more about the global FeO map derived from PLS model, we compare our work with Lucey's band-ratio result and Lunar Prospector (LP) (Figs. 1.10 and 1.11; Table 1.2). On a global view, the global mean of FeO abundance of PLS model is 7.6 wt%, and the value is 7.8 wt% for both Lucey's and LP's results. Peak values of FeO in the highland (global mode) are 5.1 wt% for PLS model and 6.4 wt% for LP, while Lucey's algorithm derives a relatively lower value less than 5 wt%. Peak distribution of FeO in the mare for the three model is

Fig. 1.9 Comparison of statistical results. (**a**) Global histogram of FeO from LP, (**b**) Lucey 2000, (**c**) PLS modeling

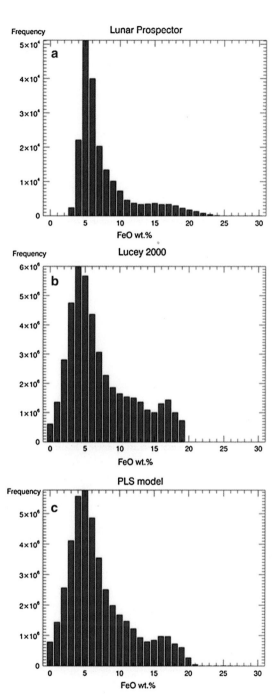

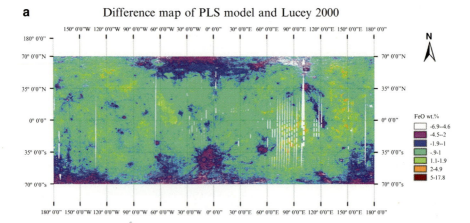

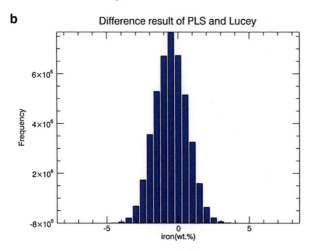

Fig. 1.10 (**a**) Difference map between iron map derived from PLS model and Lucey's algorithm (PLS minus Lucey's). (**b**) The difference of the two maps shows a Gaussian distribution with an average of 0.27 wt%, and RMS is 1.13 wt%

around 17 wt%. Comparing from the statistical results (Fig. 1.9), FeO abundance derived by PLS model in lunar highland areas is a little higher than Lucey's but is similar to LP's. Detail comparisons between PLS model, Lucey's algorithm, and LP result will be discussed in the following.

In order to show the global difference between PLS model and Lucey's method, we apply Lucey's algorithm to Clementine DIM and make a difference map (PLS FeO minus Lucey's FeO), as shown in Fig. 1.10a. Most of the difference distributes within −0.9 to 1.0 wt% which is shown in green color in the difference map. PLS model gets an even higher iron abundance than Lucey's result in lunar farside, which is consistent with the statistical result comparison (Fig. 1.9b, c). Another

1 Partial Least Squares Modeling of Lunar Surface FeO Content...

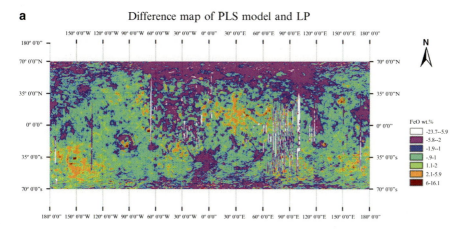

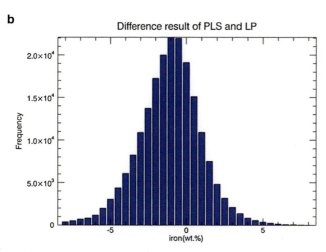

Fig. 1.11 (**a**) Difference map between iron map derived from PLS model and Lunar Prospector gamma-ray spectrometer (PLS minus Lucey's). (**b**) Statistical result of a, suggest an average of 0.9 wt% and RMS is 2.3 wt%

Table 1.2 Comparison of FeO abundance between different algorithms

FeO model	Global mean (wt%)	Global mode (wt%)	Peak value in mare (wt%)
Lucey 2000	7.8	4.7	17.1
Lunar Prospector	7.8	6.4	17
PLS model	7.6	5.1	16.9

discovery from the difference map is that iron in fresh craters derived by PLS is often lower than Lucey's result, represented by Tycho crater (located in south-southwest of the map, about 43.3°S, 11.2°W). This may be caused by different degrees of maturity suppressing between the two methods. Statistical result (Fig. 1.10b) shows a nearly Gaussian distribution of the difference, with an average value around −0.27 wt%, and root mean square error (RMS) is 1.13 wt%, indicating a relatively small difference of the global iron abundance between the two maps.

Lawrence et al. (2002) derived a global iron map from the gamma-ray counting rate of the lunar surface detected by Lunar Prospector (LP). The Fe derived by gamma-ray spectrometer is a direct measurement of Fe gamma-ray line (7.6 MeV). Moreover, gamma-ray spectrometer usually detects signal from over 10 cm below lunar regolith, so the result would be less affected by space weathering effect, making the FeO abundance more reliable than that derived by experienced algorithms. The global mode of LP is 6.4 wt%, compared to 5.1 wt% derived from PLS model. Comparing to Lucey's result, iron abundance derived by PLS modeling is more consistent with that detected by LP as far as highland regions are considered. In other words, the higher iron value that is derived from PLS model for the lunar highland regions when compared to Lucey's algorithm may suggest a more reliable result. As for iron distribution of Tycho crater, Lawrence et al. (2002) also found a large discrepancy, i.e., Lucey's results show moderate FeO abundances at 7–9 wt%, while the LP data show very low FeO abundances at 3–4 wt%. FeO content from PLS modeling is about 3.8 wt% which supports the LP result (Fig. 1.12a, b).

Considering the different spatial resolutions of the FeO maps from the spectral and gamma-ray datasets, in order to compare in detail with iron map derived by LP, we resample the Clementine iron map derived by PLS to the same spatial resolution as LP iron map (15 km/pixel) and make a global difference map (regions exceed 70°S–70°N are not included) (Fig. 1.11a). The difference distribution on the global map isn't very homogenous, but when we take a look at the global iron distribution, we find the global difference is concentrated within −0.9 to 1.0 which is colored by green, and iron abundance detected by LP in mare areas and high-latitude regions is higher than PLS model. The difference in high latitudes tends to be greater; this effect may be the influence of topographic shading or illumination conditions.

Furthermore, our iron content of PLS model for the South Pole-Aitken (SPA) basin is higher than LP's but lower than Lucey's (Fig. 1.12c). As the largest impact crater on the Moon, the SPA impact event didn't penetrate the materials from lunar mantle, which are expected to be more mafic and iron rich. The specific noritic mineralogy may account for this low FeO concentration (Lucey 2004). From the statistical result of the difference map, the global average of iron difference is within 1 wt% and the RMS is 2.3 wt%, suggesting a good consistency of PLS model and LP iron map.

In a word, we find the iron map from PLS model agrees with those from the Lucey's and LP's, though subtle difference appears for the global maps. This suggests our PLS model is a robust algorithm for the extraction of lunar iron content. Furthermore, the application of PLS method on the global lunar mosaic seems to be more consistent with LP results than those of Lucey 2000. Note that our PLS model

1 Partial Least Squares Modeling of Lunar Surface FeO Content... 17

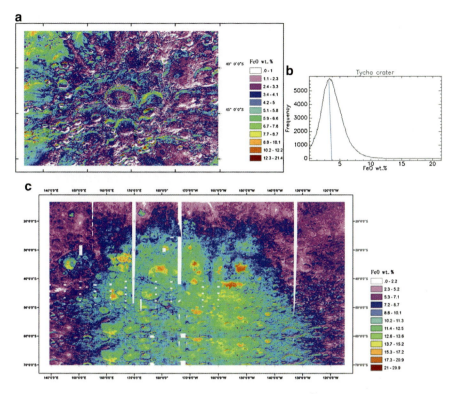

Fig. 1.12 (**a**) PLS-derived iron map of Tycho crater; (**b**) statistical result of iron distribution in Tycho crater; (**c**) PLS-derived iron map of SPA region

explored the potential for all the five available Clementine UVVIS bands and some more spectral parameter than Lucey et al. (2000), who used only Clementine's two bands (950 and 750 nm) to derive iron content. We believe our PLS model is a good test and validation for the lunar elemental mapping with available lunar spectral data and those from future lunar missions.

1.6 Indication of Lunar Magma Ocean Hypothesis

After lunar sample returned to earth, a lot of laboratory experimental analyses have been done to extract information of the lunar mineralogy and petrogenesis, which are very helpful in understanding lunar origin and evolution progress in a global or local scale. Samples from lunar highland regions contain higher plagioclase abundance and hence are rich in Al and poor in Fe compared to those from mare regions (Lucey et al. 1995). These rocks are interpreted as forming from a global circling magma ocean, and plagioclase floated in it (Wood et al. 1970; Warren and Haskin 1991). The magma ocean hypothesis was developed following the first sample return from the Moon. The crystallization of the magma ocean would result in FeO poor anorthosite rocks concentrating in the crust.

As was mentioned by Lucey et al., the key test of the magma ocean hypothesis is the abundance of anorthosite (Lucey et al. 1995). Usually, anorthosite assembles in lunar highlands, so iron abundances in these regions could represent the global anorthosite concentration. In remote detection, the global mode of FeO concentration represents iron abundance of lunar highland regions. The global mode of iron abundance derived by Lucey et al. in 1995 is about 3 wt% (in Fe, and 3.96 wt% in FeO), and he improved the algorithm in year 1998, and the new global mode of FeO is 4.8 wt% (Lucey et al. 1998). These results are consistent with our result (5.1 wt%). As far as lunar meteorites study are concerned, Korotev et al. studied eight best characterized feldspathic lunar meteorites and showed the average concentration of FeO is 4.4 ± 0.5 wt%, and the FeO range is 3–6 wt% (Korotev et al. 1996; 2003; Korotev 2005). The global mode of PLS-derived FeO map is 5.1 wt%, which also agrees well with meteorite studies and remote sensing results (Lucey et al. 1995, 1998, 2000; Lawrence et al. 2002) and thus could also support the magma ocean hypothesis of lunar crust.

1.7 Conclusions

We derived a new iron model with PLS method, which has been verified to be able to derive robust iron abundances for the Moon. We apply this model to Clementine DIM and obtain global distribution of iron. Our results show that peak distribution of iron abundance in highlands and mare regions are 5.1 wt% and 16.9 wt%, respectively. Comparing our iron map to Lucey's algorithm as well as that detected by Lunar Prospector gamma-ray spectrometer, we find the three results agree well in mare regions, while PLS model and LP iron maps show higher iron content in highlands. Local comparisons (e.g., Tycho crater and SPA basin) also suggest our PLS model is reliable and more consistent with the LP results. Besides, the PLS model-derived iron abundance peak of lunar farside is 5.1 wt%, which agrees well with the lunar meteorites that are assumed from lunar highland. Our global FeO distributions are also consistent with the lunar magma ocean hypothesis as has been presented by previous work (Lucey et al. 1995, 1998; Wood et al. 1970; Warren and Haskin 1991).

Although our PLS algorithms have already shown its potential for extraction of lunar iron abundance, it should be kept in mind that there are limitations, i.e., the exact physical significance of PLS is not as evident as experience algorithms, and PLS regression highly depends on the type of the modeling data inputs. More lunar samples and precise geographic location of them would definitely contribute to the improvement of PLS modeling for iron. Interference Imaging Spectrometer (IIM) onboard Chang'E-1 has achieved the abundance of some key elements of the Moon (Ling et al. 2011; Wu et al. 2012; Jin et al. 2013). As is known, China's new lunar lander and rover mission, "Chang'E-3" lander and "Yutu" rover, respectively, have launched in December 2013, and the rover will be released to detect mineral distribution of the lunar surface, especially in Sinus Iridum (Liu et al. 2013), which has never been set foot on by any lander or rover before. Spectral data from VIS-NIR Imaging Spectrometer (VNIS) onboard Yutu rover may provide good opportunities and more constraints for lunar compositional studies and as PLS modeling of lunar iron abundance as well.

Acknowledgements This work was supported by the National Natural Science Foundation of China (11003012, U1231103), the Natural Science Foundation of Shandong Province (ZR2011AQ001), Independent Innovation Foundation of Shandong University (2013ZRQP004), and Graduate Innovation Foundation of Shandong University at WeiHai, GIFSDUWH (yjs13026).

References

Blewett DT, Lucey PG, Hawke BR (1997) Clementine images of the lunar sample-return stations: refinement of FeO and TiO_2 mapping techniques. J Geophys Res 102(E7):16319–16325

Eliason E, Isbell C, Lee E et al (1999) The Clementine UVVIS global lunar mosaic. Cited 20 May 2013. http://www.lpi.usra.edu/lunar/tools/clementine/instructions/UVVIS_DIM_Info.html

Fischer EM, Pieters CM (1994) Remote determination of exposure degree and iron concentration of lunar soils using VIS-NIR spectroscopic methods. Icarus 111(2):475–488

Fischer EM, Pieters CM (1996) Composition and exposure age of the Apollo 16 Cayley and Descartes regions from Clementine data: normalizing the optical effects of space weathering. J Geophys Res 101(E1):2225–2234

Gillis JJ, Jolliff BL, Korotev RL (2004) Lunar surface geochemistry: global concentrations of Th, K, and FeO as derived from lunar prospector and Clementine data. Geochim Cosmochim Acta 68(18):3791–3805

Jin SG, Arivazhagan S, Araki H (2013) New results and questions of lunar exploration from SELENE, Chang'E-1, Chandrayaan-1 and LRO/LCROSS. Adv Space Res 52(2):285–305

Korotev RL (2005) Lunar geochemistry as told by lunar meteorites. Chemie der Erde 65:297–346

Korotev RL, Jolliff BL, Rockow KM (1996) Lunar meteorite Queen Alexandra Rang 93069 and the iron concentration of the lunar highlands surface. Meteorit Planet Sci 31:909–924

Korotev RL, Jolliff BL, Jolliff RA (2003) Feldspathic lunar meteorites and their implications for compositional remote sensing of the lunar surface and the composition of the lunar crust. Geochim Cosmochim Acta 67(24):4895–4923

Lawrence DJ, Feldman WC, Elphic RC (2002) Iron abundances on the lunar surface as measured by the Lunar Prospector gamma-ray and neutron spectrometers. J Geophys Res 107(E12):5130

Le Mouelic S, Lucey PG, Langevin Y (2002) Calculating iron contents of lunar highland materials surrounding Tycho crater from integrated Clementine UV-visible and near-infrared data. J Geophys Res 107:E10,5074

Li L (2006) Partial least squares modeling to quantify lunar soil composition with hyperspectral reflectance measurements. J Geophys Res 111:E04102

Li L (2008) Quantifying lunar soil composition with partial least squares modeling of reflectance. Adv Space Res 42:267–274

Li L (2011) Quantifying TiO_2 abundance of lunar soils: partial least squares and stepwise multiple regression analysis for determining causal effect. J Earth Sci 22(5):549–565

Ling Z, Zhang J, Liu J et al (2011) Preliminary results of FeO mapping using imaging interferometer data from Chang'E-1. Chin Sci Bull 56(4–5):376–379

Liu B, Liu J, Zhang G et al (2013) Reflectance conversion methods for the VIS/NIR imaging spectrometer aboard the Chang'E-3 lunar rover: based on ground validation experiment data. Res Astron Astrophys 13(7):862–874

Lucey PG (2004) Mineral maps of the moon. Geophys Res Lett 31:L08701

Lucey PG, Taylor GJ, Malaret E (1995) Abundance and distribution of iron on the moon. Science 268(5214):1150–1153

Lucey PG, Blewett DT, Hawke BR (1998) Mapping the FeO and TiO_2 content of the lunar surface with multispectral imagery. J Geophys Res 103(E3):3679–3699

Lucey PG, Blewett DT, Jolliff BL (2000) Lunar iron and titanium abundance algorithms based on final processing of Clementine ultraviolet–visible images. J Geophys Res 105(E8):20297–20305

Mckay DS, Fruland RM, Heiken GH (1974) Grain size and the evolution of lunar soils. In: Proceedings of the lunar science conference 3rd, Pergamon Press, New York, pp 983–995

Milliken RE, Mustard JF (2005) Quantifying absolute water content of minerals using near-infrared reflectance spectroscopy. J Geophys Res 110:E12001

NASA PDS Geosciences Node ftp://pds-geosciences.wustl.edu/geocopy/imaging/clem1-l-u-5-dim-uvvis-v1.0/cl_4001/catalog/

Pieters CM, Stankevich DG, Shkuratov YG et al (2002) Statistical analysis of the links among lunar mare soil mineralogy, chemistry, and reflectance spectra. Icarus 155:285–298

Pieters CM, Shkuratov Y, Kaydash V et al (2006) Lunar soil characterization consortium analysis: pyroxene and maturity estimates derived from Clementine image data. Icarus 184:83–101

Warren PH, Haskin L (1991) Lunar chemistry. In: Heiken GH et al (eds) Lunar sourcebook. Cambridge University Press, Cambridge, pp 357–474

Whiting ML, Li L, Ustin SL (2004) Predicting water content using Gaussian model on soil spectra. Remote Sens Environ 89:535–552

Wilcox BB, Lucey PG, Gillis JJ (2005) Mapping iron in the lunar mare: an improved approach. J Geophys Res 110:E1101

Wood JA, Dickey JS, Jr, Marvin UB et al (1970) Lunar anorthosites and a geophysical model of the moon. In: Proceedings of the Apollo 11 lunar science conference, Pergamon Press, New York, pp 965–988

Wu Y, Xue B, Zhao B et al (2012) Global estimates of lunar iron and titanium contents from the Chang'E-1 IIM data. J Geophys Res 117:E02001

Yen AS, Murray BC, Rossman GR (1998) Water content of the Martian soil: laboratory simulations of reflectance spectra. J Geophys Res 103:11125–11133

Chapter 2
Quantitative Characterization of Lunar Mare Orientale Basalts Detected by Moon Mineralogical Mapper on Chandrayaan-1

S. Arivazhagan

Abstract Efficient lunar resource utilization requires accurate and quantitative evaluation of mineral and glass abundances, distribution, and extraction feasibility, especially for ilmenite (TiO_2). The modal analyses have performed on lunar basaltic terrains using hyperspectral remote sensing data along with ground truth chemistry and mineralogy. The main aim of the present work is to characterize the lunar Mare Orientale basalts based on TiO_2 content and quantify the lunar surface minerals, including clinopyroxene, orthopyroxene, plagioclase, and olivine. The Orientale basin is one of the youngest impact multi-ringed basins on the Moon covering 930 km in diameter centered at 20°S 95°W. The morphological features in the Orientale basin have developed interest among geoscientist to explore further study on this region. Based on the Apollo orbital, geochemical, and Earth-based spectral data, it is concluded that the Orientale ejecta are uniformly feldspathic in composition, almost pure anorthosite with no evidence of ultramafic components (Hawke, Geophys Res Lett 18(11):2141–2144, 1991). Greeley et al. (Geophys Res 98:17183–17205, 1993) have conferred the Orientale basin bearing the low-Ti basalts by using Galileo images. In this study, parts of basaltic regions of Mare Orientale, Lacus Veris, and Lacus Autumni of the Orientale basin are investigated using Moon Mineralogical Mapper (M^3) data of onboard Chandrayaan-1 orbiter. Lucey's (1998) TiO_2 estimation method and spectral profiles and spectral unmixing techniques have been used to detect and map the minerals, including plagioclase, clinopyroxene, orthopyroxenes, olivine, and various basalts such as low-, medium-, and high-Ti basalts. The Orientale data were acquired by M^3's reduced resolution mode with 20–40 nm spectral resolution and 140 m/pixel across the 40 km field of view. The RELAB mineral spectra of plagioclase, clino/orthopyroxenes, olivine and

---

S. Arivazhagan (✉)
Department of Geology, Periyar University, Salem 636 011, Tamil Nadu, India

Department of Civil Engineering, KSR College of Engineering, Tiruchengode 637215, Tamil Nadu, India
e-mail: arivusv@gmail.com

© Springer-Verlag Berlin Heidelberg 2015
S. Jin et al. (eds.), *Planetary Exploration and Science: Recent Results and Advances*,
Springer Geophysics, DOI 10.1007/978-3-662-45052-9_2

various basaltic spectra, chemistry and mineralogy have been employed to unmixing analysis. Comparing the spectral profiles of the basaltic regions with the RELAB basaltic spectra, the distribution and nature of TiO_2 basalts in the Orientale basaltic regions have been analyzed in quantitative manner in the present research.

Keywords Mare Orientale • Lacus Veris • Lacus Autumni • Spectral unmixing • End-member

2.1 Introduction

The lunar mare basalts are concentrated on the near surface of the Moon. The low-albedo smooth plains of lunar mare basalt covered almost 20 % of the lunar surface (Head 1976). Right from the beginning of lunar observation, the distinction between the lunar highlands and maria has been recognized (Taylor 1975). The 16 % of the lunar surface is covered by about 23 lunar mares, most of which are located in the near side of the Moon and the remaining three mares such as Mare Orientale, Mare Moscoviense and Mare Ingenii are located in the far side of the Moon. The lunar rock samples collected through Apollo and "Luna" series missions indicated that the lunar mare materials are predominantly composed of basalts. The lunar surface is mainly composed of plagioclase, pyroxene, olivine, ilmenite, agglutinitic and volcanic glass. The mineralogical diversity of the mare basalts has been examined through spectroscopy and linked to the surface units of returned samples (Pieters 1978, 1993; Staid et al. 1996; Staid 2000; Lucey et al. 2006).

The major minerals among various types of mare basalt are pyroxene, Mg-rich olivine, Ca-rich feldspar, and ilmenite (McCord et al. 1972). Lunar basalt samples approximately consist of 51 % of pyroxene, 27 % plagioclase, 8 % olivine and 11 % opaques (Crown and Pieters 1987). In general, the Fe- and Mg-rich mare regions contain abundant calcic pyroxenes and the Al-rich highlands contain low-calcium pyroxenes; both exhibit diagnostic absorption features in the reflectance spectral profile (Adams 1974). The dominant pyroxene in the mare basalt is high Ca-clinopyroxene. Titanium is one of the major elements in mare rocks and accounts for 0.5–13 %. Based on titanium content, the basalts can be divided into three types: high-Ti basalts, low-Ti basalts, and high-Al, low-Ti basalts (Cloutis and Gaffey 1991). Although a variety of distinct basalt types exist, pyroxene is the most abundant mineral in these basalts, followed by plagioclase. The amount of olivine and ilmenite in the basalts varies from minor to 20 % (Pieters and Englert 1997).

Hyperspectral remote sensing or imaging spectroscopy and reflectance spectroscopy are efficient tools to detect and quantify the mineral compositions on Earth and planetary surfaces (Shkuratov et al. 2007; Chen et al. 2007; Pieters et al. 2008; Clark et al. 2008). Nowadays, although mineral species could be successfully detected by hyperspectral data (Clark et al. 2003), the mineral abundance is difficult to retrieve. Retrieval of mineral abundance is hindered by the complexities of mixture characteristics of mineral spectra.

The mineral mapping of the lunar surface can be broadly subdivided into global and regional mineral mapping. The regional mineral mapping concentrates on some important locations of mare and highland region where the drastic mineral compositional changes exist (Staid and Pieters 2000; LeMouélic et al. 1999; LeMouélic and Langevin 2001; Pieters et al. 2001). The purpose of mineral mapping is efficient lunar resource utilization, which requires accurate and quantitative evaluation of mineral and glass abundances, distribution, and extraction feasibility, especially for ilmenite. Common rock-forming minerals exhibit wavelength-dependent spectral features throughout the VNIR (0.4–1.0 μm), SWIR (1.0–2.5 μm), and TIR (5–25 μm) wavelength ranges.

The linear spectral unmixing is an efficient tool to quantify the materials distributed in the image. The spectral unmixing tool is used to decompose a reflectance (or corrected radiance) source spectrum into a set of given end-member spectra. The result of the unmixing is a measure of the membership of the individual end-member to the source spectrum. Linear unmixing has proven to be the most efficient algorithm to separate spectral fingerprints in hyperspectral images, but it requires known reference for giving input to match the spectra called end-member spectra for all of the probes present in the image. In general terms, linear unmixing is currently the most powerful technique for matching the spectral variations and establishing the quantitative mapping. It is often the case in remote sensing, that one wants to deal with identification, detection, and quantification of fractions of the target materials for each pixel for diverse coverage in a region using unmixing approaches to discern the proportion of heterogeneity (Kanniah et al. 2001). Linear mixture modeling assumes that the signal received by the satellite sensor depends on the proportion of individual surface components during the mixing process, such as soil, water, and vegetation present in a particular pixel (Abdul 2003). The unmixing decomposes a mixed pixel into a collection of end-members and a set of fractional abundances that indicate the proportion of each end-member available in the images. The contribution of each pixel is assigned to the percentage of area each ground cover class occupies in the pixel (Boardman 1989).

2.2 Study Area

In the present study, the basaltic regions of the Orientale basins such as Mare Orientale South's central part centered at 21.3′S and 265.4′E, Lacus Veris' central part located at 14°S and 273°E, and the main portion of the Lacus Autumni centered at 12°S and 277.5′E test sites are selected for carrying out the analysis. The following criteria were taken in mind to select the Orientale basin as the study area; Mare Orientale (the "eastern sea") is one of the most striking lunar features on a large scale. Since the Orientale basin has not been sampled by the Apollo program, the basin's precise age is not known. Unlike most other basins on the Moon, Orientale is relatively less flooded by mare basalts and exposes much of the basin structure to view. The Orientale basin is important to our overall understanding of the geology of large-impact basins on the lunar surface.

The Orientale basin has ~930 km in diameter and covers an area of ~700,0000 km² (Head 1974); within the basin, the largest Mare Orientale covers an area of ~52,700 km², and its volume is ~46,000 km³ (Whitten et al. 2011). Previous studies estimated that the thickness of the Mare Orientale was less than ~1–2 km (Head 1974; Solomon and Head 1980) and possibly up to ~1 km (Greeley 1976; Scott et al. 1977). With the advent of recent LRO wide-angle Camera (LROC) image, the Digital Terrain Model of the Orientale basin and its spatial profiles have been established; the depth of the Maunder crater situated in the Mare Orientale is estimated as 6.04 km. Medium-Ti basalt flood eruptions fill much of the center of the Orientale basin (3.70 Ga) (Kadel et al. 1993). The location of the Mare Orientale (A), Lacus Veris (B), and Lacus Autumni (C) are shown in Fig. 2.1, and the sites selected for the analysis are marked in the square line.

The Lacus Veris has 396 km diameter and consists of five mare ponds, the largest of which covers an area of ~8,890 km² and the smallest of which is ~145 km. Lacus Veris (also called as Spring Lake) flat extent of North South lengthened shape. The crescent-shaped Lacus Veris is situated between the two mountainous rings constituting Montes Rook to the North and the floor filled with very dark material. Medium-Ti (<4 wt% TiO$_2$) basalt signatures are present in the northern

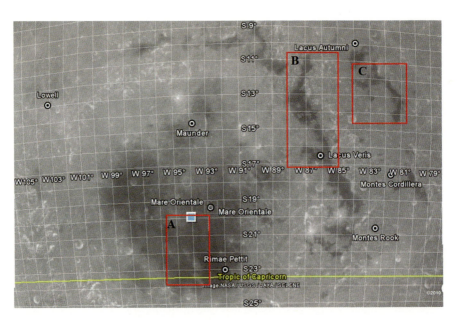

Fig. 2.1 Mare Orientale basin and test sites selected for the present study. The test sites are marked as (**A**) Mare Orientale South Central par, (**B**) Lacus Veris and (**C**) Lacus Autumni

and northeastern mare deposits of Lacus Veris, but medium–high (3 7 wt% TiO_2) basaltic signatures observed in the central and southern deposits (Kadel et al. 1993).

The Lacus Autumni is located between Montes Cordillera and Montes Rook and composed of several mare ponds. The largest of its three ponds covers an area of $\sim$2,060 km^2 and the smallest is $\sim$815 km^2 (Whitten et al. 2011). Medium–high (3–7 wt% TiO_2) basalt signatures observed for the northern mare deposits of Lacus Autumni, and medium-Ti (<4 wt% TiO_2) basalt signatures are present in the central and southern Lacus Autumni mare deposits (Kadel et al. 1993).

2.2.1 Objectives of the Study

The Mare Orientale, Lacus Veris, and Lacus Autumni of the Orientale basin basaltic regions were selected, and the following objectives were taken for the present study:

- TiO_2 estimation using Lucey's 1998 method.
- To quantify the lunar surface minerals such as clinopyroxene, orthopyroxene plagioclase and olivine using Moon Mineralogical Mapper Hyperspectral data and RELAB spectra.
- Establish the low-, medium-, and high -Ti basalt concentration map.
- Comparison of observed and RELAB basaltic spectra.
- Study of relation between TiO_2 content and age of the mare basalts.

When humans intend to establish lunar base, they will undoubtedly use the local rocks, minerals, and soils for construction materials and various chemical and metallurgical commodities (Chambers et al. 1995). The mineralogy of Orientale basin mare basalts is of particular interest, especially the question of how they compare with that on the nearside (Whitten et al. 2011). Morphological analysis result of Orientale basin basalts has been interpreted as emplacement of mare ponds during single eruptive events (Yingst and Head 1997). Several researchers have studied the Orientale basin basaltic composition, but none have been able to establish the actual mineralogy due to spatial and spectral limits of available data (Greeley et al. 1993; Kadel 1993; Staid 2000). Previous studies have found that Mare Orientale and associated mare ponds within the basin are highly contaminated with local highland material and concluded that Mare Orientale, Lacus Veris, and Lacus Autumni are probably of similar composition of lunar nearside basalts (Spudis et al. 1984; Staid 2000).

Yingst and Head (1997, 1999) found no obvious flow fronts and no distinct differences in mare composition in the individual ponds in Lacus Veris and Lacus Autumni, which assumed that individual mare ponds are the results of the single eruptive events. There is some indication of slight variation within the mare pond composition, which could either be due to lateral and vertical mixing of feldspathic material with mare soils or actual mineralogical differences (Kadel 1993; Staid 2000).

The wide range of TiO_2 contents within the returned samples has been used as a chemical property for separating them into groups including high- (>9 wt%, TiO_2), low- (1.5–9 wt% TiO_2), and very low-Ti (<1.5 wt% TiO_2) basalts, which generally cannot be related to the same source regions (Papike et al. 1998). The observed distribution of lunar basalt types suggests that mare volcanism was regionally complex with no simple correlation between composition and absolute age (e.g., Pieters 1978; Hiesinger et al. 2003, 2011).

Space weathering and mixing have altered the reflectance properties of the lunar surface over time; hence it is difficult to characterize the mineralogy of emplaced basalts from remote measurements of optically mature soils (Staid et al. 2011). In contrast, relatively crystalline and unweathered basaltic regolith associated with younger impact craters still retains diagnostic absorption features that can be interpreted based on measurements of returned lunar samples (Pieters 1977; McCord et al. 1981). M^3 data from Chandrayaan-1 have provided both the high spatial and spectral resolutions capable of investigating the detailed reflectance properties of small mare craters and allowed more direct characterization of basalt mineralogy than measurements of lunar soils that have been altered by space-weathering and non-mare contamination (Staid et al. 2011). Estimation of the relative abundance and composition of pyroxene and olivine will provide constraints on basaltic source regions, temporal evolution, and emplacement mechanisms.

2.3 Materials and Methods

India's first lunar mission Chandrayaan-1 Moon Mineralogical Mapper data, RELAB chemistry, and spectra were used in the present study. The Moon Mineralogy Mapper (M^3) is an imaging spectrometer developed by Brown University and the Jet Propulsion Laboratory, which sampled the lunar terrain from visible to near-infrared spectral region (400–3,000 nm), and provides high spatial and spectral resolution data for mineralogical study of the entire lunar surface (Goswami and Annadurai 2009). Remotely obtained reflectance spectra, combined with spectroscopic, chemical, and mineralogical data acquired in the Reflectance Experiment Laboratory (RELAB) from the lunar returned samples, can provide constraints on understanding and establishing compositional information about unexplored or unsampled planetary surfaces.

2.3.1 Methods

The methodology adopted in the study is shown in Fig. 2.2.

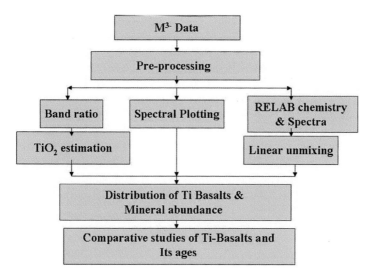

Fig. 2.2 The flow chart showing methodology adopted in the study

2.3.1.1 Data Acquisition and Preprocessing

The Orientale basin data has been downloaded from the orbital data explorer with reference to latitude and longitude information. The Chandrayaan-1 M^3 operates in two modes, including global mode and target mode. The data used here is acquired from global mode, which provides 85 bands in the spectral interval of 20–40 nm covering spectral range between 400 and 3,000 nm. The spatial resolution of global mode M^3 data is 140 m/pixel (Green et al. 2007; Pieters et al. 2009).

The preprocessing involves geo-referencing and apparent reflectance conversion from the given radiance. Using the glt file which carries information about latitude and longitude, the radiance data have been geo-referenced with ENVI software. Based on the method described by Staid et al. (2011), apparent reflectance was converted and bands were truncated at 2,500 nm, which also reduced additive contributions due to emissivity.

2.3.1.2 Spectral Profile

The mineralogy composition of the planetary surfaces has been investigated for decades using the visible and near-infrared (VISNIR) spectroscopy, owing to the abundance of diagnostic mineral absorptions in the electromagnetic spectrum (McCord et al. 1981; Pieters et al. 1996; Isaacson et al. 2011).Spectral reflectance measurements are increasingly being used to investigate the surfaces of planetary bodies throughout the solar system. The reflectance spectra of rocks exhibit characteristic signatures depending upon the mineralogy, chemical composition, and physical properties. Remotely obtained spectra in the visible and near-infrared wavelength regions can provide information on the mineralogy and modal abundance of a planetary surface (Adams 1974, 1975). The minerals can be identified based on spectral features, including absorptions pattern, the parameter structures, absorption band positions, band strength, and band depth, which are directly related to mineral chemistry (Hunt and Salisbury 1970, 1971). The same principle can be adopted to map the mineralogical compositions of objects in the outer space. The visible and near-infrared reflectance spectra of mature lunar soils have been used to classify the mare basalts from the nearside of the moon (McCord et al. 1976; Pieters 1978; BVSP 1981).

The spectral profiles of the Mare Orientale, Lacus Veris, and Autumni were collected randomly from the preprocessed M^3 data. From the collected spectral profiles, the average spectrum has been taken for further analysis. From the spectra, it is observed that, there are two broad major absorption features at 950 and 2,100 nm due to dominance of clinopyroxene and minor amount of olivine content. Another insignificant absorption is located at 1,260 nm due to plagioclase content of the basaltic terrain. The spectra of the three test sites are shown in Fig. 2.3. To avoid the maturity effect of the mare soil in the spectra, the spectral profiles were collected in the fresh shallow impact craters. There is much variation observed in between the spectra from fresh shallow impact craters and that from the dark color matured mare soils. From the matured mare soil spectra, it is difficult to delineate the basaltic mineralogy.

Based on the morphological features, the spectral profiles have drawn for the preliminary understating of the rock types and their mineralogy. From this spectral profile, it has been understood that the dominant minerals of this region are

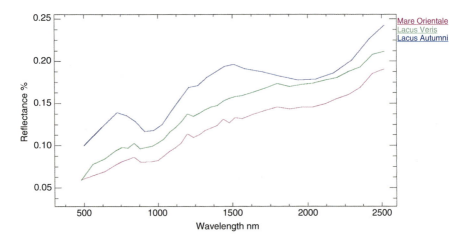

Fig. 2.3 Mare Orientale, Lacus Veris and Lacus Autumni spectra

2 Quantitative Characterization of Lunar Mare Orientale Basalts Detected...

plagioclase and clinopyroxene and the dominant rock types in this region arc gabbroic anorthosite and anorthosite gabbro. These results have been validated through spectral mixing and unmixing through the available Apollo sample data.

2.3.1.3 TiO$_2$ Estimation

High-Ti mare soils are attractive resources for lunar liquid oxygen (LLOX) production because of their unconsolidated nature, high ilmenite abundance, and widespread occurrence (Chambers et al. 1995). There is much debate over whether to use mare soils or basalts as raw materials for mineral extraction, especially ilmenite (Heiken and Vaniman 1990; Taylor and McKay 1992). Ilmenite abundance in mare basalts and soils are an important resource in the lunar exploration projects. Chemically, mare basalts can be divided into two broad groups: (1) the older high-Ti group (age 3,550–3,850 Ma, TiO$_2$ content 9–13 %) and (2) the younger, low-Ti group (age 3,150–3,450 Ma, TiO$_2$ content 1–5 %). Experimental studies show that, the low-Ti basalts could have been derived from an olivine-pyroxene source rock at depths ranging from 200 to 500 km, while the high-Ti basalts could have been derived from olivine-pyroxene-ilmenite cumulates in the outer 150 km of the Moon (Papike et al. 1976).

The following algorithm derived by Lucey et al (1998) for estimation of TiO$_2$ weight percentage is used:

$$\theta_{Ti} = \arctan\left(\frac{(R_{415}/R_{750}) - 0.45}{R_{750} - 0.05}\right) \tag{2.1}$$

$$TiO_2 = \left(\theta_{Ti}^2 \times 20.79\right) - \left(\theta_{Ti} \times 22.928\right) + 5.909 \tag{2.2}$$

Based on this analysis, it is observed that the percentage of TiO$_2$ weight in most of the Orientale basin basalts never varies more than 15 %.

Medium-Ti basalt flood eruptions have filled much of the center of the Orientale Basin (>3.70 Ga) and much of the Orientale is again flooded (<3.45 Ga) with medium–high-Ti basalts. The older medium-Ti basalts are exposed only in west-central Mare Orientale. The last eruptions in southern-east region of Mare Orientale are of high-Ti basalts (Kadel et al. 1993). Mare Orientale basalts bear high-Ti concentration and it's calculated through Lucey's 1998 algorithm and M^3 data. The south central portion's eastern region has TiO$_2$ in the range of 5–10 wt%, and the central region basalts have 10–12 wt % of TiO$_2$. In few places, the TiO$_2$ concentration exceeds up to 15 wt%. That most of the places in the Mare Orientale have >10 wt% of the TiO$_2$ indicates this region bearing high-Ti basalts. The spectral profile of the Mare Orientale basalts matching with RELAB's high-Ti mare soils spectra has confirmed the high-Ti basalts. The TiO$_2$ concentration map of the Mare Orientale is shown in Fig. 2.4a.

Hawke et al. (1991) calculated that Lacus Autumni and Lacus Veris' southern part bears lower-TiO$_2$ concentration, while the northern part of the Lacus Veris bears

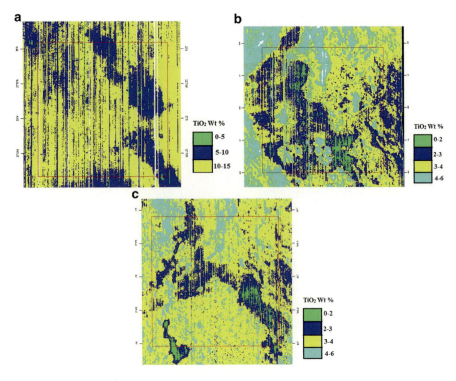

Fig. 2.4 TiO$_2$ wt % of (**a**). Mare Orientale, (**b**) Lacus Veris and (**c**) Lacus Autumni

intermediate-TiO$_2$ concentration. TiO$_2$ analysis of the center portion of the Lacus Veris indicates that, the region is dominated by the medium-Ti basalts in the range of <6 wt%. This is shown in Fig. 2.4b.

The Lacus Autumni also bears the medium-Ti basalts which concentration range is <6 wt%. The TiO$_2$ wt% of the Lacus Autumni is shown in Fig. 2.4c. The dominant basaltic region bears TiO$_2$ concentration range between 2–3 wt%. The variation of the Ti content observed in the spectral profile of the Lacus Autumni doesn't exceed 6 %.

2.3.1.4 Spectral Mixing and End-Member Selection

Manned and unmanned missions to the Moon have returned about 382 kg of lunar rocks and soils (Vaniman et al. 1991). These were collected from the lunar nearside's typical region, within and around the Procellarum KREEP Terrane (PKT) (Jolliff et al. 2000), or from equatorial latitudes on the eastern limb. Linear spectral mixture analysis can be used to model the spectral variability in multispectral or hyperspectral images and to establish the relation of its results and the physical abundance of surface constituents represented by the spectral end-members (Tompkins et al. 1997). In order to use a linear mixture model, the spectral reflectance of the "pure" end-members should be measured. Ideally, ground-based reflectance spectra would be acquired from the known minerals and rocks to produce accurate end-members. Since we are concentrating the lunar surface, we are not able to provide the specific high-precision ground truth information/spectra at this present situation. Spectral mixture analysis (SMA), a technique based on modeling image spectra as the linear combination of end-members, has been used to derive the fractional contribution of end-member materials to image spectra in a wide variety of applications. An end-member is a "pure" spectrum of a material or target and has a unique spectral signature. Image end-members are pure pixels from image itself. The end-member selection included the following components: plagioclase, ortho-/clinopyroxene, and olivine in the mineral class and low-, medium-, and high-Ti basalts in rock classes. Performing spectral unmixing requires appropriate end-member selection, since unsuitable end-member may lead to meaningless fraction mapping. In this paper, seven end-members have been identified and extracted from the image data for spectral unmixing analysis.

The bidirectional reflectance spectra of low-, medium-, and high-Ti basaltic soils acquired from various grain sizes such as <10, 10–20, 20–45, and <45 μm. Based on the grain size variation, there is not much variation observed in the spectral curves apart from the broad weak absorption near 1,000 nm for the grain size of 20–45 μm sample. To establish the basaltic end-member's spectra, the RELAB spectra of low-, medium-, and high-Ti basalts of various grain sizes have been taken and averaged. The spectral values are averaged for the various grain sizes, and one spectral end-member is derived for each basaltic soil. The average of the bulk chemical composition in oxides and in modal abundance of minerals in the various grain sizes of low Ti (15071-52), medium Ti (12030-14), and high Ti (71501-35) is given in Tables 2.1 and 2.2, respectively. The spectral profiles of RELAB mineral end-member (a) and basaltic end-members (b) are shown in Fig. 2.5. The spectra from 450 to 2,500 nm are used as end-member to estimate the abundance of the various basalts in the Orientale basin basaltic terrain. Apart from the basalts, the plagioclase, ortho-/clinopyroxene, and olivine spectra from RELAB are also utilized in the present study.

2.3.1.5 Linear Spectral Unmixing

Spectral unmixing methods have been proven useful in the application of imaging spectroscopy for geological studies. It is understand that, grain size effects intimate mixtures and mineral coatings, so other spectroscopic spectral unmixing methods have proven useful in the application of imaging spectroscopy to geological studies. Linear spectral unmixing (LSU) has been proposed for the analysis of hyperspectral images to compute the fractional contribution of the detected end-members to each pixel in the image. Linear spectral unmixing is used to discriminate one type of material from another. It provides an estimation of how much material available in a pixel and a method of classifying broad categories of materials in a pixel.

Table 2.1 The average bulk chemistry of low, medium, and high-Ti basalts of various grain sizes such as <45, 20–45, 10–20, and <10 μm

Bulk chemistry (oxides wt %)	Low-Ti basalt (15071)-52	Medium-Ti basalt (12030)-14	High-Ti basalt (71501)-35
SiO_2	46.07	46.25	31.87
TiO_2	1.89	3.32	9.52
$Al2O_3$	13.87	11.70	11.83
$Cr2O_3$	0.44	0.43	0.43
MgO	10.88	9.42	9.49
CaO	10.52	9.78	10.36
MnO	0.19	0.20	0.22
FeO	13.87	16.27	16.05
Na_2O	0.40	0.46	0.38
K_2O	0.16	0.29	0.09
P_2O_5	0.15	0.25	0.06
SO_2	0.11	0.12	0.19

Source: RELAB

Table 2.2 The average modal abundance of minerals for low-, medium-, and high-Ti basalts of various grain sizes such as <45, 20–45, 10–20, and <10 μm

Modal abundance of minerals (wt %)	Low-Ti basalt (15071)-52	Medium-Ti basalt (12030)-14	High-Ti basalt (71501)-35
Ilmenite	1.63	2.93	9.86
Plagioclase	19.10	15.76	18.76
Pyroxene	16.56	23.50	14.60
Olivine	2.86	3.50	3.40
Agglutinitic glass	52.16	48.06	45.40
Volcanic glass	3.90	1.43	6.70
Others	3.76	4.80	1.30

Source: RELAB

A specific problem of LSU is the determination of the end-members; to this end we employ two approaches: the Convex Cone Analysis and another one based on the detection of morphological independence. The planetary surface materials have specific spectral features, which are related to their composition. Each image pixel is assumed to be some mixture of various component materials, but regular algorithms for classification cannot identify more than one class within one pixel. Hence, the spectral unmixing is used to deal with this case.

Quantitatively retrieving mineral abundances from hyperspectral data is one of the promising and challenging geological application fields of hyperspectral remote sensing. However, its mixture characteristic of mineral spectra and deconvolution method of mixture spectra are the most basic obstacles in using hyperspectral data

2 Quantitative Characterization of Lunar Mare Orientale Basalts Detected...

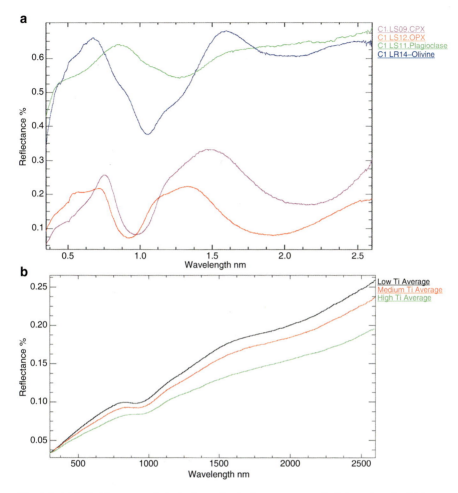

Fig. 2.5 (**a**) RELAB spectra of plagioclase, clino/ortho pyroxene and olivine minerals. (**b**) Low-, medium- and high-Ti basaltic end member spectra from RELAB

(Bokun et al. 2008). Pure materials can be measured under controlled conditions in the laboratory, but for remotely sensed data, multiple components possibly will be included in one measurement (pixel) causing spectral mixing.

2.4 Results and Discussion

Based on the spectral profiles, we are able to discriminate the difference between the ortho- and clinopyroxene such as Mg, Fe, and Ca pyroxenes. Based on the TiO_2 estimation, spectral profiles, and linear spectral unmixing, the following results were obtained and discussed herewith.

2.4.1 Mare Orientale

The averaged low-, medium-, and high-Ti basaltic spectra have been derived from the RELAB mare soil spectra. The bulk chemistry and modal abundance of the minerals have been listed in Table 2.1 and 2.2, respectively. Based on the RELAB basaltic spectra, the concentration of the different types of basalts is mapped in the Mare Orientale, Lacus Autumni, and Lacus Veris basaltic regions. The south central part of the Mare Orientale is comprised of high-Ti basalts ranges around 21 % and medium-Ti basalts dominant in 40 %, and in few craters it leads upto 60 % and low-Ti basalts ranges around 6 %. The comparative analysis of the different types of basalts is given in Table 2.3. The concentration of the high-Ti basalts and medium-Ti basalts is shown in Fig. 2.6a, b.

Compared with the Lacus Autumni and Veris, the south central part of the Orientale basalts bears high-Ti basalts, which ranges up to 15 %. Hence, based on the TiO_2 concentration, the Lacus Autumni and Veris might be of the same origin and similar ages.

In the south central part of the Mare Orientale basaltic region, the soil consists of <20 % of plagioclase. Some impact craters located in the inner side of the

Table 2.3 The observed Modal abundance of minerals for Mare Orientale, Lacus Autumni and Lacus Veries from M^3 data

Modal abundance of minerals and Ti-basalts (values are in wt %)	Mare Orientale	Lacus Autumni	Lacus Veris
Plagioclase	20	21	12
Pyroxene (CPX +OPX)	21	32	21
Olivine	4	4	2.5
Ilmenite	–	–	–
Agglutinitic glass	–	–	–
Volcanic glass	–	–	–
Others	–	–	–
Basalts			
High-Ti basalts	21	10	40
Medium-Ti-basalts	60	–	–
Low-Ti basalts	6	21	>60
Modal abundance of minerals (wt %)	Mare Orientale	Lacus Autumni	Lacus Veries
Ilmenite	–	–	–
Plagioclase	20	21	12
Pyroxene	21	32	21
Olivine	4	4	2.5
Agglutinitic glass	–	–	–
Volcanic glass	–	–	–
Others	–	–	–

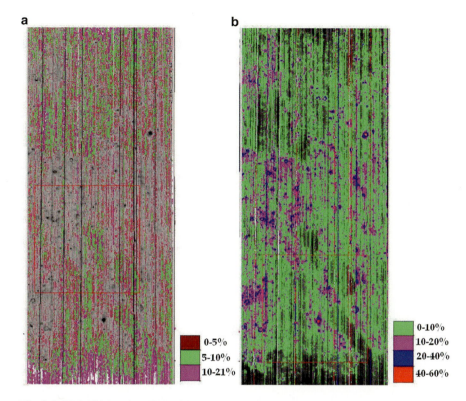

Fig. 2.6 High Ti (**a**) and medium Ti (**b**) basalts distribution map of South Central region of Mare Orientale

basaltic regions contain plagioclase in the range of up to 40 %, which excavated the highland material from deeper crust. In the central part of the Mare Orientale region, clinopyroxene is the dominant mineral which ranges from 8–12 % to 22 %, and this almost consistent with pyroxene content of the high-Ti basalts of RELAB samples. There is no measurable amount of orthopyroxene observed in the south central part of the Mare Orientale. Olivine content of the south central part of Mare Orientale varies from 2 to 4 %.

2.4.2 Lacus Veris

The high-Ti-bearing basalt concentration range is up to 40 % in the Lacus Veris region. The region is dominated by high-Ti basalts which accounts for 27 %. Like Lacus Autumni, this region also does not have much medium-Ti basaltic concentration. The entire western region of the Lacus Veris is dominated by low-Ti-bearing basalts. The distribution of the high- and low-Ti basalts is shown in Fig. 2.7a, b, respectively.

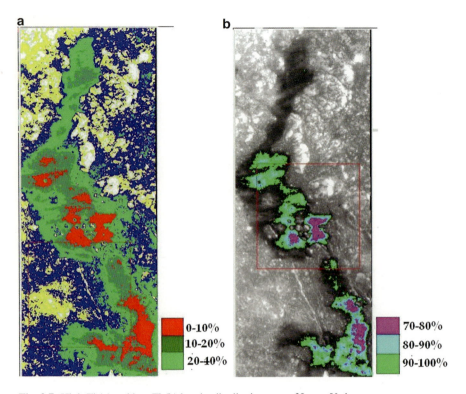

Fig. 2.7 High-Ti (**a**) and low-Ti (**b**) basalts distribution map of Lacus Veris

The concentration of the CPX is 15 % in the Lacus Veris basaltic terrain. In some places, it exceeds up to 21 %. This is almost similar with the high-Ti basaltic chemistry of the RELAB sample no. 71501-35 in the grain size of 20–45 μm. Like previous test sites, the Lacus Veris also does not have much amount of OPX. The plagioclase content of this region is comparatively lesser than the Mare Orientale and Lacus Autumni. The plagioclase and olivine content in the Lacus Veries basaltic region is around 12 % and 2.5 % respectively. The olivine distribution is observed only in the small areas of the Lacus Veries basaltic region. The concentration of the olivine is in and around 2.5 % in the Lacus Veris region. That is also presented in small areas and not distributed in the all basaltic region.

2.4.3 Lacus Autumni

The high-Ti basaltic concentration of Lacus Autumni region is shown in Fig. 2.8a. The Lacus Autumni consists of only 10 % of the high-Ti basalts, however in some specific regions, the content of the high-Ti basalts is increased up to 80 %, which is marked in magenta color shown in the Fig. 2.8a. Low-Ti basalts concentrated in

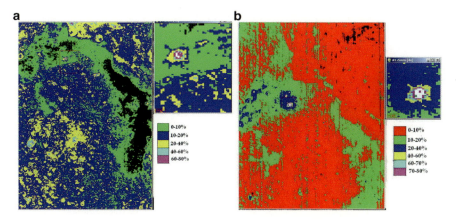

Fig. 2.8 High-Ti (**a**) and low-Ti (**b**) basalts distribution map of Lacus Autumni

the range of 21 % in most of the Lacus Autumni, but in some specific locations, the content of the Low-Ti basalts exceeds up to 70–80 %, which is shown in Fig. 2.8b in the colors of cyan and magenta. There is no much amount of medium-Ti basalt distribution in the Lacus Autumni region.

The clinopyroxene (CPX) in the central part of the Lacus Autumni is around 32 %. The outside of the Lacus Autumni, CPX concentration varies in the range from 32 to 42 %. The southern-east portion of the Lacus Autumni has slightly lower CPX with the amount of 26 %. There is no much amount of OPX measured in the Lacus Autumni region. The plagioclase content of the Lacus Autumni is 27 %, 21 % in central part and 30 % in external side of the Lacus Autumni. The olivine concentration of the Lacus Autumni is nearly 4 %. The western portion of the Lacus Autumni has lower olivine content compared to the central portion. The impact craters on the Lacus Autumni have slightly higher olivine content that is nearly up to 5 %. It could be due to excavation of deeper material due to impact.

2.4.4 Age of Mare Orientale, Lacus Veris, and Lacus Autumni

Compared with lunar highland, mare basalts occupy lesser area, but they contain much information about the thermal history of the moon and the nature of the lunar interior (Head 1976). The age of mare basalts studied is between 3.15 and 3.96 Gy. But recent photogeologic (crater-counting) studies indicate that there are 2.5-Gy-old younger basalts on the moon, which yet to be sampled. The returned samples were classified into two broad categories based on the TiO_2 concentration: high-titanium group (ages ~3.55–3.85 Gy; TiO_2, 9–14 wt%) and the low-titanium group (ages 3.15–3.45 Gy; TiO_2, 1–5 wt%). Basalts returned by Apollo 11 and 17 are dominated by high-titanium group including olivine and ilmenite, whereas Apollo 12 and 15 and Luna 16 returned samples dominated by low-titanium group which is almost olivine (Head 1976).

M³ images with high spatial and spectral resolution can be used to count the craters for given size on the Mare Orientale, Lacus Veris, and Lacus Autumni and to produce the reliable model. In mare Orientale, craters >5 km were counted on the Hevelius formation and craters >0.75 km were counted on the Maunder formation. This crater count method calculated that ages are ~3.68 Ga for the ejecta and ~3.64 Ga for the melt sheet. Orientale was emplaced shortly after the Maunder formation in ~3.58 Ga. This delay of ~60–100 Ma between basin formation and volcanism in three study areas argues against Orientale impact-pressure-release melting ages (Whitten et al. 2011).

In the Lacus Veris, the crater count chronology deduced the age of five largest ponds in the range of ~3.20 to ~3.69 Ga (Whitten et al. 2011), which is consistent with the previous studies conducted by Greeley et al. (1993) and Kadel (1993).

In the Lacus Autumni, crater counts give an age range between ~3.47 and ~1.66 Ga (Whitten et al. 2011). The young model age of ~1.66 Ga falls well within the calculated time range of mare volcanism occurring on the lunar nearside (Hiesinger et al. 2000). This discrepancy in the model ages could be the result of sizeable differences in the definition of the count area or count statistics used. The detailed discussion can be found in Whitten et al. (2011) and relevant references.

2.4.5 Validation

The results revealed from the RELAB spectra and chemical data have been used to validate the result of this present research. Apart from the RELAB data, the results have been compared with the previous Clementine UVVIS data global mineral mapping study conducted by Bokun et al 2010. The RELAB low-, medium-, and high-Ti basaltic spectra were compared with Mare Orientale, Lacus Veris, and Autumni spectra established from M³ data, which is shown in Fig. 2.9. In this,

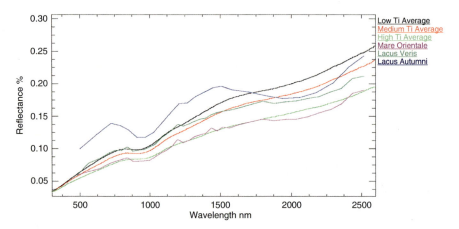

Fig. 2.9 Low-, medium- and high-Ti basalts spectra with Mare Orientale, Lacus Veris and Lacus Autumni spectra

the Mare Orientale spectra almost match with the RELAB high-Ti basaltic spectra. The Lacus Veris spectrum is plotted in between the low- and medium-Ti RELAB basaltic spectra. The Lacus Autumni spectra were not matched with this group of basaltic spectra, and it indicates admixing and contamination of the maria regolith with materials excavated or ejected from underlying and neighboring rocks in the Lacus Autumni region.

2.5 Conclusion and Future Work

Since the RELAB basaltic samples were brought from the nearside of the moon, there may be slight variation in the Orientale basin basalts. Apollo geochemical ground tracks and sample collection did not cover this region; hence, the relative end-member spectral library and chemistry such as plagioclase, ortho- and clinopyroxenes, and olivine and low-, medium-, and high-Ti basalts are only used. The more precise results can be provided once we get samples from Orientale basalts.

The spectral mixtures were derived from RELAB spectra, based on lunar rock classification, and they are considered as end-members for estimating the abundance of the rock content in the specified area using M^3 data. If there are similar kinds of studies for the other basaltic terrain on the lunar surface, the results can be used as validation. However, only the global lunar mineral mapping maps using the UVVIS multispectral Clementine data are available at present, so the current results have been validated with RELAB chemistry.

To improve this work, it is important to compare the fraction covers with field data in order to verify the data obtained through the images. In addition, field spectral data is necessary for end-member selection, because a good choice of end-members is the most important step for a good spectral unmixing. Mixed pixels may cover a region containing different components, and therefore, traditional image classification approach should be altered, which assigns a particular class of ground cover to each pixel (Kanniah et al. 2001).

Low-Ti basalts are younger than the high-Ti basalts. But remote sensing data based studies suggests that high-Ti basalts are younger. Finally, there is no obvious correlation between the ages and low-Ti and high-Ti concentrations (e.g., Pieters 1978; Hiesinger et al. 2003, 2011). Based on the TiO_2 analysis, the Mare Orientale basalts have high-TiO_2 ($\sim$15 %) content compared with the Lacus Veris and Lacus Autumni ($<$6 %). Papike et al. (1976) has concluded that the Mare Orientale age is older than the Lacus Veris and Lacus Autumni, which supports the result of crater counting given by Whitten et al. (2011). In the future, if we obtain sample from the Orientale basin, with radiometric dating and geochemical measurements, we will get precise chronology of the Orientale basin and clarify the relation between the TiO_2 content and ages. In mineral mapping, all basaltic terrains from the Orientale basin are dominant with clinopyroxene followed by plagioclase.

The future work will focus on the thermal and photometrically corrected data, which would be used to estimate the minerals and mineral mixtures using Hapke's single-scattering albedo (SSA) modeling and spectral unmixing. The error of mineral abundances derived from reflectance spectra and single-scattering albedo is 20.05 % and 5.03 %, respectively (Bokun et al. 2008). Since the significant variation between the linear spectral unmixing and single-scattering albedo, SSA should be employed for getting more accuracy in quantitative mapping. Once corrections are done, the Modified Gaussian Model (MGM) could perform to derive the modal percentage of the subset of Mg-Fe and Ca pyroxenes and olivine, which will obtain the better results.

Acknowledgement The author acknowledges PLANEX, Physical Research Laboratory (ISRO), Ahmedabad, and Council of Scientific and Industrial Research (CSIR), New Delhi, for Postdoctoral Research Fellowship. Moreover, the author thanks the anonymous reviewer's critical review which helps to improve the paper.

References

Abdul S (2003) Subpixel classification of ground surface features. GIS Dev 7(11):20–24

Adams JB (1974) Visible and near-infrared diffuse reflectance spectra of pyroxenes as applied to remote sensing of solid objects in the solar system. J Geophys Res 79:4829–4836

Adams JB (1975) Interpretation of visible and near-infrared diffuse reflectance spectra of pyroxenes and other rock-forming minerals. In: Karr C Jr (ed) Infrared spectroscopy of lunar and terrestrial minerals. Academic, New York, pp 91–116

Basaltic Volcanism Study Project (BVSP) (1981) Basaltic volcanism on the terrestrial planets. Pergamon Press, Oxford, p 1286 (Heiken GH, Vaniman DT, French BM (eds) (1991) Lunar sourcebook. Cambridge University Press, Cambridge, p 736)

Boardman JW (1989) Spectral and spatial unmixing: applications of singular value decomposition. In: Proceedings of image processing, Reno, 1989

Bokun Y, Shengweia L, Runsheng W, Xiaofanga G, Weidongb S (2008) Experiment study on quantitative retrieval of mineral abundances from reflectance spectra, remote sensing of the environment: 16th national symposium on remote sensing of China (eds) Qingxi Tong Proceedings of SPIE, vol 7123, p 712303. doi:10.1117/12.815549

Bokun Y, Runsheng W, Fuping G, Zhenchao W (2010) Minerals mapping of the lunar surface with Clementine UVVIS/NIR data based on spectra unmixing method and Hapke model. ICARUS 208:11–19

Chambers JG, Taylor LA, Patchen A (1995) Quantitative mineralogical characterization of lunar high-TI mare basalts and soils for oxygen production. J Geophys Res 100(E7):14391–14401

Chen X, Warner T, Campagna D (2007) Integrating visible, near-infrared and short-wave infrared hyperspectral and multispectral thermal imagery for geological mapping at Cuprite, Nevada [J]. Remote Sens Environ 110:344–356

Clark R, Swayze G, Livo R, et al (2003) Imaging spectroscopy: Earth and planetary remote sensing with the USGS Tetracorder and expert systems [J]. J Geophys Res 108(E2):5–1 to 5–44

Clark RN et al (2008) Compositional mapping of Saturn's satellite Dione with Cassini VIMS and implications of dark material in the Saturn system. Icarus 193:372–386

Cloutis EA, Gaffey MJ (1991) Pyroxene spectroscopy revisited: spectral compositional correlations and relationships to geothermometry. J Geophys Res 96:22809–22826

Crown DA, Pieters CM (1987) Spectral properties of plagioclase and pyroxene mixtures and the interpretation of lunar soil spectra. ICARUS 72:492–506

Goswami JN, Annadurai M (2009) Chandrayaan-1: India's first planetary science mission to the Moon. Curr Sci 96(4):486–491

Greeley R (1976) Modes of emplacement of basalt terrains and an analysis of mare volcanism in the Orientale Basin. In: Proceedings of the 7th lunar science conference, Houston, TX, 15–19 March 1976, vol 3 (A77-34651 15-91). Pergamon Press, Inc., New York, pp 2747–2759

Greeley R et al (1993) Galileo imaging observations of lunar maria and related deposits. J Geophys Res 98:17183–17205. doi:10.1029/93JE01000

Green RO, Pieters CM, Mouroulis P, Sellars G, Eastwood M, Geier S, Shea J (2007) The Moon Mineralogy Mapper: characteristics and early laboratory calibration results. Lunar Planet Sci [CD-ROM] 38, Abstract 2354

Hawke BR, Lucey PG, Taylor GJ, Bell JF, Peterson CA, Blewett DT, Horton K, Smith GA, Spudis PD (1991) Remote sensing studies of the Orientale region of the Moon: a pre-Galileo view. Geophys Res Lett 18(11):2141–2144

Head JW (1974) Orientale multi-ringed basin interior and implications for the petrogenesis of lunar highland samples. Moon 11:327–356. doi:10.1007/BF00589168

Head JW (1976) Lunar volcanism in space and time. Rev Geophys 14(4):265–300. doi:10.1029/RG014i002p00265

Heiken GH, Vaniman DT (1990) Characterization of lunar ilmenite resources. In: Proceedings of the 20th lunar and planetary science conference, Houston, TX, 13–17 March 1989 (A90-33456 14-91). Lunar and Planetary Institute, Houston, pp 239–247

Hiesinger H, Jaumann R, Neukum G, Head J (2000) Ages of mare basalts on the lunar nearside. J Geophys Res 105(E12):29239–29275

Hiesinger H, Head JW, Wolf U, Jaumann R, Neukum G (2003) Ages of lunar mare basalts in Mare Frigoris and other nearside maria, Lunar Planet Sci [CD-ROM], 34, Abstract 1257

Hiesinger H, Head III JW, Wolf U, Jaumann R, Neukum G (2011) Ages and stratigraphy of lunar mare basalts: a synthesis. In: Ambrose WA, Williams DA (eds) Recent advances and current research issues in lunar stratigraphy, Special paper, vol 447. Geological Society of America, pp 1–51. doi:10.1130/2011.2477(01)

Hunt GR, Salisbury JW (1970) Visible and near infrared spectra of minerals and rocks. I. Silicate minerals. Mod Geol 1:283–300

Hunt GR, Salisbury JW (1971) Visible and near infrared spectra of minerals and rocks. II. Carbonates. Mod Geol 2:23–30

Isaacson PJ, Pieters CM, Besse S, Clark RN, Head JW, Klima RL, Mustard JF, Petro NE, Staid MI, Sunshine JM, Taylor LA, Thaisen KG, Tompkins S (2011) Remote compositional analysis of lunar olivine-rich lithologies with Moon Mineralogy Mapper (M^3) spectra. J Geophys Res 116:E00G1. doi:10.1029/2010JE003731

Jolliff BL, Gillis JJ, Haskin LA, Korotev RL, Wieczorek MA (2000) Major lunar crustal terranes: surface expressions and crust-mantle origins. J Geophys Res 105:4197–4216

Kadel SD (1993) Multispectral and morphologic analyses of lunar mare basalts in the Orientale Basin, M.Sc., Arizona State University, Phoenix

Kadel SD, Greeley R, Neukum G, Wagner R (1993) The history of mare volcanism in the Orientale Basin; mare deposit ages, compositions and morphologies, Lunar Planet Sci [CD ROM], 24, Abstract 1374

Kanniah KD, Ng Su Wai, Alvin Lau Meng Shin, Abd Wahid Rasib (2001) Linear mixture modelling applied to IKONOS data for mangrove mapping. http://www.a-a-r-s.org/acrs/proceeding/ACRS2005/Papers/FRR2-2.pdf

LeMouélic SL, Langevin Y (2001) The olivine at the lunar crater Copernicus as seen by Clementine NIR data. Planet Space Sci 49:65–70

LeMouélic SL, Langevin Y, Erard S (1999) The distribution of olivine in the crater Aristarchus inferred from Clementine NIR data. Geophys Res Lett 26:1195–1198

Lucey PG, Blewett DT, Hawke BR (1998) Mapping the FeO and TiO_2 content of the lunar surface with multispectral imagery. J Geophys Res 103(E2):3679–3699

Lucey PG et al (2006) Understanding the lunar surface and space – Moon interactions. Rev Miner Geochem 60:83–219. doi:10.2138/rmg.2006.60.2

McCord TB, Charette MP, Johnson TV, Lebofsky LA, Pieters C, Adams JB (1972) Lunar spectral types. J Geophys Res 77:1349–1359

McCord TB, Pieters CM, Feierberg MA (1976) Multispectral mapping of the lunar surface using ground based telescopes. Icarus 29:1–34

McCord TB, Clark RN, Hawke BR, Mcfadden LA, Owensby PD, Pieters CM, Adams JB (1981) Moon: near-infrared spectral reflectance, a first good look. J Geophys Res 86(B11):10883–10892

Papike JJ, Hodges FN, Bence AE, Cameron M, Rhodes JM (1976) Mare Basalts' crystal chemistry, mineralogy, and petrology. Rev Geophys 14(4):475–540

Papike JJ, Ryder G, Schearer CK (1998) Lunar samples. In: Papike JJ (ed) Planetary materials. Mineralogical Society of America, Washington, DC

Pieters CM (1977) Characterization of lunar mare basalt types-II: spectral classification of fresh mare craters. In: Proceedings of the 8th lunar science conference, Houston, TX, 14–18 March 1977, vol 1 (A78-41551 18-91). Pergamon Press, Inc., New York, pp 1037–1048

Pieters CM (1978) Mare basalt types on the front side of the Moon: a summary of spectral reflectance data. In: Proceedings of the ninth lunar and planetary science conference, pp 2825–2849

Pieters CM, Englert PA (eds) (1997) Remote geochemical analysis: Elemental and mineralogical composition. Cambridge University Press, New York, p 467

Pieters CM, Head JW, Sunshine JM, Fischer EM, Murchie SL, Belton M, McEven A, Gaddis L, Greeley R, Neukum G, Jaumann R, Hoffmann H (1993) Crustal diversity of the Moon: compositional analyses of Galileo solid state imaging data. J Geophys Res 98:17127–17148. doi:10.1029/93JE01221

Pieters CM, Mustard JF, Sunshine JM (1996) Quantitative mineral analyses of planetary surfaces using reflectance spectroscopy. Spec Publ Geochem Soc 5:307–325

Pieters CM, Head JW, Gaddis L, Duke M (2001) Rock types of South Pole-Aitken Basin and extent of basaltic volcanism. J Geophys Res 106(E11):28001–28022

Pieters CM, Klima RL, Hiroi T, Dyar MD, Lane MD, Treiman AH, Noble SK, Sunshine JM, Bishop JL (2008) Martian dunite NWA 2737: integrated spectroscopic analyses of brown olivine. J Geophys Res 113(1–12):E06004

Pieters CM, Boardman J, Buratti B, Chatterjee A, Clark R, Glavich T, Green R, Head J, Isaacson P, Malaret E, McCord T, Mustard J, Petro N, Runyon C, Staid M, Sunshine J, Taylor L, Tompkins S, Varanasi P, White M (2009) The Moon Mineralogy Mapper (M^3) on Chandrayaan-1, Indian Academy of Sciences. Curr Sci 96(4):500–505

Scott DH, McCauley JF, West MN (1977) Geologic map of the west side of the moon, Miscellaneous investigations series map I-1034. U.S. Geological Survey, Reston

Shkuratov Y, Opanasenko N, Zubko E, Grynko Y, Korokhin V, Pieters C, Videen G, Mall U, Opanasenko A (2007) Multispectral polarimetry as a tool to investigate texture and chemistry of lunar regolith particles. Icarus 187:406–416

Solomon SC, Head JW (1980) Lunar mascon basins: Lava filling, tectonics, and evolution of the lithosphere. Rev Geophys Space Phys 18:107–141. doi:10.1029/RG018i001p00107

Spudis PD, Hawke BR, Lucey P (1984) Composition of Orientale Basin deposits and implications for the lunar basin-forming process. In: Proceedings of the 15th lunar planetary science conference, Part 1, J Geophys Res 89(Suppl):C197–C210

Staid MI (2000) Remote determination of the mineralogy and optical alteration of lunar basalts using clementine multispectral images; global comparisons of mare volcanism. Ph.D. thesis, Brown University, Providence, RI

Staid MI, Pieters CM (2000) Integrated spectral analysis of mare soils and craters: applications to eastern nearside basalts. Icarus 145:122–139. doi:10.1006/icar.1999.6319

Staid MI, Pieters CM, Head JW (1996) Mare Tranquillitatis: Basalt emplacement history and relation to lunar samples. J Geophys Res 101:23213–23228. doi:10.1029/96JE02436

Staid MI et al (2011) The mineralogy of late stage lunar volcanism as observed by the Moon Mineralogy Mapper on Chandrayaan-1. J Geophys Res. doi:10.1029/2010JE003735

Taylor SR (1975) Lunar science: a post-Apollo view. Pergamon, New York, p 372

Taylor LA, McKay DS (1992) An ilmenite feedstock on the moon; beneficiation of rocks versus soils In: Proceedings of the 23rd lunar and planetary science conference, League City, TX, 16–20 March 1992, Lunar and Planetary Institute, Houston, TX, pp 1411–1412

Tompkins S, Mustard JF, Pieters CM, Forsyth DW (1997) Optimization of endmembers for spectral mixture analysis. Remote Sens Environ 59(3):472–489

Vaniman D, Dietrich J, Taylor GJ, Heiken G (1991) Exploration, samples, and recent concepts of the moon. In: Heiken GH, Vaniman DT, French BM (eds) Lunar sourcebook: a user's guide to the moon. Cambridge University Press/Lunar and Planetary Institute, New York

Whitten J, Head JW, Staid M, Pieters CM, Mustard J, Clark R, Nettles J, Klima RL, Taylor L (2011) Lunar mare deposits associated with the Orientale impact basin: new insights into mineralogy, history, mode of emplacement, and relation to Orientale Basin evolution from Moon Mineralogy Mapper (M^3) data from Chandrayaan-1. J Geophys Res 116:E00G09. doi:10.1029/2010JE003736

Yingst RA, Head JW (1997) Volumes of lunar lava ponds in South Pole–Aitken and Orientale basins: implications for eruption conditions, transport mechanisms and magma source regions. J Geophys Res 102:10909–10931. doi:10.1029/97JE00717

Yingst RA, Head JW (1999) Geology of mare deposits in South Pole–Aitken Basin as seen by Clementine UVVIS data. J Geophys Res 104:18957–18979. doi:10.1029/1999JE900016

Chapter 3
Gravity Changes over Russian River Basins from GRACE

Leonid V. Zotov, C.K. Shum, and Natalya L. Frolova

Abstract Gravity Recovery and Climate Experiment (GRACE) twin satellites have been observing the mass transports of the Earth inferred by the monthly gravity field solutions in terms of spherical harmonic coefficients since 2002. In particular, GRACE temporal gravity field observations revolutionize the study of basin-scale hydrology, because gravity data reflect mass changes related to ground and surface water redistribution, ice melting, and precipitation accumulation over large scales. However, to use the GRACE data products, de-striping/filtering is required. We applied the multichannel singular spectrum analysis (MSSA) technique to filter GRACE data and separate its principal components (PCs) at different periodicities. Data averaging over the 15 largest river basins of Russia was performed. Spring 2013 can be characterized by the extremely large snow accumulation occurred in Russia. Melting of this snow induced large floods and abrupt increase of river levels. The exceptional maxima are evident from GRACE observations, which can be compared to the hydrological models, such as Global Land Data Assimilation System (GLDAS) or WaterGAP Global Hydrology Model (WGHM), and gauge data. Long-periodic climate-related changes were separated into PC 2. Finally, it was observed that there were mass increases in Siberia and decreases around the Caspian Sea. Overall trend over Russia demonstrates mass increase until 2009, when it had a maximum, followed by the decrease.

Keywords Earth's gravity field • GRACE • Hydrological changes • MSSA

L.V. Zotov (✉)
Sternberg Astronomical Institute, Lomonosov Moscow State University, Universitetsky pr., 13, 119991 Moscow, Russia

National Research University Higher School of Economics, Moscow, Russia
e-mail: wolftempus@gmail.com

C.K. Shum
Division of Geodetic Science, School of Earth Sciences, The Ohio State University, Columbus, OH, USA

Institute of Geodesy & Geophysics, Chinese Academy of Sciences, Wuhan, China

N.L. Frolova
Department of Hydrology, Faculty of Geography, Lomonosov Moscow State University, Moscow, Russia

© Springer-Verlag Berlin Heidelberg 2015
S. Jin et al. (eds.), *Planetary Exploration and Science: Recent Results and Advances*,
Springer Geophysics, DOI 10.1007/978-3-662-45052-9_3

3.1 Introduction

Space-based Earth Observing Systems provided a substantially large amount of information to the scientific community in the recent decades. One of the most important contributions of these data sets is their use to study climate change. Cumulative effects of redistribution of masses in the Earth system can be seen in the changes of the gravity field of the Earth. Gravimetry is a science with a long history. Gravity measurement techniques for land and ocean have been developing all over the twentieth century. But only the space era opened the possibility to study global gravity field and its changes over a planetary scale, including inaccessible distant regions.

Technological achievements of our epoch – the NASA/DLR Gravity Recovery and Climate Experiment (GRACE) twin satellites mission was launched on 17.03.2002 from Plesetsk kosmodrom. It allows the observations of monthly changes in Earth's gravity field with unprecedented accuracy, working already for 11 years by the time of our study, which is twice more than expected. Though the battery power is ten times less than at a launch time, there is a possibility that the mission period could be extended till 2017, when the GRACE Follow-on Mission is anticipated to be launched.

GRACE satellites fly in a near-polar orbit at $\sim$500 km altitude following one another at a distance of $\sim$220 km. Accelerations of each satellite, occurring during the flight above the Earth's mass anomalies, influence the range between the two GRACE satellites. Microwave K-band range measurements represent the fundamental observational data, containing information about the gravity field. Data centers located at GeoForschungsZentrum (GFZ) (Potsdam), Center for Space Research (CSR) (Austin), and Jet Propulsion Laboratory (JPL) (Pasadena) process these data, taking into account onboard GPS, accelerometers, star cameras, etc, to produce the level one (L1) data products (Case et al. 2004). Then through sophisticated gravity field inversion techniques with regularization (Tikhonov et al. 1998; Wang et al. 2012), correcting the aliasing effects of the atmospheric pressure changes over land and over ocean, applying solid Earth, ocean and pole tides, and other corrections, level 2 (L2) data product (Bettadpur 2007) is obtained, representing monthly gravity field in form of Stokes coefficients (3.1) of spherical decomposition on the surface of Earth's mean radius (Panteleev 2000; Sagitov 1979).

Modeling of the mean gravitational field is the primary goal of the space gravity missions (Kenyon et al. 2007). Contemporary models incorporate information, obtained from CHAllenging Minisatellite Payload (CHAMP), GRACE, and Gravity field and steady-state Ocean Circulation Explorer (GOCE) satellites. But GRACE also provides monthly anomalies (1 month is required to cover the Earth). If the mean model is subtracted from the GRACE monthly Stokes coefficient, it is possible to see month-to-month changes with several microgals accuracy (1 Gal $=$ 0.01 m/s^2) and spatial resolution of $\sim$300 km. Monthly L2 files are accessible from GFZ, CSR, and JPL archives, but one needs to mitigate, for example, removing or filtering, the meridional correlated high-frequency noise patterns, called stripes.

The source of these noises is the same polar orbit of the satellites, imperfect observability of the gravity signals, etc. Scientific teams are developing various methods of stripe filtering, discussed below.

GRACE data can be used in geophysical, glaciological, oceanographic, and hydrological studies. Earth rotation research, geodynamics, and climate change studies also benefit from this mission. A wide range of GRACE science and applications have been extensively published in journals or presented in international assemblies.

For the largest country in the world, Russia, not fully covered by meteorological and hydrological observational networks, GRACE data are especially useful. By spring 2013 it became clear that Russia has gone through a very snowy winter. Near Moscow the thickness of snow exceeded up to twice the nominal value and was more than 70 cm. The questions aroused: can it be detected by GRACE?

In this work the study of regional changes of the gravity field from GRACE over Russia will be presented. A novel data processing technique – multichannel singular spectrum analysis (MSSA) – will be applied for the GRACE data filtering toward improved separation of various signal components, related to hydrological, seasonal, and climatological changes. Initial data and method will be presented in the next section, and results of processing will be presented in Sect. 3.3, followed by discussions and conclusion.

3.2 Data Processing

3.2.1 Initial Data Preparation

We used JPL Level-2 RL05 GRACE monthly geopotential field data from 01.2003 through 06.2013 with the set of Stokes coefficients complete to degree 60. Release 5 (RL05) of the L2 data product is more accurate than the previous version (RL04) primarily because of GRACE data and model improvement. Six months of data (06.03, 01.11, 06.11, 05.12, 10.12, 03.13) were linearly interpolated (overall, $N = 126$ files were used). Absence of some of these monthly solutions in the recent years is caused by the difficulties in battery power maintenance on board.

The spherical harmonic decomposition of the gravity field is given by

$$V(\varphi, \lambda, r) = \frac{GM}{r} \sum_{n=2}^{\infty} \sum_{m=0}^{n} \left(\frac{a}{r}\right)^n (C_{nm} \cos m\lambda + S_{nm} \sin m\lambda) P_n^m(\sin \varphi), \quad (3.1)$$

here C_{nm}, S_{nm} are normalized Stokes coefficients, representing the exterior geopotential spherical harmonic expansion, n is the degree, m is the order of the spherical harmonics, P_n^m are the fully normalized associated Legendre polynomials, a is the mean equatorial radius of the Earth, and the arguments φ, λ, r are latitude, longitude, and radius, respectively (Panteleev 2000).

The coefficients of zero and first degree are set to zero due to the choice of the coordinate system. GRACE is insensitive to degree-one coefficients (geocenter). Estimates of C_{20} (oblateness) coefficients from GRACE are not very reliable; they were replaced by SLR-derived solutions. Since we are interested in the monthly changes, the averaged field over 10 years was subtracted from the GRACE monthly Stokes coefficients. GIA[1] correction was also applied according to Paulson et al. (2007) model. Finally, the results were converted to the surface mass changes in terms of equivalent water height (EWH) level (cm), according to Wahr et al. (1998)

$$\Delta h(\varphi, \lambda, t) = \frac{2\pi a p_{ave}}{3 p_w} \sum_{n=2}^{60} \sum_{m=0}^{n} \frac{2n+1}{1+k_n} W_n(\Delta C_{nm}(t) \cos m\lambda$$

$$+ \Delta S_{nm}(t) \sin m\lambda) P_n^m(\sin \varphi), \tag{3.2}$$

here $\Delta C_{nm}(t)$, $\Delta S_{nm}(t)$ are normalized Stokes coefficient differences with respect to the mean (model); p_{ave} and p_w are average densities of the Earth and seawater, respectively; k_n is the load Love number of degree n; and W_n is a filter coefficient. All spectral filter coefficients in Eq. (3.2) were set to one, so we did not apply any filtering except MSSA.

3.2.2 Multichannel Singular Spectrum Analysis Method

Filtering of GRACE data is needed, because they contain meridional error – stripes. Such errors are primarily due to the fact that GRACE satellites are in the same polar orbit and satellite-to-satellite tracking only observes along-track direction, causing the gravity field inversion problem to be near singular.

The orbital and instrument errors are correlated in the resonant orders of the spherical harmonic Stokes coefficients, causing the so-called stripes high-spatial-frequency errors. An unfiltered map, obtained as the difference between January 2003 and January 2013 initial data, is shown in Fig. 3.1.

Different authors use a variety of filtering methods to minimize stripes and reduce noises in the GRACE monthly gravity field solutions. Among them are Gaussian filtering with symmetric and asymmetric Gaussian function (Han et al. 2005), Wiener (Klees et al. 2007) and regularizing (Kusche et al. 2009) filters, whose coefficients depend on degree and order, and de-striping/smoothing (Duan et al. 2009; Guo et al. 2010; Swenson and Wahr 2006) filters, operating to remove the anomalously large values from the resonant orders of the Stokes coefficients.

---

[1]Glacial isostatic adjustment (GIA), related to the restoration of isostatic equilibrium, is especially large in Canada and Scandinavia due to the release of massive ice sheets located there 20,000 years ago.

3 Gravity Changes over Russian River Basins from GRACE

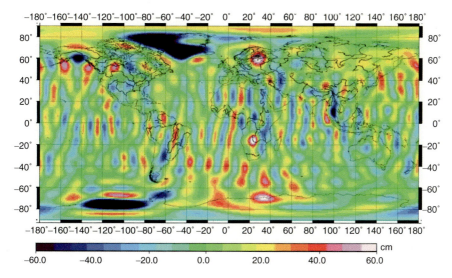

Fig. 3.1 Unfiltered GRACE EWH difference between 01.2003 and 01.2013 (2013–2003). Vertical stripes distort the signal

Filtering methods based on principal component analysis (PCA), empirical orthogonal functions (EOF), singular spectrum analysis (SSA), and independent component analysis (ICA) were also proposed. PCA by the name of EOF analysis was applied to the GRACE data in works by Rangelova et al. (2007), Schrama et al. (2007), and Wouters and Schrama (2007). In Rangelova et al. (2007), SSA was also tested. The rotation of PCA components to increase their meaningfulness was recommended in Rangelova and Sideris (2008). In Han et al. (2005), non-isotropic filtering was used. It is a kind of nonlinear modification of EOF analysis for nonstationary time series, where PCs are obtained by means of time series envelope calculation and orthogonalization. Good review of EOF-based methods of GRACE data filtering can be found in Boergens et al. (2014). All these methods are quite close to MSSA, but we find that MSSA is more flexible, despite its mathematical complexity, and is thus preferred in this study. For the first time, we applied MSSA for the filtering of GRACE observations in Zotov and Shum (2009). In Rangelova et al. (2010), MSSA was also applied to regional GRACE data, but the length of time series was yet too short to choose parameter L appropriately. Here we will demonstrate the abilities of MSSA on GRACE data of 11-year extent. Multichannel singular spectrum analysis, also called extended EOF, is a generalization of singular spectrum analysis (SSA) for the multidimensional (multichannel) time series (Ghil et al. 2002; Jollife 2001). SSA, in its turn, is based on PCA, generalized for the time series in such way that instead of the simple correlation matrix, the trajectory matrix is analyzed. It is obtained through the time series embedding into the L-dimensional space. Parameter L is called lag or "caterpillar" length. When $L = 1$, SSA becomes PCA (trajectory matrix becomes non-lagged signal covariance matrix). SSA algorithm has four stages: (a) formation of trajectory matrix, (b) its singular value decomposition (SVD), (c) singular numbers grouping, and (d) principal components recovering through Hankelization. SSA algorithm is described in details in Golyandina et al. (2001) and Golyandina (2004). MSSA includes similar iterations.

Firstly (a), we select lag parameter L. For every Stokes coefficient, we have time series of length N. The trajectory matrix can be built for one time series component (channel); let's say for the Stokes coefficient $C_{ij}(t_k)$, $k = 0, \ldots, N - 1$, as follows:

$$\mathbf{X}_{C_{ij}} = \begin{pmatrix} \Delta C_{ij}(t_0) & \Delta C_{ij}(t_1) & \ldots & \Delta C_{ij}(t_{K-1}) \\ \Delta C_{ij}(t_1) & \Delta C_{ij}(t_2) & \ldots & \Delta C_{ij}(t_K) \\ \vdots & \vdots & \ddots & \vdots \\ \Delta C_{ij}(t_{L-1}) & \Delta C_{ij}(t_L) & \ldots & \Delta C_{ij}(t_{N-1}) \end{pmatrix}, \tag{3.3}$$

here $K = N - L + 1$.

The trajectory matrices $\mathbf{X}_{C_{ij}}, \mathbf{X}_{S_{ij}}$ for every Stokes coefficient C_{ij} and S_{ij} should be incorporated into large block matrix $\mathbf{X}$ as follows:

$$\mathbf{X} = [\mathbf{X}_{C_{2,0}}, \mathbf{X}_{S_{2,0}} \ldots, , \mathbf{X}_{C_{ij}}, \mathbf{X}_{S_{ij}}, \ldots, \mathbf{X}_{C_{60,60}} \mathbf{X}_{S_{60,60}}]^T. \tag{3.4}$$

Thus, in our realization, we put blocks one behind another. This multichannel trajectory matrix, composed of blocks for every channel, can be used to calculate the lagged covariance matrix $\mathbf{A} = \mathbf{X}^T\mathbf{X}$.

At the second stage (b), SVD should be applied to $\mathbf{X}$.

$$\mathbf{X} = \mathbf{U}\mathbf{S}\mathbf{V}^T.$$

As a result, a sequence of singular numbers (SNs) s_i standing along the diagonal of matrix $\mathbf{S}$ in order of decreasing values (see Fig. 3.2) and the corresponding eigenvectors $\mathbf{v}_i$ (left) and $\mathbf{u}_i$ (right) are obtained. If to solve the eigenvalue problem for $\mathbf{A} = \mathbf{V}\mathbf{S}^T\mathbf{S}\mathbf{V}^T$, then the eigenvalues will be squared singular numbers $\lambda_i = s_i^2$ and left eigenvectors $\mathbf{v}_i$, included as columns in $\mathbf{V}$, form empirical orthogonal functions (EOFs).

The ith component corresponds to the matrix

$$\mathbf{X}^i = s_i \mathbf{u}_i \mathbf{v}_i^T.$$

In MSSA we reconstruct the vectorial PCs from this matrix, knowing its structure, similar to the structure of $\mathbf{X}$. It is done through Hankelization (d), allowing to reconstruct every channel of ith component from the corresponding blocks of matrix $\mathbf{X}^i$, organized as in (3.4). Suppose we need to reconstruct the C_{lm} channel. Then each kth count can be obtained from the averaging along the side diagonals of the corresponding matrix block $\mathbf{Y} = \mathbf{X}^i_{C_{lm}}$. The first and the last L elements are

3 Gravity Changes over Russian River Basins from GRACE

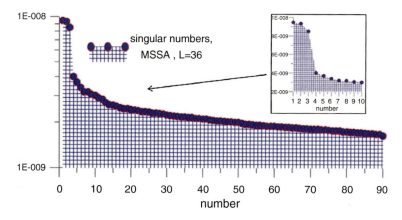

Fig. 3.2 Distribution of singular numbers, defining MSSA components energy

calculated from the fewer number of **Y** values, so the first and the last points of PCs will be less consistent. It is supposed that elements on the side diagonals of matrix **Y** are approximately equal and it is almost Hankel. In case when it doesn't hold strictly, some kind of edge effect appears.

Grouping (c) of components is required, when some of SNs are related to one and the same PC and have similar behavior that could be detected with the use of ω correlations and other techniques (Golyandina et al. 2001). Then, SNs (s_i) should be grouped together and reconstructed as one PC before stage (d). It can be done after Hankelization (d) by simple item-by-item summation of components. Details on grouping and theorems about separability of components can be found in (Golyandina 2004).

As a result, the set of PCs with decreasing amplitudes representing different modes of time series variability are obtained.

The main parameter of the algorithm – the time lag L, which determines the dimensionality of the time series embedding space, should be chosen heuristically, using recommendations given in Golyandina (2004). It should not be larger than $N/2$, and it is better to choose it as a multiplier of periods, expected in the time series. In earlier works, we used $L = 24$ (Zotov and Shum 2009; Zotov 2012). But with extend of period of observations, we have chosen $L = 36$ months (3 years) that allows to better separate the components. No other filters like Gaussian smoothing were used, though it is possible; see Eq. (3.2).

In Zotov and Shum (2009) and Zotov (2012), we found MSSA more flexible than simple EOF for recognition of trend, modulated oscillations of different periods, and denoising of multidimensional time series. Different channels "help" each other to capture spatiotemporal correlation patterns. Lagged matrix **X** allows to find them in L-dimensional space. The obtained PCs extract correlations, which present in all the channels simultaneously.

3.3 Results of Processing

We applied MSSA in the spectral domain to the Stokes coefficients. The distribution of singular numbers is represented in Fig. 3.2. SNs were grouped into several PCs which were converted into spatial maps of EWH. The first two largest SNs were grouped into PC 1 capturing annual cycle, the next two SNs into PC 2, representing trend (slow changes). The sum of MSSA SNs 1–10 represents the largest part of signal variability (energy). Higher-order PCs (SN > 10) contain high-frequency components, such as noises related to the stripes and some part of the signal from transient events, such as coseismic deformation after earthquakes. Detailed analysis of MSSA PCs for global maps was given in Zotov and Shum (2009).

Simulated Topological Networks (STN-30p, http://www.wsag.unh.edu/Stn-30/stn-30.html) database was used to constrain the region of study to the basins of 15 large Russian rivers (Fig. 3.3, left). Table 3.1 contains information about these basins. The map of the sum of SNs 1–10 for the last month (06.2013) of the data span in the constrained area is represented in Fig. 3.3, right. This map includes contribution from annual PC 1, long periodic PC 2, and other components except the stripes, which are mostly removed (they go to SNs > 10). The animated maps of all the obtained PCs are accessible on the website http://lnfm1.sai.msu.ru/~tempus/GRACE/index.htm.

The signal was averaged over the territory constrained by the basins of 15 large Russian rivers. Results are shown in Fig. 3.4. On top the black curve represents the mean sum of SNs 1–10. The purple curve represents the initial data (sum of all PCs) before MSSA. It is seen that SNs 1–10 sum includes almost all the variability of the initial data. The trend (PC 2) is shown in blue. It has a maximum in 2009, then decreases. This trend is defined mostly by Siberian river basins (Fig. 3.6). The red curve depicts the prediction made in February 2013 by neural network (NN), containing nine neurons in three layers (Zotov 2005). Prediction was made when the data for spring months were not yet available. Later, when they were obtained, we found out that the prediction was inappropriate (NN was too simple). The observed

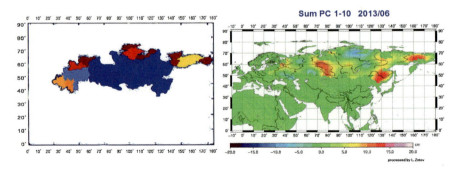

Fig. 3.3 Drainage basins of 15 Russian rivers and the sum of SNs 1–10 over these basins for 06.2013

3 Gravity Changes over Russian River Basins from GRACE 53

Table 3.1 Information about 15 Russian river basins used in this study according to STN-30p

Basin name	Basin length, km	Basin area, km²	Basin name	Basin length, km	Basin area, km²
Rivers of Arctic basin (Asia)			Rivers of Arctic basin (European part)		
Ob	4,257.41	3,025,923.25	Dvina	1,414.96	360,944.03
Yenisei	4,898.93	2,578,730.25	Pechora	1,491.99	314,291.81
Lena	4,365.71	2,441,815.75	River of Pacific Ocean basin (Far East)		
Kolyma	1,971.77	665,648.06	Amur	3,644.61	1,754,681.0
Khatanga	1,370.25	370,352.91	Rivers of European part		
Indigirka	1,451.13	334,126.38	Volga	2,785.36	1,476,411.38
Anadyr	1,011.96	225,847.92	Dnepr	1,543.60	508,839.19
Yana	997.73	224,992.69	Don	1,400.51	423,038.44
Olenek	1,644.33	223,189.27			

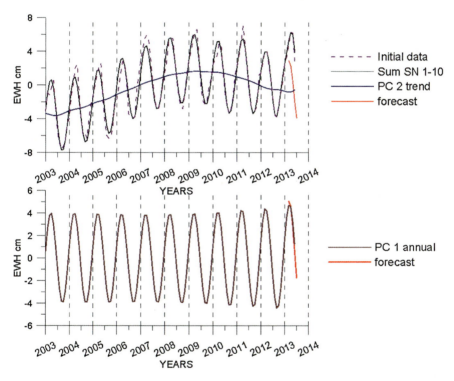

Fig. 3.4 Average mass changes in the basins of 15 large Russian rivers for the sum of SNs 1–10, trend PC 2 (*top*), and annual PC 1 (*bottom*), together with a forecast by neural network

level of mass anomaly sufficiently surpasses the prediction. Unprecedented maxima can be seen in April 2013 that can be attributed to huge snow accumulation over Russia.

Figure 3.4, bottom, depicts the average plot only for the annual oscillation captured by PC 1. Prediction for this annual term worked better. Annual term

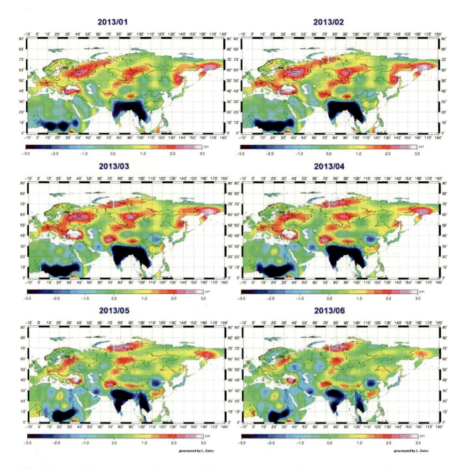

Fig. 3.5 The differences for the annual PC 1 between monthly (January–June 2013) maps and previous 10-year averages for the selected months

demonstrates the tendency of amplitude increase since 2009. Some part of spring maxima signal was also assigned to PC 2, probably, as a result of the edge effect.

To see the anomalous snow accumulation, we calculated the differences for the annual PC 1 between monthly (January–June 2013) maps and average maps over 10 previous years (2003–2012) for the corresponding months (monthly anomalies). The maps are shown in Fig. 3.5 for 6 months. All Eurasia is shown; however, this study focuses on Russia. It is seen that snow accumulation was very large over the European part of Russia, West Siberia, Chukotka, and the Far East with respect to the average for the selected months. Positive anomalies start to grow in January, reach maximum in April, and disappear in June.

Weather conditions in the basins of Russian rivers in winter and spring are quite cold, temperatures are below zero, and most of the rivers are frozen. Groundwater mass is not changing during winter. We thus concluded that the increase of mass

3 Gravity Changes over Russian River Basins from GRACE

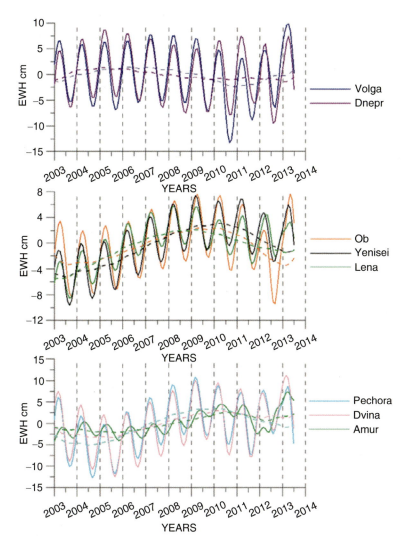

Fig. 3.6 Average mass changes captured by the sum of SNs 1–10 and trends (PC 2) for particular river basins

observed by GRACE can be related mostly to the snow accumulation. According to (Report on Climate 2013), the amount of water stored in snow in winter 2012–2013 was the largest in Russia since 1967. When snow melted in April–June, it caused floods. Fortunately, the ground still had capacity to absorb water. That helped to avoid extreme floods and state of emergency. Still, the occurrence of flood on many rivers was less than 2 % (event happens once in 50 years); their water levels increased in spring by several meters, which was detected by the river gauges.

We calculated the average plots for particular rivers of Siberia, European part, Russian North, and the Far East (Fig. 3.6). Our MSSA curves were found to be consistent with CNES RL03-v1 solutions for particular river basins (www.thegraceplotter.com). Curves for all the rivers confirm the height maxima in spring 2013. Different amplitudes of seasonal cycle and different trend behaviors for European and Siberian rivers can be seen. Trends for European rivers (Fig. 3.6, top) are mainly decreasing, while for Siberian rivers (Fig. 3.6, middle), they demonstrate maxima in 2009 like the overall trend (PC 2) in Fig. 3.3, top. Since the Siberian river basins are very large (Table 3.1), they dominate the mass changes within Russia. These river discharges are an important driver for Arctic climate change. If to multiply the basin area from Table 3.1 by average mass change, the total water storage anomaly can be calculated for each river.

The curve for Amur river in the Far East (Fig. 3.6, bottom) is quite different from others. The amplitude of its annual cycle is small, but since 2012, observed mass in the basin of Amur is quickly increasing. In August 2013, huge flood occurred there, caused by heavy precipitation. Recent studies (Reager et al. 2014) have shown that GRACE data incorporation sufficiently improves flood forecast.

Minimum in summer 2010 for Volga river (Fig. 3.6, top) is a footprint of the heat wave that occurred in the European part of Russia, accompanied by fires and aerosol pollution, producing smoke and causing great difficulties for the habitants of Moscow (Barriopedro et al. 2010).

Let us look at the map of a climatologically driven trend, captured by PC 2, in Fig. 3.7. The map of difference for PC 2 between 2003 and 2013 years showing the changes of gravity that occurred since 2003 is much cleaner than the unfiltered map in Fig. 3.1. Himalayas glaciers melting, pattern of Sumatra earthquake coseismic deformation, and changes in China, India, and Africa seen on the map are out of

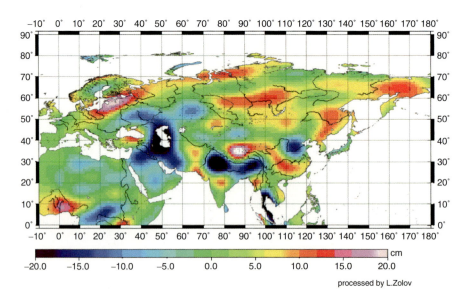

Fig. 3.7 Difference between 2003 and 2013 years (2013–2003) for the trend component (PC 2)

the scope of this study. Mass increase has been observed in some Siberian regions, such as sources of Lena and Irtysh, which may be related to the degradation of permafrost (Frappart et al. 2010; Landerer et al. 2010). Potentially as a result of global warming, melted ground ice is replaced by water, which increases the density, mass, and, consequently, gravity field. Negative anomaly over the Caspian Sea can be related to its level decrease, as reported in Zonn et al. (2010).

3.4 Discussion

For large territories like Russia, satellite gravity field data represent an important source of hydrologic information. In this study monthly gravity field solutions from GRACE in terms of mass (EWH) changes were processed by MSSA and averaged over the 15 largest Russian river basins. Annual component (PC 1) shows amplitude increase since 2009 (Figs. 3.4 and 3.6). Unprecedented maximum in spring 2013 is caused by the huge snow accumulation over the territory of Russia (Figs. 3.5 and 3.6). Trend component shows increase since 2003, maximum in 2009, followed by the decrease. This behavior is mostly defined by Siberian river basins. Map for the trend (Fig. 3.7) shows mass anomalies increase in Siberia and decrease over the Caspian Sea.

We cannot answer the question why precipitation increased in winter-spring 2013 in Russia, causing unprecedented snow accumulation. It could be related to the anomalies of atmospheric mass transfer from Atlantic, Gulf Stream circulation, El Niño/La Niña, or conditions in the Arctic. There are some evidences in favor of global warming pause (hiatus). Such questions could be answered only after extended interdisciplinary research, involving climatology, meteorology, and other sciences. Some issues and projections for precipitation and anomalous weather conditions into the future could be found in IPCC Fifth Assessment Report (2013) or the report of Hydrometeorological Center of Russia (Report on Climate 2013).

In this work we did not plan to find explanation of the causes of meteorological and hydrological changes over Russia. Our goal was to present observational data, its processing technique, and show its usefulness for hydrological and climatological studies, exploration of our planet.

We used MSSA to filter data and separate meaningful components, founding it a promising method for GRACE data processing. MSSA method has greater flexibility than simple EOF and could be useful in the analysis of other satellite observations, such as altimetry, water vapor, and precipitation data (Zotov 2012).

Exact physical interpretation of the obtained signals requires comparison to hydrological models (GLDAS, WGHM) and ground-based observations. The remaining questions are (i) what is the useful part of the signal in PC 1 and PC 2, (ii) how to reduce boundary effects for the first and the last points of PCs, and (iii) how to better separate secular change from annual and, probably, other periodic signals? As for the last question, fortunately, we already have 11 years of GRACE data, and quite a good separation can be achieved by the appropriate choice of L.

Gravity field can also reveal or constrain the planet's internal structure. In the recent decades, gravity remote sensing has given impressing results for the Earth (GRACE, GOCE) and for the Moon (Gravity Recovery and Interior Laboratory (GRAIL)). Space missions to Mercury, Venus, Saturn, Jupiter, Enceladus, Titan, and Pluto would benefit from dedicated gravimetry or gradiometry sensors in addition to other geodetic instruments to constrain internal structures of terrestrial planets.

Acknowledgements This work is partially sponsored by RFBR grants N 12-02-31184, 15-05-02340, and N 13-05-00113. Paris observatory 2-month position was allocated for the first author. The Ohio State University (OSU) component of the research was partially supported by grants by NSF/IGFA Belmont Forum Project (Grant No. ICER-1342644) and by the Chinese Academy of Sciences/SAFEA International Partnership Program for Creative Research Teams (Grant No. KZZD-EW-TZ-05).

References

IPCC Fifth Assessment Report (2013) Climate Change 2013: The Physical Science Basis. http://www.climatechange2013.org/
Barriopedro D, Fischer EM, Luterbacher J, Trigo RM, Garcia-Herrera R (2011) The hot summer of 2010: redrawing the temperature record map of Europe. Science. doi:10.1126/science.1201224
Bettadpur S (2007) Level-2 gravity field product user handbook. ftp://podaac.jpl.nasa.gov/pub/grace/doc/L2-UserHandbook_v2.3.pdf
Boergens E, Rangelova E, Sideris MG, Kusche J (2014) Assessment of the capabilities of the temporal and spatiotemporal ICA method for geophysical signal separation in GRACE data. JGR 119:4429–4447. doi:10.1002/2013JB010452
Case K, Kruizinga G, Sien-Chong Wu (2004) GRACE level 1B data product user handbook. ftp://podaac.jpl.nasa.gov/pub/grace/doc/Handbook_1B_v1.2.pdf
Duan X, Guo J, Shum C, van der Wal W (2009) Towards an optimal scheme for removing correlated errors in GRACE data. J Geodesy 83:1095–1106. doi:10.1007/s00190-009-0327-0
Guo J, Duan X, Shum C (2010) Non-isotropic filtering and leakage reduction for determining mass changes over land and ocean using GRACE data. Geophys J Int 181:290–302. doi:10.1111/j.1365-246X.2010.04534.x
Frappart F, Papa F, Guntner A, Ramillien G, Prigent C, Rossow W, Bonnet M (2010) Interannual variations of the terrestrial water storage in the Lower Ob' Basin from a multisatellite approach. Hydrol Earth Syst Sci 14:2443–2453
Ghil M, Allen RM, Dettinger MD et al (2002) Advanced spectral methods for climatic time series. Rev Geophys 40(1):3.1–3.41
Golyandina N,Nekrutkin V, Zhigljavskyet A (2001) Analysis of time series structure: SSA and related techniques. Chapman & Hall/CRC, New York/London
Golyandina N (2004) Method "Caterpillar-SSA": analysis of the time series. (In Russian) SPB., BBM
Han S-C, Shum CK, Jekeli Ch et al (2005) Non-isotropic filtering of GRACE temporal gravity for geophysical signal enhancement. Geophys J Int 163(1):18–25
Jollife IT (2001) Principal component analysis. Springer, New York
Kenyon S et al (2007) Toward the next Earth gravitational model. In: SEG annual meeting, San-Antonio. http://library.seg.org/doi/abs/10.1190/1.2792518
Klees R, Revtova E, Gunter B et al (2008) The design of an optimal filter for monthly GRACE gravity models. Geophys J Int 175(5768):417–432

3 Gravity Changes over Russian River Basins from GRACE

Kusche J, Schmidt R, Petrovic S et al (2009) Decorrelated GRACE time-variable gravity solutions by GFZ and their validation using a hydrological model. J Geodesy 83:903–913

Landerer F, Dickey J, Zlotnicki V (2010) Terrestrial water budget of the Eurasian pan-Arctic from GRACE satellite measurements during 2003–2009. J Geophys Res Atmos, D 23115. doi:10.1029/2010JD014584

Panteleev VL (2000) Theory of figure of the Earth. Lectures in Russian. http://lnfm1.sai.msu.ru/grav/russian/lecture/tfe/index.html

Paulson A, Zhong S, Wahr J (2007) Inference of mantle viscosity from GRACE and relative sea level data. Geophys J Int 171:497

Report on the Climate Peculiarities on the Territory of Russian Federation (2013). Hydrometeorological Center of Russia. http://www.meteoinfo.ru/climate

Rangelova E, Wal W, Braun A et al (2007) Analysis of GRACE time-variable mass redistribution signals over North America by means of principal components analysis. J Geophys Res 112, F03002

Rangelova E, Sideris M (2008) Contributions of terrestrial and GRACE data to the study of the secular geoid changes in North America. J Geodyn 46:131–143

Rangelova E et al (2010) Spatiotemporal analysis of the GRACE-derived mass variations in North America by means of multi-channel singular spectrum analysis. In: Mertikas SP (ed) Gravity, geoid and earth observation. IAG Symposia 135, Springer, Berlin/Heidelberg

Reager JT, Thomas BF, Famiglietti JS (2014) River basin flood potential inferred using GRACE gravity observations at several months lead time. Nat Geosci. doi:10.1038/ngeo2203

Sagitov MU (1979) Lunar gravimetry. Nauka, Moscow

Schrama E, Wouters B, Lavallee D (2007) Signal and noise in gravity recovery and climate experiment (GRACE) observed surface mass variations. J Geophys Res 112, B08407. doi:10.1029/2006JB004882

Swenson S, Wahr J (2006) Post-processing removal of correlated errors in GRACE data. Geophys Res Lett 33, L08402. doi:10.1029/2005GL025285

Tikhonov AN, Leonov AS, Yagola, AG (1998) Nonlinear ill-posed problems. Chapman and Hall, London/New York

Wahr J, Molenaar M, Bryan F (1998) Time variability of the earth's gravity field: hydrological and oceanic effects and their possible detection using grace. J Geophys Res Solid Earth 103(B12):30,205–30,229. doi:10.1029/98JB02844

Wang YF, Yagola AG, Yang CC (2012) Computational methods for applied inverse problems, De Gruyter & Higher, Berlin/Education Press, Beijing

Wouters B, Schrama E (2007) Improved accuracy of GRACE gravity solution through empirical orthogonal function filtering of spherical harmonics. Geophys Res Lett 34, L23711. doi:10.1029/2007GL032098

Zonn IS, Glantz MH, Kostianoy AG, Kosarev AN (2010) The Caspian Sea encyclopedia. Springer, Berlin/London

Zotov LV (2005) Regression methods of Earth rotation prediction. Mosc Univ Phys Bull 5, 64–68 (in Russian)

Zotov L, Shum CK (2009) Singular spectrum analysis of GRACE observations. In: AIP proceedings of the 9th Gamow summer school, Odessa

Zotov L (2012) Application of multichannel singular spectrum analysis to geophysical fields and astronomical images. Adv Astron Space Phys 2:82–84

Chapter 4
Gravimetric Forward and Inverse Modeling Methods of the Crustal Density Structures and the Crust-Mantle Interface

Robert Tenzer and Wenjin Chen

Abstract The numerical models and results of the gravimetric interpretation of the crustal density structures and the Moho geometry are presented. The numerical scheme applied utilizes the gravimetric forward and inverse modeling derived in a frequency domain. Methods for a spectral analysis and synthesis of the gravity and crustal structure models are applied in the gravimetric forward modeling of the gravity field generated by the major known crustal density structures. The gravimetric inversion scheme is formulated by means of a linearized Fredholm integral equation of the first kind. In numerical results we show the gravitational contributions of crustal density structures and the refined gravity field quantities, which have a minimum as well as maximum correlation with the Moho geometry. The resulting gravimetric Moho model is finally presented.

Keywords Crust • Density • Gravity • Isostasy • Moho

4.1 Introduction

Seismic data are primarily used in geophysical studies investigating the Moho geometry. In the absence or a low coverage of seismic data, gravimetric or combined gravimetric-seismic methods can be applied. Several different gravimetric methods of the Moho depth determination have been developed and applied in global and regional studies. For more details we refer readers to articles, for instance, by Čadek and Martinec (1991), Braitenberg and Zadro (1999), Arabelos et al. (2007), Sjöberg (2009), Braitenberg et al. (2010), Sampietro (2011), Eshagh et al. (2011), Sampietro et al. (2013), Bagherbandi (2012), Bagherbandi et al. (2013), Bagherbandi and Tenzer (2013), and Tenzer et al. (2013). The gravimetric methods are formulated for a chosen isostatic scheme. The Pratt-Hayford isostatic model is

---

R. Tenzer (✉) • W. Chen
The Key Laboratory of Geospace Environment and Geodesy, School of Geodesy and Geomatics, Chinese Ministry of Education, Wuhan University, 129 Luoyu Road, Wuhan, China
e-mail: rtenzer@sgg.whu.edu.cn

© Springer-Verlag Berlin Heidelberg 2015
S. Jin et al. (eds.), *Planetary Exploration and Science: Recent Results and Advances*,
Springer Geophysics, DOI 10.1007/978-3-662-45052-9_4

based on the assumption of a variable compensation density (Pratt 1855; Hayford 1909; Hayford and Bowie 1912), while a variable compensation depth is considered in the Airy-Heiskanen isostatic model (Airy 1855; Heiskanen and Vening Meinesz 1958). Vening Meinesz (1931) modified the Airy-Heiskanen theory by introducing a regional isostatic compensation based on a thin plate lithospheric flexure model (cf. Watts 2001). The regional compensation model was later adopted in the Parker-Oldenburg isostatic method (Oldenburg 1974). Moritz (1990) utilized the Vening Meinesz inverse problem of isostasy for the Moho depth estimation. Sjöberg (2009) reformulated Moritz's problem, called herein the Vening Meinesz-Moritz (VMM) problem of isostasy, as that of solving a nonlinear Fredholm integral equation of the first kind. Sjöberg and Bagherbandi (2011) developed and applied a least-squares approach, which combined seismic and gravity data in the VMM isostatic inverse scheme for a simultaneous estimation of the Moho depth and density contrast. They also presented and applied the non-isostatic correction to model for discrepancies between isostatic and seismic models (cf. Bagherbandi and Sjöberg 2012).

In this study, we present a numerical scheme of the Moho depth determination using gravity and crustal density structure models. This numerical scheme utilizes expressions for the gravimetric forward and inverse modeling in a frequency domain. The functional model for the gravimetric inverse problem (in terms of a nonlinear Fredholm integral equation of the first kind) is established based on the assumption that the refined gravity disturbances have a maximum correlation with the Moho geometry. This maximum correlation can be attained by applying the topographic and crust components stripping gravity corrections to gravity disturbances, yielding the consolidated crust-stripped gravity disturbances. However, these gravity data still comprise a long-wavelength signal from unmodeled mantle heterogeneities and uncertainties of a crustal structure model. A linearization of the Fredholm integral equation of the first kind is applied, which incorporates the isostatic compensation scheme based on minimizing the correlation between the isostatic gravity and (a priori) Moho depth data. The resulting complete crust-stripped isostatic gravity disturbances are then used to solve the gravimetric inverse problem for finding the Moho depths. These input gravity data are computed by applying the gravimetric forward modeling.

4.2 Gravimetric Forward Modeling

The consolidated crust-stripped gravity disturbances δg^{cs} are obtained from the corresponding gravity disturbances δg after applying the topographic and crust density contrasts stripping gravity corrections. The computation is realized according to the following scheme (Tenzer et al. 2012a):

$$\delta g^{cs} = \delta g - g^t - g^b - g^i - g^s - g^c, \tag{4.1}$$

4 Gravimetric Forward and Inverse Modeling Methods of the Crustal Density... 63

where g^t, g^b, g^i, g^s, and g^c are, respectively, the gravitational attractions generated by the topography and density contrasts of the ocean (bathymetry), ice, sediments, and remaining anomalous density structures within the consolidated (crystalline) crust.

The gravity disturbance δg at a point (r, Ω) is computed according to the following expression (e.g., Heiskanen and Moritz 1967):

$$\delta g\,(r, \Omega) = \frac{GM}{R^2} \sum_{n=0}^{\bar{n}} \sum_{m=-n}^{n} \left(\frac{R}{r}\right)^{n+2} (n+1)\, T_{n,m} Y_{n,m}\,(\Omega), \tag{4.2}$$

where $GM = 3986005 \times 10^8\,\mathrm{m^3 s^{-2}}$ is the geocentric gravitational constant, $R = 6{,}371 \times 10^3\,\mathrm{m}$ is the Earth's mean radius, $Y_{n,m}$ are the (fully normalized) surface spherical harmonic functions of degree n and order m, $T_{n,m}$ are the (fully normalized) numerical coefficients which describe the disturbing gravity potential T (i.e., the difference between the Earth's and normal gravity potentials), and $\bar{n}$ is the maximum degree of spherical harmonics. The coefficients $T_{n,m}$ are obtained from the coefficients of a global geopotential model (describing the Earth's gravity field) after subtracting the spherical harmonic coefficients of the normal gravity field. The 3-D position is defined in the spherical coordinate system (r, Ω), where r is the spherical radius and $\Omega = (\varphi, \lambda)$ is the spherical direction with the spherical latitude φ and longitude λ. The full spatial angle is denoted as $\Phi = \{\Omega' = (\varphi', \lambda') : \varphi' \in [-\pi/2, \pi/2] \wedge \lambda' \in [0, 2\pi)\}$. For gravity points situated at the Earth's surface, the geocentric radius r is calculated as $r \cong R + H$, where H is the topographic height.

Tenzer et al. (2012a) developed and applied a uniform mathematical formalism for computing the topographic and stripping gravity corrections of the Earth's inner density structures. This numerical scheme utilizes the expression for the gravitational attraction g (defined as a negative radial derivative of the respective gravitational potential V; i.e., $g = -\partial V/\partial r$) generated by an arbitrary volumetric mass layer with a variable depth and thickness while having laterally distributed vertical mass density variations. The gravity correction g at a point (r, Ω) is computed using the following expression:

$$g\,(r, \Omega) = \frac{GM}{R^2} \sum_{n=0}^{\bar{n}} \sum_{m=-n}^{n} \left(\frac{R}{r}\right)^{n+2} (n+1)\, V_{n,m} Y_{n,m}\,(\Omega), \tag{4.3}$$

where the potential coefficients $V_{n,m}$ read

$$V_{n,m} = \frac{3}{2n+1} \frac{1}{\rho^{\mathrm{Earth}}} \sum_{i=0}^{I} \left(\mathrm{Fl}_{n,m}^{(i)} - \mathrm{Fu}_{n,m}^{(i)}\right). \tag{4.4}$$

The numerical coefficients $\{Fl^{(i)}_{n,m}, Fu^{(i)}_{n,m} : i = 0, 1, \ldots, I\}$ in Eq. (4.4) are given by

$$Fl^{(i)}_{n,m} = \sum_{k=0}^{n+2} \binom{n+2}{k} \frac{(-1)^k}{k+1+i} \frac{L^{(k+1+i)}_{n,m}}{R^{k+1}}, \qquad (4.5)$$

and

$$Fu^{(i)}_{n,m} = \sum_{k=0}^{n+2} \binom{n+2}{k} \frac{(-1)^k}{k+1+i} \frac{U^{(k+1+i)}_{n,m}}{R^{k+1}}. \qquad (4.6)$$

The terms $\sum_{m=-n}^{n} L_{n,m} Y_{n,m}$ and $\sum_{m=-n}^{n} U_{n,m} Y_{n,m}$ in Eqs. (4.5) and (4.6) define the spherical lower-bound and upper-bound laterally distributed radial density variation functions L_n and U_n of degree n, respectively. These spherical functions and their higher-order terms $\{L^{(k+1+i)}_n, U^{(k+1+i)}_n : \quad k = 0, 1, \ldots; \quad i = 1, 2, \ldots, I\}$ are defined by

$$L^{(k+1+i)}_n (\Omega) = \begin{cases} \dfrac{4\pi}{2n+1} \iint\limits_{\Phi} \rho\left(D_U, \Omega'\right) D^{k+1}_L\left(\Omega'\right) P_n(t)d\Omega' \\[2mm] \qquad = \displaystyle\sum_{m=-n}^{n} L^{(k+1)}_{n,m} Y_{n,m}\left(\Omega\right) \quad i = 0 \\[4mm] \dfrac{4\pi}{2n+1} \iint\limits_{\Phi} \beta\left(\Omega'\right) a_i\left(\Omega'\right) D^{k+1+i}_L\left(\Omega'\right) P_n(t)d\Omega' \\[2mm] \qquad = \displaystyle\sum_{m=-n}^{n} L^{(k+1+i)}_{n,m} Y_{n,m}\left(\Omega\right) \quad i = 1,\, 2, \ldots I \end{cases} \qquad (4.7)$$

and

$$U^{(k+1+i)}_n (\Omega) = \begin{cases} \dfrac{4\pi}{2n+1} \iint\limits_{\Phi} \rho\left(D_U, \Omega'\right) D^{k+1}_U\left(\Omega'\right) P_n(t)d\Omega' \\[2mm] \qquad = \displaystyle\sum_{m=-n}^{n} U^{(k+1)}_{n,m} Y_{n,m}\left(\Omega\right) \quad i = 0 \\[4mm] \dfrac{4\pi}{2n+1} \iint\limits_{\Phi} \beta\left(\Omega'\right) a_i\left(\Omega'\right) D^{k+1+i}_U\left(\Omega'\right) P_n(t)d\Omega' \\[2mm] \qquad = \displaystyle\sum_{m=-n}^{n} U^{(k+1+i)}_{n,m} Y_{n,m}\left(\Omega\right) \quad i = 1, 2, \ldots, I \end{cases} \qquad (4.8)$$

4 Gravimetric Forward and Inverse Modeling Methods of the Crustal Density... 65

For a specific volumetric layer, the mass density ρ is either constant ρ, laterally varying $\rho(\Omega')$, or – in the most general case – approximated by the laterally distributed radial density variation model using the following polynomial function (for each lateral column):

$$\rho\left(r', \Omega'\right) = \rho\left(D_U, \Omega'\right) + \beta\left(\Omega'\right) \sum_{i=1}^{I} a_i\left(\Omega'\right)\left(R - r'\right)^i,$$

$$\text{for} \quad R - D_U\left(\Omega'\right) \geq r' > R - D_L\left(\Omega'\right), \tag{4.9}$$

where a nominal value of the lateral density $\rho(D_U, \Omega')$ is stipulated at the depth D_U. This density distribution model describes the radial density variation by means of the coefficients $\{a : i = 1, 2, \ldots, I\}$ and β within a volumetric mass layer at a location Ω'. Alternatively, when modeling the gravitational field of anomalous mass density structures, the density contrast $\Delta\rho(r', \Omega')$ of a volumetric mass layer respective to the reference crust density ρ^c is defined as

$$\Delta\rho\left(r', \Omega'\right) = \rho\left(r', \Omega'\right) - \rho^c$$

$$= \Delta\rho\left(D_U, \Omega'\right) + \beta\left(\Omega'\right) \sum_{i=1}^{I} a_i\left(\Omega'\right)\left(R - r'\right)^i,$$

$$\text{for} \quad R - D_U\left(\Omega'\right) \geq r' > R - D_L\left(\Omega'\right), \tag{4.10}$$

where $\Delta\rho(D_U, \Omega')$ is a nominal value of the lateral density contrast at the depth D_U.

The coefficients $L_{n,m}$ and $U_{n,m}$ combine the information on the geometry and density (or density contrast) distribution of a volumetric layer. These coefficients are generated to a certain degree of spherical harmonics using discrete data of the spatial density distribution (i.e., typically provided by means of density, depth, and thickness data) of a particular structural component of the Earth's interior.

From Eqs. (4.1) and (4.3), the spectral representation of the consolidated crust-stripped gravity disturbance δg^{cs} is written as

$$\delta g^{cs}\left(r, \Omega\right) = \frac{GM}{R^2} \sum_{n=0}^{\bar{n}} \sum_{m=-n}^{n} \left(\frac{R}{r}\right)^{n+2} (n + 1)\, T_{n,m}^{cs} Y_{n,m}\left(\Omega\right), \tag{4.11}$$

where the potential coefficients $T_{n,m}^{cs}$ consist of the following components:

$$T_{n,m}^{cs} = T_{n,m} - V_{n,m}^{t} - V_{n,m}^{b} - V_{n,m}^{i} - V_{n,m}^{s} - V_{n,m}^{c}. \tag{4.12}$$

Tenzer et al. (2009a, b) and Tenzer et al. (2012b) demonstrated that the consolidated crust-stripped gravity disturbances δg^{cs} have a maximum correlation with the

CRUST2.0 Moho geometry. Tenzer and Chen (2014) proposed an isostatic compensation scheme based on minimizing a spatial correlation between the isostatic gravity disturbances and the Moho geometry. According to this scheme, the isostatic compensation attraction g^i defines the stripping gravity correction of a homogenous crust layer (below the geoid surface) respective to the upper mantle density. For the adopted constant densities of the reference crust and upper mantle, the stripping density contrast is defined as a constant value of the Moho density contrast $\Delta\rho^{c/m}$. The isostatic compensation attraction g^i reads

$$g^i(r, \Omega) = \frac{GM}{R^2} \sum_{n=0}^{\bar{n}} \left(\frac{R}{r}\right)^{n+2} \sum_{m=-n}^{n} F_{n,m}^D Y_{n,m}(\Omega), \tag{4.13}$$

where the numerical coefficients $F_{n,m}^D$ are given by

$$F_{n,m}^D = \frac{3}{2n+1} \frac{n+1}{n+3} \frac{\Delta\rho^{c/m}}{\bar{\rho}^{\text{Earth}}} \sum_{k=1}^{n+3} \binom{n+3}{k} \frac{(-1)^k}{R^{k+1}} D_{n,m}^{(k)}. \tag{4.14}$$

The Moho depth coefficients $D_{n,m}$ and their higher-order terms $\{D_{n,m}^{(k)} : k = 2, 3, 4, \dots\}$ are computed from discrete values D'_0 of the a priori Moho model using the following expression:

$$D_n^{(k)}(\Omega) = \frac{2n+1}{4\pi} \iint_{\Omega' \in \Phi} D'^k_0 P_n(t) d\Omega'$$

$$= \sum_{m=-n}^{n} D_{n,m}^{(k)} Y_{n,m}(\Omega) \quad (k = 1, 2, 3, 4, \dots). \tag{4.15}$$

The complete crust-stripped isostatic gravity disturbance δg^m is obtained from the consolidated crust-stripped gravity disturbance δg^{cs} after subtracting the isostatic compensation attraction g^i. Hence

$$\delta g^m(r, \Omega) = \delta g^{cs}(r, \Omega) - g^i(r, \Omega). \tag{4.16}$$

The values of δg^m are used as the input data to determine the Moho depths in the gravimetric inverse scheme.

4.3 Gravimetric Inverse Problem

Tenzer and Chen (2014) formulated a relation between the complete crust-stripped isostatic gravity disturbances δg^m and the Moho depth corrections $\delta D'$ by means of a linearized Fredholm integral of the first kind:

$$\delta g^m (r, \Omega) \cong G \Delta \rho^{c/m} \iint_{\Omega' \in \Phi} T\left(r, \psi, D_0'\right) \delta D' d\,\Omega', \tag{4.17}$$

where T is the integral kernel. They gave the solution of Eq. (4.17) in the following spectral form:

$$\delta g^m (r, \Omega) = \frac{GM}{R^2} \sum_{n=0}^{\bar{n}} \left(\frac{R}{r}\right)^{n+2} (n+1) \sum_{m=-n}^{n} F_{n,m}^{\delta D} Y_{n,m} (\Omega), \tag{4.18}$$

where the numerical coefficients $F_{n,m}^{\delta D}$ read

$$F_{n,m}^{\delta D} = \frac{3}{2n+1} \frac{\Delta \rho^{c/m}}{\rho^{\text{Earth}}} \sum_{k=0}^{n+2} \binom{n+2}{k} \frac{(-1)^k}{R^{k+1}} \delta D_{n,m}^{(k)}. \tag{4.19}$$

The inverse solution to the system of linearized observation equations (in Eq. (4.19)) yields the Moho depth correction coefficients $\delta D_{n,m}^{(k)}$. These coefficients are converted into the Moho depth corrections $\delta D'$ using the following expression:

$$\delta D_n^{(k)} (\Omega) = \frac{2n+1}{4\pi} \iint_{\Omega' \in \Phi} D_0'^k P_n(t) \delta D' d\Omega'$$

$$= \sum_{m=-n}^{n} \delta D_{n,m}^{(k)} Y_{n,m} (\Omega) \quad (k = 1, 2, 3, 4, \ldots). \tag{4.20}$$

4.4 Numerical Realization and Results

The gravity disturbances were generated using the GOCO03S coefficients (Mayer-Guerr et al. 2012) with a spectral resolution complete to degree 180 of spherical harmonics. The spherical harmonic terms of the normal gravity field were computed according to the parameters of GRS-80 (Moritz 2000). The same spectral resolution was used to compute the topographic and bathymetric (ocean density contrast) stripping gravity corrections. These two gravity corrections were computed from the DTM2006.0 coefficients of the solid topography (Pavlis et al. 2007). The average density of the upper continental crust 2,670 kg m^{-3} (cf. Hinze 2003) was adopted for defining the topographic and reference crustal densities. The bathymetric stripping gravity correction was computed using the depth-dependent seawater density model (see Tenzer et al. 2012c). For the reference crustal density of 2,670 kg m^{-3} and the surface seawater density of 1,027.91 kg m^{-3} (cf. Gladkikh and Tenzer 2011), the nominal ocean density contrast (at zero depth) equals 1642.09 kg m^{-3}.

The parameters of the depth-dependent seawater density model in Eq. (4.10) are $\beta = 0.00637\,\text{kg m}^{-3}$, $a_1 = 0.7595\,\text{m}^{-1}$, and $a_2 = -4.3984 \times 10^{-6}\,\text{m}^{-2}$ (cf. Tenzer et al. 2012b). The 5×5 arc-min continental ice-thickness data derived from Kort and Matrikelstyrelsen (KMS) dataset for Greenland (Ekholm 1996) and from the updated data for Antarctica assembled by the BEDMAP project (Lythe et al. 2001) were used to generate the coefficients of a global ice-thickness model. These coefficients combined with the DTM2006.0 topographic coefficients were then used to compute the ice (density contrast) stripping gravity correction with a spectral resolution complete to degree 180 of spherical harmonics. For the density of the glacial ice 917 kg m^{-3} (cf. Cutnell and Kenneth 1995), the ice density contrast equals 1,753 kg m^{-3}. The sediment and consolidated crust (density contrasts) stripping gravity corrections were computed using the 2×2 arc-deg data of the global crustal model CRUST2.0 (Bassin et al. 2000) with a spectral resolution complete to degree/order 90.

The stepwise consolidated crust-stripped gravity disturbances are shown in Fig. 4.1. The corresponding statistics are summarized in Table 4.1. As seen from a comparison of the GOCE03S gravity disturbances (in Fig. 4.1a) and the consolidated crust-stripped gravity disturbances (in Fig. 4.1f), the application of the topographic and stripping gravity corrections due to the bathymetry, ice, sediments, and consolidated crust changed the gravity map considerably. Whereas the GOCE03S gravity disturbances are globally mostly within ±250 mGal, the consolidated crust-stripped gravity disturbances vary between −1,416 and 473 mGal. These gravity disturbances are mostly positive over oceans, while negative over continents. The gravity maxima are situated along the oceanic subduction zones. The most pronounced feature in the gravity map is the global tectonic configuration of the boundaries between the oceanic and continental lithospheric plates, which is distinctively marked by the absolute gravity minima. The extreme gravity minima apply over significant orogens of the Tibetan Plateau, Himalayas, and Andes. As mentioned before, these refined gravity disturbances have a maximum correlation with the (a priori) Moho model.

The 2×2 arc-deg discrete data of the CRUST2.0 Moho depths were used to calculate the compensation attraction according to Eqs. (4.13), (4.14), and (4.15) using the Moho density contrast of 485 kg m^{-3} (Tenzer et al. 2012b). This correction was further applied to the consolidated crust-stripped gravity disturbances. The complete crust-stripped isostatic gravity disturbances are shown in Fig. 4.2. These gravity disturbances are everywhere positive and vary between 451 and 1,171 mGal, with a mean of 752 mGal and a standard deviation of 67 mGal. The spatial pattern of the isostatic gravity disturbances reveals the signature of the mantle lithosphere density structure.

The complete crust-stripped isostatic gravity disturbances computed on a 1×1 arc-deg surface grid were used to determine the Moho depths globally using the same grid-sampling interval. The system of observation equations was formed

4 Gravimetric Forward and Inverse Modeling Methods of the Crustal Density... 69

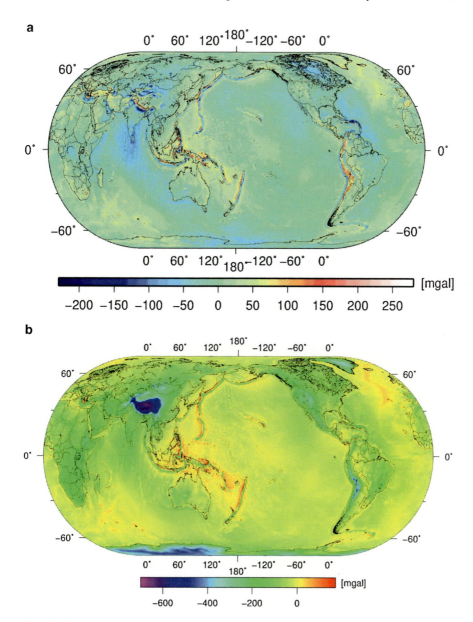

Fig. 4.1 The stepwise consolidated crust-stripped gravity disturbances computed globally on a 1×1 arc-deg surface grid: (**a**) GOCO03S δg; (**b**) topography-corrected δg^{T}; (**c**) topography-corrected and bathymetry-stripped δg^{TB}; (**d**) topography-corrected and bathymetry- and ice-stripped δg^{TBI}; (**e**) topography-corrected and bathymetry-, ice-, and sediment-stripped δg^{TBIS}; (**f**) consolidated crust-stripped δg^{cs}

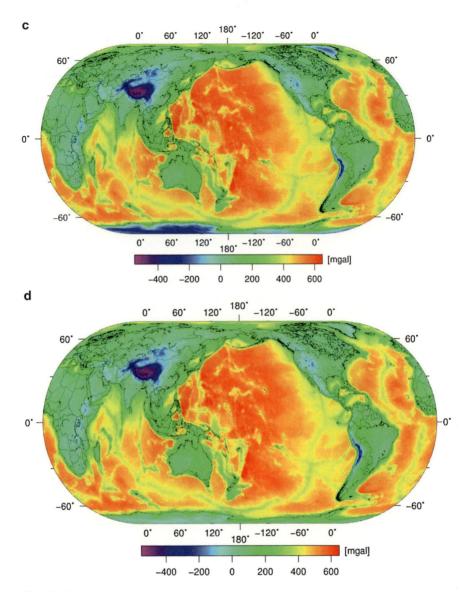

Fig. 4.1 (continued)

4 Gravimetric Forward and Inverse Modeling Methods of the Crustal Density...

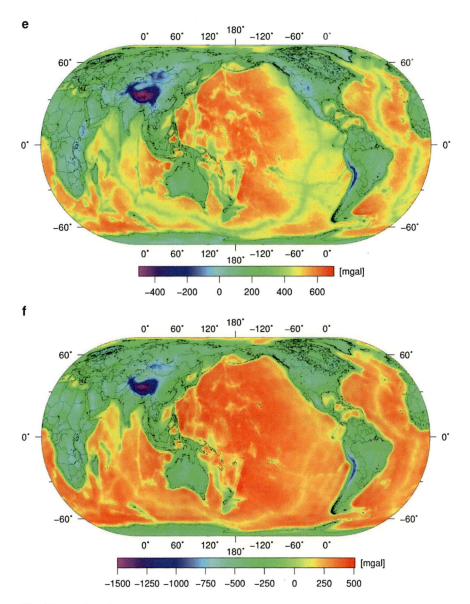

Fig. 4.1 (continued)

Table 4.1 Statistics of the stepwise consolidated crust-stripped gravity disturbances (as shown in Fig. 4.1)

Gravity disturbances	Min [mGal]	Max [mGal]	Mean [mGal]	STD [mGal]
δg	−225	257	1	29
δg^T	−647	156	−72	107
δg^{TB}	−511	636	260	233
δg^{TBI}	−508	639	285	202
δg^{TBIS}	−492	667	327	192
δg^{cs}	−1,416	473	18	332

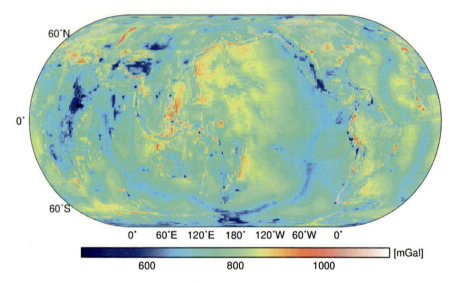

Fig. 4.2 The complete crust-stripped isostatic gravity disturbances computed globally on a 1 × 1 arc-deg surface grid

according to the expression in Eq. (4.18). The inverse solution was carried out iteratively using a Gauss-Seidel scheme (e.g., Young 1971). The regularization was applied to stabilize the ill-posed solution. The gravimetric Moho solution is shown in Fig. 4.3. The Moho depths vary globally between 3.6 and 85.1 km, with a mean of 13.6 km and a standard deviation of 13.4 km. The global map revealed a typical pattern of the Moho geometry with the enhanced contrast between the thick continental crust and thinner oceanic crust. The largest Moho depths were confirmed under Andes and Himalayas with extension beneath the Tibetan Plateau, where the maximum Moho depths exceed 70 km.

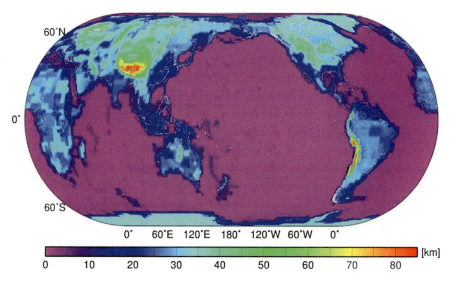

Fig. 4.3 The gravimetric Moho solution computed globally on a 1 × 1 arc-deg grid

4.5 Discussion and Concluding Remarks

The consolidated crust-stripped gravity disturbances have a maximum correlation with the Moho geometry. However, uncertainties of a crustal structure model contribute to the total error budget. Tenzer et al. (2012a) estimated that relative errors in computed values of the consolidated crust-stripped gravity data can reach about 10 %. Moreover, these gravity data still comprise a long-wavelength signal from unmodeled mantle heterogeneities. Bagherbandi and Sjöberg (2012) proposed a procedure of treating the long-wavelength gravity signal of the mantle in solving the VMM inverse problem of isostasy. They applied the method of Eckhardt (1983) to estimate the maximum degree of long-wavelength spherical harmonic terms, which should be removed from the gravity field in prior of solving the VMM problem. The principle of this procedure was based on finding the representative depth of gravity signal attributed to each spherical harmonic degree term. The spherical harmonics, which have the depth below a certain limit (chosen, for instance, as the maximum Moho depth), are then removed from the gravity field.

The estimation of the Moho depth uncertainties is not simple, because there is not enough information on the accuracy of input data. The expected largest uncertainties in the estimated Moho depths are mainly due to the inaccuracies of crustal models currently available and unmodeled mantle heterogeneities. The relative errors in computed values of the consolidated crust-stripped gravity data of about 10 % yield similar relative uncertainties in the estimated Moho depths, provided that the gravity errors propagate almost linearly into the Moho errors. Most of these errors are due to large uncertainties in the CRUST2.0 sediment and consolidated crust data, while the global models of the Earth's gravity field, topography, ice, and bathymetry have a relatively high resolution and accuracy. Similar to uncertainties in the input gravity data, the errors in the Moho density contrast propagate proportionally to the Moho depth errors. In addition to these uncertainties, we disregarded the mantle density heterogeneities and the core-mantle geometry. The effect of mantle density structure on the gravity field and the Moho geometry is not yet sufficiently known, because of the absence of a reliable mantle density model.

References

Airy GB (1855) On the computations of the effect of the attraction of the mountain masses as disturbing the apparent astronomical latitude of stations in geodetic surveys. Roy Soc (Lond) Ser B 145:101–104

Arabelos D, Mantzios G, Tsoulis D (2007) Moho depths in the Indian ocean based on the inversion of satellite gravity data. In: Huen W, Chen YT (eds) Solid Earth, ocean science and atmospheric science, vol 9, Advances in geosciences. World Scientific Publishing, Hackensack, pp 41–52

Bagherbandi M (2012) A comparison of three gravity inversion methods for crustal thickness modelling in Tibet plateau. Asian J Earth Sci 43(1):89–97

Bagherbandi M, Sjöberg LE (2012) Non-isostatic effects on crustal thickness: a study using CRUST2.0 in Fennoscandia. Phys Earth Planet Inter 200–201:37–44

Bagherbandi M, Tenzer R (2013) Comparative analysis of Vening Meinesz-Moritz isostatic models using the constant and variable crust-mantle density contrast – a case study of Zealandia. J Earth Syst Sci 122(2):339–348

Bagherbandi M, Tenzer R, Sjöberg LE, Novák P (2013) Improved global crustal thickness modeling based on the VMM isostatic model and non-isostatic gravity correction. J Geodyn 66:25–37

Bassin C, Laske G, Masters TG (2000) The current limits of resolution for surface wave tomography in North America. Eos Trans AGU 81:F897

Braitenberg C, Zadro M (1999) Iterative 3D gravity inversion with integration of seismologic data. Boll Geofis Teor Appl 40(3/4):469–476

Braitenberg C, Mariani P, Reguzzoni M, Ussami N (2010) GOCE observations for detecting unknown tectonic features. In: Proceedings of the ESA living planet symposium, 28 June–2, July 2010, Bergen, Norway, ESA SP-686

Čadek O, Martinec Z (1991) Spherical harmonic expansion of the earth's crustal thickness up to degree and order 30. Stud Geophs Geod 35:151–165

Cutnell JD, Kenneth WJ (1995) Physics, 3rd edn. Wiley, New York

Eckhardt DH (1983) The gains of small circular, square and rectangular filters for surface waves on a sphere. Bull Geodyn 57:394–409

Ekholm S (1996) A full coverage, high-resolution, topographic model of Greenland, computed from a variety of digital elevation data. J Geophys Res B10(21):961–972

Eshagh M, Bagherbandi M, Sjöberg LE (2011) A combined global Moho model based on seismic and gravimetric data. Acta Geodaetica et Geophysica Hungarica 46(1):25–38

Gladkikh V, Tenzer R (2011) A mathematical model of the global ocean saltwater density distribution. Pure Appl Geophys 169(1–2):249–257

Hayford JF (1909) The figure of the earth and isostasy from measurements in the United States. USCGS, Washington, DC

Hayford JF, Bowie W (1912) The effect of topography and isostatic compensation upon the intensity of gravity, USCGS, special publication, no. 10. G.P.O., Washington, DC

Heiskanen WA, Moritz H (1967) Physical geodesy. Freeman W.H., New York

4 Gravimetric Forward and Inverse Modeling Methods of the Crustal Density... 75

Heiskanen WA, Vening Meinesz FA (1958) The Earth and its gravity field. McGraw-Hill Book Company, Inc., New York

Hinze WJ (2003) Bouguer reduction density, why 2.67. Geophysics 68(5):1559–1560

Lythe MB, Vaughan DG (2001) BEDMAP consortium: BEDMAP; a new ice thickness and subglacial topographic model of Antarctica. J Geophys Res B Solid Earth Planets 106(6): 11335–11351

Mayer-Guerr T, Rieser D, Höck E, Brockmann JM, Schuh W-D, Krasbutter I, Kusche J, Maier A, Krauss S, Hausleitner W, Baur O, Jäggi A, Meyer U, Prange L, Pail R, Fecher T, Gruber T (2012) The new combined satellite only model GOCO03s. Presented at GGHS2012, Venice, October

Moritz H (1990) The figure of the Earth. Wichmann H, Karlsruhe

Moritz H (2000) Geodetic reference system 1980. J Geodyn 74:128–162

Oldenburg DW (1974) The inversion and interpretation of gravity anomalies. Geophysics 39: 526–536

Pavlis NK, Factor JK, Holmes SA (2007) Terrain-related gravimetric quantities computed for the next EGM. In: Kiliçoglu A, Forsberg R (eds) Gravity field of the Earth. Proceedings of the 1st international symposium of the International Gravity Field Service (IGFS), Harita Dergisi, Special issue no. 18, General Command of Mapping, Ankara, Turkey

Pratt JH (1855) On the attraction of the Himalaya mountains and of the elevated regions beyond upon the plumb-line in India. Trans Roy Soc (Lond) Ser B 145:53–100

Sampietro D (2011) GOCE exploitation for Moho modeling and applications. In: Proceedings of the 4th international GOCE user workshop, 31 March–1 April 2011, Munich, Germany

Sampietro D, Reguzzoni M, Braitenberg C (2013) The GOCE estimated Moho beneath the Tibetan Plateau and Himalaya. In: Rizos C, Willis P (eds) International association of geodesy symposia, Earth on the edge: science for a sustainable planet. Proceedings of the IAG general assembly, 28 June–2 July 2011, Melbourne, Australia, vol 139. Springer, Berlin

Sjöberg LE (2009) Solving Vening Meinesz-Moritz Inverse Problem in Isostasy. Geophys J Int 179(3):1527–1536

Sjöberg LE, Bagherbandi M (2011) A method of estimating the Moho density contrast with a tentative application by EGM08 and CRUST2.0. Acta Geophys 58:1–24

Tenzer R, Chen W (2014) Expressions for the global gravimetric Moho modeling in spectral domain. Pure Appl Geophys 171(8):1877–1896

Tenzer R, Hamayun K, Vajda P (2009a) Global maps of the CRUST2.0 crustal components stripped gravity disturbances. J Geophys Res 114(B):05408

Tenzer R, Hamayun K, Vajda P (2009b) A global correlation of the step-wise consolidated crust-stripped gravity field quantities with the topography, bathymetry, and the CRUST2.0 Moho boundary. Contrib Geophys Geodesy 39(2):133–147

Tenzer R, Gladkikh V, Vajda P, Novák P (2012a) Spatial and spectral analysis of refined gravity data for modelling the crust-mantle interface and mantle-lithosphere structure. Surv Geophys 33(5):817–839

Tenzer R, Hamayun K, Novák P, Gladkikh V, Vajda P (2012b) Global crust-mantle density contrast estimated from EGM2008, DTM2008, CRUST2.0, and ICE-5G. Pure Appl Geophys 169(9):1663–1678

Tenzer R, Novák P, Gladkikh V (2012c) The bathymetric stripping corrections to gravity field quantities for a depth-dependant model of the seawater density. Mar Geod 35:198–220

Tenzer R, Bagherbandi M, Hwang C, Chang ETY (2013) Moho interface modeling beneath Himalayas, Tibet and central Siberia using GOCO02S and DTM2006.0. Special issue on geophysical and climate change studies in Tibet, Xinjiang, and Siberia from satellite geodesy. Terres Atmos Ocean Sci 24(4):581–590

Vening Meinesz FA (1931) Une nouvelle méthode pour la réduction isostatique régionale de l'intensité de la pesanteur. Bull Geod 29:33–51

Watts AB (2001) Isostasy and flexure of the lithosphere. Cambridge University Press, Cambridge

Young D (1971) Iterative solutions of large linear systems. Academic, New York

Chapter 5
Radar Exploration of Mars: Recent Results and Progresses

Stefano Giuppi

Abstract Radar is the acronym for RAdio Detection And Ranging. It is an object-detection system, which principles consist basically in the transmission, propagation, and reflection of radio waves. After the first exploitations in the military field, radar evolved as a useful device also in the civil field, widely extending its applications. After having been tested on Earth, radar capabilities to penetrate a planet surface have been applied on Mars exploration. MARSIS, part of the payload of ESA Mars Express mission, and SHARAD, embarked on board NASA MRO spacecraft, are two nadir-looking radar sounders which use synthetic aperture radar (SAR) techniques. The two instruments are complementary: MARSIS is able to detect subsurface interfaces at great depth, while SHARAD can better discriminate subsurface interfaces close to the surface. The two radars achieved information on Martian craters, both exposed and buried, provided geophysical evidences for the former existence of an ocean in the Martian northern hemisphere, investigated Martian pedestal craters, provided a useful contribution to analyze the nature of the Medusae Fossae Formation, and probed the ice-rich polar layered deposits of Mars, detected a boundary in many areas of plains off the south polar layered deposits. The analysis of MARSIS data enables to study also the Martian ionosphere and to estimate its TEC producing the related maps. Through volumetric (3D) study, the two radars provide an opportunity to extend our knowledge of a planetary body to the third dimension, allowing to detect features that are difficult to investigate in vertical profiles. Since Martian polar terrains are considered a close analogue to the material forming the crusts of Jovian satellites Europa and Ganymede, a radar sounder, RIME, has been selected as part of ESA's first large-class mission in Cosmic Vision Program, JUICE, the first orbiter on an icy moon which will investigate the emergence of habitable worlds around gas giants, characterizing Ganymede, Europa, and Callisto as planetary objects and potential habitats.

Keywords Radar • Mars • Subsurface interfaces • MARSIS • SHARAD

S. Giuppi (✉)
Istituto di Astrofisica e Planetologia Spaziali – Istituto Nazionale di Astrofisica, Rome, Italy
e-mail: Stefano.giuppi@iaps.inaf.it

© Springer-Verlag Berlin Heidelberg 2015
S. Jin et al. (eds.), *Planetary Exploration and Science: Recent Results and Advances*,
Springer Geophysics, DOI 10.1007/978-3-662-45052-9_5

5.1 Introduction

Starting with a brief historical summary of the radar genesis, the radar operating principles are discussed. Next in Sect. 5.4, the two radars, which are investigating Mars surface, subsurface, and atmosphere, are described, including a discussion about the different capabilities of the two radars in terms of surface depth penetration, subsurface detection, and resolution, along with the space missions they are part of. Their main scientific accomplishments are described in Sect. 5.5. Section 5.6 provides a brief presentation of a radar involved in a new coming space mission.

The term radar, currently part of the vocabulary of many languages as a common noun, is the acronym for RAdio Detection And Ranging. It is an object-detection system which uses radio waves in order to determine the range, altitude, direction, or speed of objects. Its uses are very versatile; it can be used to detect aircraft, ships, vehicles, missiles, spacecrafts, and weather formations. Its working principle, described more in detail in Sect. 5.3, is based on the transmission of pulses of radio waves or microwaves through a dish or an antenna. When the waves encounter an object (usually called target), a part of the wave energy gets reflected and received by a dish or an antenna usually located at the same site as the transmitter.

5.2 Early Radar History

The history of radar is highly linked to the history of electromagnetism, and it does not have a well-defined origin because the vicissitudes that led to its creation and use in the military and then civil fields have followed different paths in different countries.

In 1886, German physicist Heinrich Hertz showed that radio waves could be reflected from solid objects. In 1904, German engineer Christian Hülsmeyer, through experiments on the detection of electromagnetic waves reflected from ships, developed and patented an apparatus capable of detecting the presence of distant metallic obstacles thanks to returning echoes. The apparatus, referred to as "telemobiloskop," was tested as an anticollision device and can be considered the true forerunner of the modern radar.

The telemobiloskop was publicly presented in Cologne on 18th May 1904. The system was composed of a directional antenna, a receiver, and a coherer transmitter. The device allowed the detection of a metal object in a certain direction with a range of about 3 km, without specifying the distance, and operated with $40 \div 50$ cm wavelength waves. This device could work even in adverse weather conditions, demonstrating the feasibility of detecting a ship in dense fog, and was judged favorably by the press, but it did not arouse the interest of naval authorities nor industry, and it was not put into production.

5 Radar Exploration of Mars: Recent Results and Progresses

The principle of radar, namely, the possibility of target detection offered by shortwaves, was for the first time announced in June 1922 by Italian physicist Guglielmo Marconi in a famous speech at the Institute of Radio Engineers (USA): "As it was for the first time demonstrated by Hertz, electric waves can be completely reflected by conducting bodies. In my experience, I found reflection and detection effects of electric waves by metallic objects at a distance of miles. I think it should be possible to design a device by means of which a ship can radiate a wave beam in a desired direction in order that, if those waves encounter a metal object, such as another ship, they will be reflected to a receiver, shielded with respect to the transmitter of the transmitting ship, immediately indicating the presence and bearing of the other ship in fog or bad weather."

In 1922, engineer Albert Taylor of the Naval Research Laboratory (NRL) in Washington, DC, made the first observation of the radar effect. He observed that a ship passing between a transmitter and a receiver reflected some of the radio waves back to the transmitter. In 1930, further tests at the NRL showed a fluctuation in the signal when a plane flied through a beam from a transmitting antenna. Finally, when scientists and engineers learned how to use a single antenna for both transmitting and receiving, the interest for radar and its capabilities of tracking aircraft and ships grew worldwide.

The technologies that led to the modern version of radar were developed independently and in great secrecy before the Second World War especially in France, Germany, Italy, Japan, the Netherlands, the Soviet Union, the United Kingdom, and the United States.

Though the first studies on radar devices have been developed in the United States, the British were the first to investigate the feasibility of aircraft detection by electromagnetic waves propagation and to fully exploit radar as a defense system against aircraft attack. Those studies let sir Robert Watson-Watt to the production of a first prototype in 1935, which he called RADAR. The most notable result of British efforts was the Chain Home network of radars which defended Great Britain by detecting approaching German aircrafts in the Battle of Britain in 1940.

After those first exploitations in the military field, radar evolved as a useful device also in the civil field. A natural evolution of military radar was to equip aircrafts with radar device warning of obstacles in their path, displaying weather information and giving accurate altitude readings, while in the marine field, radars are used to measure the distance between ships to prevent collision, to navigate, and to fix position at sea when within range of shore or other fixed references such as islands. In the course of time, radar extended its application to air traffic control, vessel traffic service, meteorological monitoring, radar astronomy, antimissile systems, and outer space surveillance.

In geology, specialized ground-penetrating radars are used to map the composition of Earth crust. At the end of the twentieth century, the feasibility of using such radars to investigate Mars subsurface has been studied. And at the beginning of the twenty-first century, two different space missions to the Red Planet carried a radar as part of their payloads.

5.3 Radar Operating Principles

The basic principles of a radar consist in the transmission, propagation, and reflection of radio waves. Figure 5.1 shows a functional diagram of a radar system: the transmitter generates a pulse of electromagnetic energy which is routed through a transmit-receive switch, called duplexer, to an antenna, which radiates it into free space. The transmit-receive switch has the function of protecting the sensitive receiver from the high-power transmitted pulse. When the electromagnetic pulse encounters objects along its way, it is scattered (reradiated), and part of the scattered signal returns to the radar where it is collected by the antenna and routed through the transmit-receive switch to the receiver. The received signal is enhanced, interfering signals are reduced, and measurements of the object are made by signal processing and presented on displays.

The purpose of radar instruments is not just to detect objects but mainly to measure the range between the radar and the object. When a pulse is transmitted, a clock is started and stopped as soon as a received signal is detected. The range between the radar and the detected object is expressed by $R = c\Delta t/2$, where R is the range from the radar to the object in m, c is the speed of light in the vacuum $= 299{,}792{,}458$ m/s, and Δt is the elapsed time in s.

As mentioned before, the radar transmits pulsed signals. The characteristics of a radar waveform are the wavelength of the propagated energy $\lambda = c/f_c$ in m, where c is the speed of light in the vacuum and f_c is the radar carrier frequency in Hz; the angular frequency $\omega = 2\pi f_c$ in rad/s; the pulse repetition interval (PRI) in s, which is the time between radar pulses; and its reciprocal the pulse repetition frequency (PRF) in Hz, which is the number of pulses sent per time interval, with PRF $= 1/$PRI (Fig. 5.2).

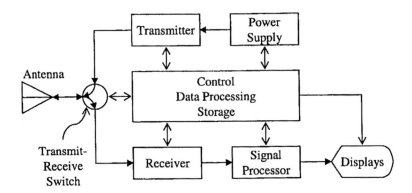

Fig. 5.1 Radar block diagram

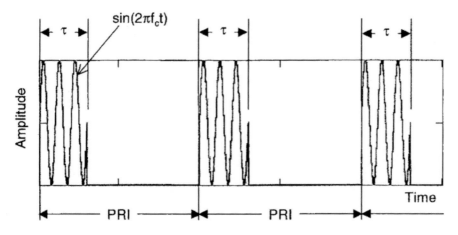

Fig. 5.2 Characteristics of a radar waveform

5.3.1 The Radar Equation

In the radar systems analysis, it is fundamental to know the maximum distance at which it is possible to detect a target once the power of the transmitted signal is fixed. This information, which is the basis of radar performance, is expressed in the radar range equation:

$$P_r = \frac{P_p \cdot G^2 \cdot \lambda^2 \cdot \sigma}{(4\pi)^3 \cdot R^4 \cdot L} \quad (5.1)$$

where P_r is the power of the signal sensed by the receiver in W; P_p is the peak transmitted power in W; G is the radar antenna gain, which is the ratio of focused signal over isotropic radiation; λ is the wavelength of the transmitted signal in m; σ is the target radar cross section in m^2; R is the radar to object range in m; and L is a loss factor due mainly to imperfect dielectric medium (Skolnik 1980).

5.4 Two Radars on Mars Exploration: MARSIS and SHARAD

Radar capabilities to penetrate a planet surface, tested on Earth to map the crust composition, suggested to study the feasibility to use specialized radar devices on Mars exploration. Two instruments were developed for this purpose by the University of Rome "La Sapienza," Italy, in partnership with NASA's Jet Propulsion Laboratory (JPL) in Pasadena, California, USA: MARSIS and SHARAD.

5.4.1 ESA's Mars Express Mission: MARSIS

The Mars Advanced Radar for Subsurface and Ionosphere Sounding (MARSIS) (Picardi et al. 2000; Jordan et al. 2009) is part of the payload of Mars Express (MEX), one of the most challenging missions ever attempted by European Space Agency (ESA). In fact, MEX is Europe's first spacecraft sent to the Red Planet. It is so-called because it has been built more quickly than any other comparable planetary mission. It carries a suite of instruments that are investigating many scientific aspects of the Red Planet in unprecedented detail. The observations are particularly focused on Martian atmosphere, surface, and subsurface. The spacecraft orbits around Mars are highly elliptic having the pericenter (closest point to Mars) at about 250 km from the planet surface and the apocenter (furthest point from Mars) at about 11,000 km from the planet surface. This peculiarity limits data acquisition, which are possible only from altitudes lower than 1,200 km, but gives the chance to periodically perform additional observations of Phobos, one of Martian moons, when MEX passes near this natural satellite of Mars.

MEX began its journey to the Red Planet from the Baikonur launch pad in Kazakhstan on a Soyuz-Fregat launcher on 2nd June 2003 and arrived in Mars orbit on 25th December 2003, after almost 7 months.

MARSIS is a low-frequency nadir-looking radar sounder which transmits radio frequency pulses through a 40 m dipole antenna. It uses synthetic aperture radar (SAR) techniques, has the capability to map the distribution of water both solid and liquid on the Martian crust, and is able to detect what lies beneath the surface of Mars with a penetration capability up to about 5 km. It can operate as a subsurface sounder or as an ionospheric sounder. When operating in subsurface mode, MARSIS carrier frequency can be set to 1.8, 3, 4, or 5 MHz, with a 1 MHz bandwidth which enables a vertical resolution varying between 50 and 100 m in the subsurface depending on the medium characteristics. MARSIS operates in ionospheric mode at altitudes greater than 800 km sweeping the entire 0.1–5.5 MHz frequency range.

MARSIS performs on board a complex scientific processing, including azimuth and range compression, tracking and a particular algorithm used to remove iono-spheric effects, and the Contrast Method (CM) which showed to be properly applicable also to other tasks, as described in Sect. 5.5.2.

5.4.2 NASA's Mars Reconnaissance Orbiter Mission: SHARAD

The Mars Shallow Radar Sounder (SHARAD) (Seu et al. 2007) is an instrument embarked on board the NASA Mars Reconnaissance Orbiter (MRO) spacecraft.

5 Radar Exploration of Mars: Recent Results and Progresses

The spacecraft altitude varies approximately between 250 km at the pericenter, placed over the south pole, and 320 km at the apocenter. This characteristic, which precludes the opportunity to observe Phobos, allows, for each orbit, to acquire observations on both Martian hemispheres.

MRO was launched from Cape Canaveral Air Force Station on 12th August 2005, with an Atlas V-Centaur launch vehicle, and reached Mars orbit on 10th March 2006.

SHARAD transmits pulses through a 10 m dipole antenna operating exclusively as a subsurface sounder with a carrier frequency of 20 MHz and a bandwidth of 10 MHz, parameters which allow a vertical resolution varying between 10 and 20 m. SHARAD uses, as MARSIS, synthetic aperture radar (SAR) techniques; but, while MARSIS is an unfocused SAR with a best along-track resolution of 2 km, SHARAD data through a processing with a focusing algorithm (Chirp Scaling Algorithm) can achieve a best along-track resolution of 300 m.

5.4.3 MARSIS and SHARAD Performances Comparison

One of the main quests in the exploration of the Red Planet is the identification of water deposits on or below the surface of Mars, which are supposed to last as a relic of ancient Martian oceans. The capability of penetrating deep into a planet surface makes low-frequency radars the perfect candidates for detecting such deposits.

The operating principle of a subsurface sounder is extremely simple: the radar antenna transmits a pulse of electromagnetic energy which, as it reaches the planet surface, produces a first reflection echo propagating back to the radar. Due to the low frequencies adopted, a large fraction of the electromagnetic energy propagates through the crust generating additional reflections as it encounters subsurface discontinuities. Also those secondary echoes, much weaker than the front surface signal, propagate back to the receiver. The radar, through SAR processing, produces sets of synthetic apertures, called frames, adjacent to each other, which altogether form a radargram, which basically is a cross section of the planet surface.

MARSIS and SHARAD are complementary instruments: MARSIS is able to detect subsurface interfaces at depth of more than 3 km, but due to the limited bandwidth adopted, it has a limited range resolution. SHARAD, on the contrary, has a penetration capability of approximately 1,500 m, but can discriminate subsurface interfaces close to the surface, allowing to analyze the upper portions of the Martian crust.

Figure 5.3 shows two radargrams produced with data acquired over Martian south pole: MARSIS is capable to detect the basal reflector at a depth of 3.7 km, while SHARAD, which signal cannot reach the basal unit, manages to generate a radargram with a finer resolution, both horizontal and vertical, providing unique information on Martian stratigraphy.

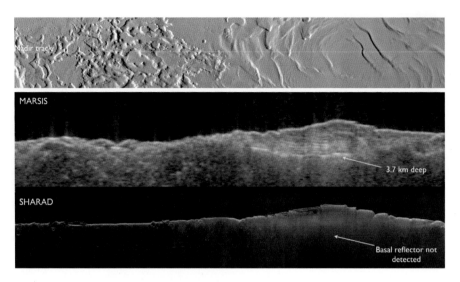

Fig. 5.3 Comparison of MARSIS and SHARAD radargrams on Mars south pole (Image credit: NASA/ESA/JPL-Caltech/University of Rome/Washington University in St. Louis)

5.5 MARSIS and SHARAD Scientific Accomplishments

MARSIS and SHARAD were originally designed with the primary scientific task of investigating the presence of water, both solid and liquid, on the Martian surface and subsurface. But the information on Mars achieved using their data exceeded by far the initial expectations.

MARSIS was the first instrument to show the cross section of Mars craters, highly contributing to their structure analysis. It was also capable to identify buried craters not detectable with other instruments. Data from Chryse Planitia, a mid-latitude northern lowland region covered by sediments delivered from the highlands, revealed the presence of a buried quasi-circular structure with a diameter of 250 km which is probably an impact basin. MARSIS data also show a planar structure which suggests the presence of a deposit more than 1 km thick (Picardi et al. 2005). On the basis of the analysis of comparable exposed basins on Mars, the data leads to the conclusion that the feature could be the basin floor or a boundary between layers of basin fill. Figure 5.4 shows Martian craters both exposed and buried.

MARSIS data gave evidence of other buried impact basins in the northern lowlands of Mars. Before Mars Express mission, the Mars Orbiter Laser Altimeter (MOLA) on Mars Global Surveyor revealed quasi-circular depressions interpreted as ancient impact craters filled by materials in later eras. Analyzing MARSIS data, Watters et al. (2006) identified 11 buried basins, many of which have not been detected by imaging instruments. Though, on the basis of MARSIS data, it will not be possible to discover all the buried basins. In fact, in order to be detected, a buried crater must be filled with materials having a dielectric constant sufficiently

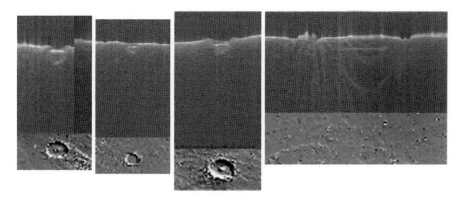

Fig. 5.4 Martian craters (exposed and buried)

different from that of the crustal material. The analysis of number and dimensions of the buried basins provided an improvement in the study of the strong contrast between the northern lowland plains and the southern highlands. In fact, the number of detected buried craters >200 km in diameter allowed to date the lowland crust back to the Early Noachian epoch and to deduce that the northern lowland crust is at least as old as the oldest exposed highland crust, suggesting that the crustal dichotomy arose early in the geologic evolution of Mars.

MARSIS, probing the physical properties of the subsurface to unprecedented depth, provided geophysical evidences for the former existence of an ocean in the Martian northern hemisphere, already suggested by previous observations. On the basis of Martian surface geology, it is known that in the past, there was abundant water flowing, as attested by dendritic valley systems (Carr 1996; Luo and Stepinski 2009; Di Achille and Hynek 2010) and wide outflow channels (Carr 1987; Baker et al. 1991). The ancient presence of Oceanus Borealis, an ocean that would have covered more than 60 % of Mars north hemisphere has been postulated (Baker et al. 1991).

Studying MARSIS data, Mouginot et al. (2012) produced a dielectric map of the Martian subsurface. Due to the radar characteristics, the map represents the volume averaged properties of the first $60 \div 80$ m below the surface. The data evidenced significant differences between the two hemispheres: while in the southern hemisphere, the area characterized by the lowest dielectric values delimits a quasi-circular area from the pole to 60°S latitude, in the northern hemisphere, the area distinguishing by low dielectric values extends with three lobes toward lower latitudes in the regions of Amazonis Planitia (150°W), Chryse Planitia (30°W), and Utopia Planitia (100°E). The region matches with substantial accuracy the shoreline which encloses the Vastitas Borealis formation. On the basis of dielectric constants measured in the region, considerably lower than that of typical volcanic materials, the presence of low-density sedimentary deposits or huge ground-ice masses or probably a combination of both has been proven, suggesting the former presence of an ancient ocean filled by aqueous sediments.

SHARAD data contributed to investigate Martian pedestal craters, being able to detect the base of their deposits. Pedestal craters, discovered by Mariner 9 mission, are craters characterized by ejecta material more resistant to erosion compared to the surrounding terrains. They distinguish from craters detected on the Moon by sitting on a raised platform making a sort of pedestal. The erosion mechanism is still debated: McCauley (1973) proposed eolian erosion as origin of the pedestal craters, while Arvidson et al. (1976) suggested that the different erosion between the pedestal and its perimetric scarp is due to different wind drag velocity threshold for erosion. On the contrary, Head and Roth (1976) on the basis of the observed azimuthal symmetry of pedestals, certainly not caused by the action of prevailing winds, claimed that the mechanisms which control the pedestals origin are mostly due to the ejecta themselves. Pedestals diameter varies between less than a kilometer to hundreds of kilometers, while their height is generally less than 500 m.

Nunes et al. (2011), using SHARAD data, analyzed almost a hundred pedestal craters. Due to the radar characteristics, radargrams of craters with diameters <30 km are affected by clutter and, therefore, show basal reflection only in few cases. Conversely, through SHARAD data it was possible to detect the base of larger craters and to measure an average dielectric constant of $\sim$4, compatible with either a mixture of ice and silicates or with a porous silicate, possibly ash deposits.

Both MARSIS and SHARAD radars, with their radargrams, provided a useful contribution to analyze the nature of the Medusae Fossae Formation (MFF). The MFF is believed to be one of the youngest geologic deposits on Mars and has unique characteristics. There is a special interest in investigating this region with a radar sounder, since the eastern part of MFF correspond to the "Stealth" region on Mars, so named because no echo was detected by radar operating from Earth, suggesting a fine-grained composition of MFF deposits, such as volcanic ash or eolian sediments. As Mars Express flied over MFF, MARSIS detected secondary echoes interpreted as a subsurface interface between the MFF material and the basal substrate. On the basis of MARSIS data, Watters et al. (2007) estimated for MFF deposits materials a dielectric constant of about 2.9 ± 0.4, which suggests that MFF deposits could either consist of dry highly porous or unconsolidated materials or of components with higher dielectric constant mixed to ice. Carter et al. (2009), studying SHARAD data, confirmed Watters results.

One of the main radar objectives is to probe the ice-rich polar layered deposits of Mars on both south and north pole. On the south pole, only MARSIS has the capabilities to penetrate deep into the deposits, which extends for few kilometers beneath the surface. MARSIS data covering Martian south pole generally show, in addition to the surface echo, a very bright secondary echo starting as the spacecraft passes over the margin of south polar layered deposits (SPLD) (Fig. 5.5).

The area marked by the lower bright reflections extends from the margin of the layered deposit to its thickest part ($\sim$3.7 km), i.e., from 310° to 0° east longitude. The interface is interpreted (Plaut et al. 2007) as the boundary between the base of the ice-rich south polar layered deposits materials and the basal substrate.

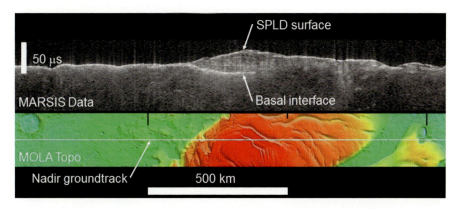

Fig. 5.5 MARSIS south pole radargram along with Mars topography from MOLA data

An evidence of deposits consisting of nearly pure water ice is the slight attenuation of the power of the signal of the lower echo. The analysis of radar data over the south pole provides also additional information about Mars interior; in fact, the basal substrate beneath the layered deposit materials shows a continuation respect to the surrounding surface topography, suggesting a thick elastic lithosphere, sign of a cold Mars interior.

Data collected over south pole allowed the generation of a map of the interface topography providing a new estimation of the thickness and volume of the south polar layered deposits. Since the vertical resolution of MARSIS data is about 100 m in ice, assuming that the signal passed through a medium with a dielectric constant equal to 3 (pure ice) and taking into account every uncertainty, the volume of the south polar layered deposits has been estimated to be 1.6×10^6 km^3, which could cover Mars whole surface with a water layer of 11 m.

Over Mars north pole, both MARSIS and SHARAD have the capabilities to detect the basal substrate. The two radars are complementary: while MARSIS can give evidence of deeper structures, SHARAD radargrams can reveal subsurface interfaces at a depth of $100 \div 200$ m which cannot be discriminated by MARSIS. Figure 5.6 shows two north pole radargrams obtained by the two radars along with the footprints on Martian surface of the spacecraft orbits. The arrows show features detected only by MARSIS. But SHARAD can identify subsurface layers in the upper portion of Mars crust providing information about Martian surface stratigraphy at scale comparable to optical images.

In order to complete the description of radar observations over Mars poles, it is worth noting that MARSIS detected a boundary, few hundred meters deep, in many areas of plains off the south polar layered deposits. Figure 5.7 shows one of those boundaries. The location of these buried interfaces closely match the Hesperian Dorsa Argentea Formation, a surface around the southern ice cap that is characterized by a group of sinuous, branched ridges that resemble eskers which form when streams are under glaciers. Radar data suggest that it could be an ancient ice-rich deposit.

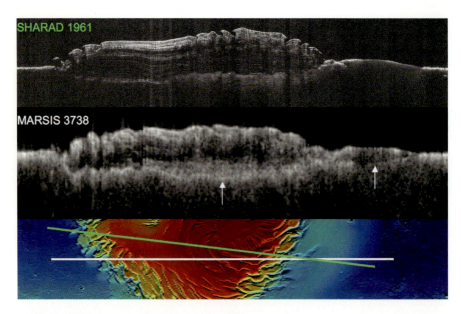

Fig. 5.6 MARSIS and SHARAD north pole radargrams

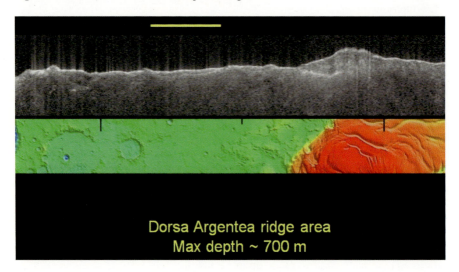

Fig. 5.7 Boundary, 700 m deep, in Dorsa Argentea ridge area

As mentioned before, Mars Express orbits are highly elliptic, and MARSIS operative window extends at each pass for about 40 min as the spacecraft altitude is less than 1,200 km. This characteristic gives the opportunity to study not only the subsurface of the Red Planet but also its ionosphere. Radar sounders already proved to be useful in the ionospheric physics study when applied to Earth ionosphere.

5 Radar Exploration of Mars: Recent Results and Progresses

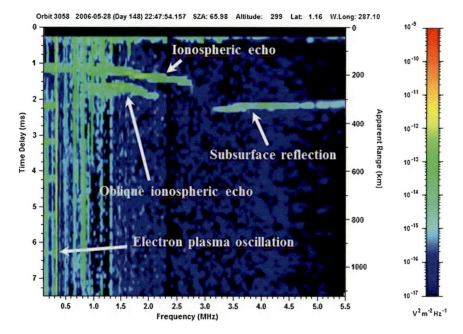

Fig. 5.8 MARSIS ionogram

Before radar observations on Mars, most of the Martian ionosphere knowledge derived from radio occultation measurements. MARSIS provided better spatial resolution measurements and extended the study to regions where radio occultation measurements cannot be made. Martian ionosphere is a perfectly reflecting surface for radio echo sounding at frequencies below the electron plasma frequency $f_p(z) = 8.98\sqrt{N_e(z)}$ Hz, which is a function of the altitude, where N_e is the electron density in m^{-3} (Gurnett and Bhattacharjee 2005).

MARSIS data are presented as an ionogram, which is a plot of the echo strength versus frequency and time delay. In a typical ionogram (Fig. 5.8), it is possible to identify three types of echoes: a very strong spike at the local electron plasma frequency $f_{p\text{-local}}$, caused by excitation of electrostatic oscillations at the electron plasma frequency; a second echo from $f_{p\text{-local}}$ to the maximum plasma frequency in the ionosphere $f_{p\text{-max}}$, due to a reflection by the ionosphere; and a third echo from $f_{p\text{-max}}$ to the maximum sounding frequency, due to a surface reflection. In regions characterized by strong crustal magnetic fields, a fourth oblique echo occurs.

MARSIS data are providing an essential contribution in the characterization of the response of Martian ionosphere to various inputs such as solar EUV flux, energetic particles, crustal magnetic fields, and solar wind. They are also analyzed to study the effects due to seasons, latitude, and local time. An example of MARSIS data contribution to the study of a peculiar aspect of Martian ionosphere will be presented in Sect. 5.5.2.

Hereafter, two MARSIS scientific accomplishments, which involve completely different Martian aspects, will be described more in detail with the purpose of pointing out the versatility of radar data on planetary investigation.

5.5.1 3D Structure of the Mars North Pole (Ice Cap and Basal Unit)

As mentioned before, the primary scientific task of MARSIS and SHARAD is the detection of water on Martian surface and subsurface. The observations of both radars are particularly devoted to Martian poles study. The two radars provide an opportunity to extend our knowledge of a planetary body to the third dimension. In fact, planetary subsurface radar sounders represent a new generation of remote sensing instruments which, through volumetric (3D) study, allow to detect features that are difficult to investigate in vertical profiles.

In order to complete the description of radar observations over Martian poles, it is worth noting that, due to flight-dynamics aspects of spacecraft orbits, there is a lack in the coverage of about 7° around the two poles, but, due to the limited extension of those areas, it is possible to acquire an accurate information of the overall Martian pole structure.

Polar campaigns have been planned periodically for both MARSIS and SHARAD instruments. MARSIS campaigns are focused alternatively on north or south pole, due to Mars Express highly elliptic orbit, while SHARAD campaigns can be performed simultaneously on both poles. Figure 5.9 shows the traces on ground of Mars Express orbits during the MARSIS north polar campaign performed on 2011.

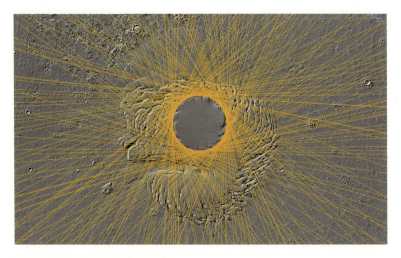

Fig. 5.9 MARSIS 2011 north polar campaign

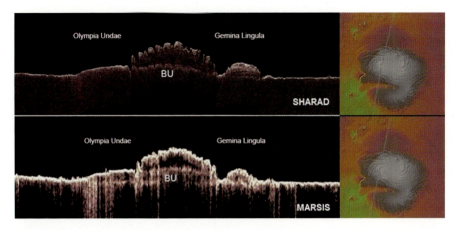

Fig. 5.10 MARSIS and SHARAD Mars north pole radargrams

For each orbit, it has been produced a radargram. Figure 5.10 shows a radargram generated with MARSIS data along with a radargram generated with SHARAD data for comparison. As mentioned before, SHARAD has the capability to better discriminate subsurface interfaces, revealing Martian superficial stratigraphy, while MARSIS signal can penetrate deeper into Martian surface achieving information even of the basal unit (BU) structure. This peculiarity of MARSIS data suggested to study the feasibility to produce a three-dimensional representation of north pole ice cap along with the underlying basal unit.

Each radargram can be considered as a cross section of the Martian crust (Fig. 5.11). Composing the radargrams in a three-dimensional structure, Frigeri et al. (2012) generated three-dimensional images in which there are easily recognizable Olympia Undae, Gemina Lingula, and Chasma Boreale (Fig. 5.12).

The peculiarity of this approach is the possibility to study Martian north pole from different perspectives and even to discriminate the ice polar cap (in cyan) and the basal unit (in yellow) (Fig. 5.13). This study revealed a cavity under Olympia Cavi, 80 km wide and 400 km long, buried under upper layered deposits (Fig. 5.14).

5.5.2 Mars Total Electron Content Evaluation

Mars has strong local magnetic fields (up to 1,600 nT at ∼90 km of altitude), connected to characteristics of the Martian crust (Acuna et al. 1999; Nielsen et al. 2006), as the data collected by Magnetometer/Electron Reflectometer (MAG/ER) onboard Mars Global Surveyor (MGS) proved. But the Red Planet is not characterized by an appreciable global magnetic field. Therefore, its ionosphere is not protected from direct interaction with the solar wind which can directly interact with it altering its local properties.

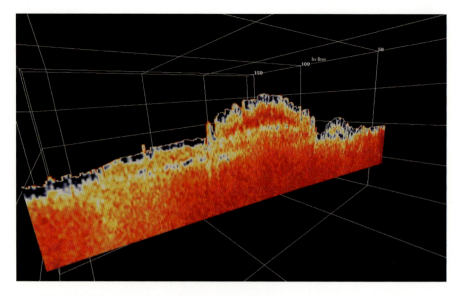

Fig. 5.11 MARSIS radargram considered as a 3D unit

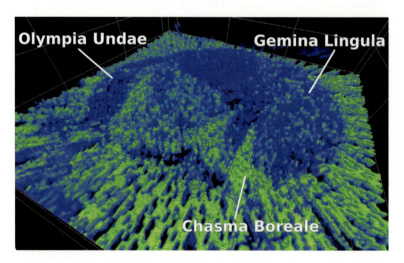

Fig. 5.12 Three-dimensional representation of Martian north pole

Consequently, the ionosphere of Mars varies considerably from the day side to the night side. The day side is also dependent on photoionization of atmospheric neutrals by the solar EUV flux, solar cycle, solar rotation, solar flares, cosmic rays, and gamma ray bursts, while the night side depends on neutral density, day-night plasma transport, and recombination rates (Lillis et al. 2009). All these factors combined with the Martian crustal magnetic field can produce some areas where

5 Radar Exploration of Mars: Recent Results and Progresses

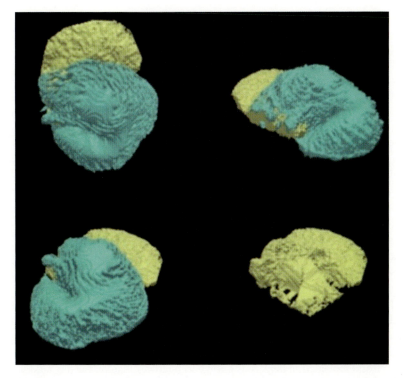

Fig. 5.13 Martian north pole from different perspectives: ice polar cap (*cyan*) and basal unit (*yellow*)

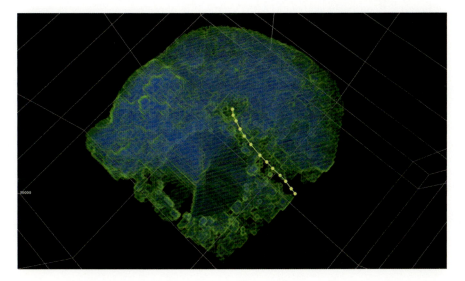

Fig. 5.14 The cavity under Olympia Cavi, buried under upper layered deposits

the ionization of the atmosphere is higher and the recombination is lower; as a consequence, high values of the electron density are present even on the night side (Safaeinili et al. 2007).

Many aspects of Martian ionosphere have been investigated in the past (Gurnett et al. 2008; Leblanc et al. 2008; Němec et al. 2010). In particular, it was proved that the total electron content (TEC) is higher where the local magnetic field is nearly vertical (Safaeinili et al. 2007), and it was observed that during solar particle events, the TEC increases by a factor of 2 or more (Lillis et al. 2010). Recently, it has been proved that MARSIS data in subsurface mode, usually used to detect subsurface features, can be processed in order to obtain reliable estimates of the TEC of the Martian ionosphere (Cartacci et al. 2012). This task was pursued using the Contrast Method (CM). CM is a tool developed to compensate distortion effects on radar signals due to the Martian ionosphere in order to analyze Martian surface and subsurface more efficiently.

Radar signals have a penetration depth in the subsurface approximately proportional to their wavelength. But, in order to propagate through the Martian ionosphere, MARSIS needs to operate with frequencies above the maximum plasma frequency, f_p, given by $f_p(z) = 8.98\sqrt{N_e}(z)$, where N_e, the electron density in m^{-3}, is a function of the altitude z (Gurnett and Bhattacharjee 2005). For this reason, Martian areas to be observed by the radar are selected mainly on the night side where the values of the sun elevation angle (SEA), which is the angle between the direction of the geometric center of the Sun apparent disk and the horizon, are negative and the electron density is known to be lower than on the day side. In fact, all frequencies lower than f_p are reflected. Frequencies higher than f_p are attenuated, delayed, and are subject to phase distortion proportionally to the electron density of the ionosphere.

In order to properly analyze MARSIS data, it is, therefore, necessary to correct the phase distortion of the signal or, at least, to reduce it. Many methods have been suggested (Safaeinili et al. 2003; Ilyushin and Kunitsyn 2004; Mouginot et al. 2008; Zhang et al. 2009) through the years; compared to them, CM is a simpler algorithm which proved to be a very reliable method.

The correctness of the application of CM to Mars TEC estimation is also proved by matching, with remarkable approximation, the theoretical model proposed by Chapman in 1931 (Fig. 5.15).

From the analysis of MARSIS data in subsurface mode, it has been possible to estimate the TEC of the Martian ionosphere and to produce the related maps, to evaluate its dependence by solar activities, and to confirm the result of Safaeinili et al. (2007) about a connection between the orientation of the crustal magnetic field and the measured TEC, which appears to be higher where the magnetic field direction is quasi-vertical and lower where the magnetic field direction is quasi-horizontal (Fig. 5.16).

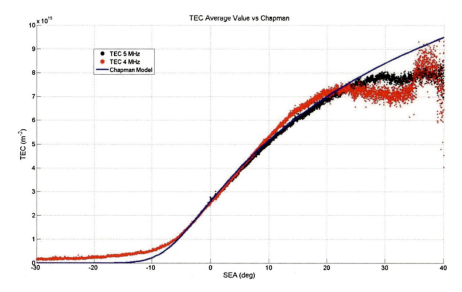

Fig. 5.15 Average TEC measured with MARSIS for 4 and 5 MHz, compared to TEC Chapman model (Reproduced from Cartacci et al. 2012)

5.6 New Radar Frontiers in Solar System Exploration

The successful appliance of radar sounders in planetary exploration stimulated the analysis of possible addition of radars as part of the payload in upcoming space missions dedicated to planetary object investigations. Since the MARSIS and SHARAD subsurface radar sounders have been observing the Martian polar terrains, which are considered a close analogue to the material forming the crusts of Jovian satellites Europa and Ganymede, it has been assessed that a radar sounder could provide unique information in the study of the geological and geophysical evolution of icy satellites.

A radar sounder, Radar for Icy Moons Exploration (RIME), will be part of ESA's first large-class mission in Cosmic Vision Program, JUICE. The JUpiter ICy moons Orbiter Mission (JUICE) will be the first orbiter on an icy moon and will investigate the emergence of habitable worlds around gas giants, characterizing Ganymede, Europa, and Callisto as planetary objects and potential habitats, and will also explore the Jupiter system as an archetype for gas giants. JUICE will first orbit Jupiter for ~2.5 years, providing 13 flybys of Callisto and 2 of Europa, and then will orbit Ganymede for 9 months. Launch is scheduled for 2022 with Jupiter arrival in 2030 and Ganymede orbit insertion in 2032.

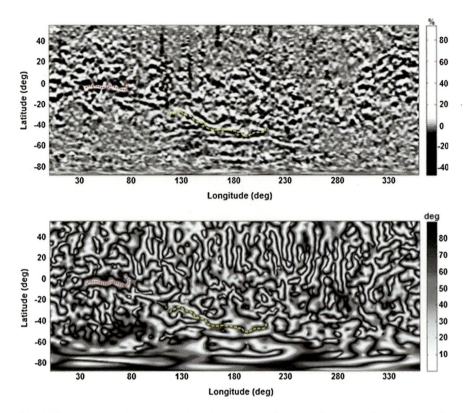

Fig. 5.16 Latitude-longitude maps of TEC (*upper panel*) and α (*lower panel*). α is the angle between the ambient magnetic field vector of internal origin and the local vertical direction. *Yellow-dashed* (*red dotted*) lines highlight quasi-vertical (*horizontal*) field (Reproduced from Cartacci et al. 2012)

Ganymede is the largest satellite in the solar system. It is known to have a deep ocean, an internal dynamo, and an induced magnetic field which makes it unique among Jupiter icy moons, and it is the richest in crater morphologies. It is an archetype of water worlds, and it is the best example of liquid environment trapped between icy layers.

Callisto is the best place to study the impactor history. It is the only known example of non-active but ocean-bearing world, and it is the witness of early ages.

Europa has a deep ocean and Juice will investigate if it is an active world. It is also the best example of liquid environment in contact with silicates.

RIME mission will be focused on subsurface exploration of Jupiter icy moons. It is based on a solid heritage from successful radar sounders operating at Mars. RIME will address the following JUICE mission goals: characterize Ganymede as a planetary object and possible habitat, explore Europa's recently active zones, and study Callisto as a remnant of the early Jovian system.

RIME is a synthetic aperture (SAR) subsurface sounding radar with a central frequency of 9 MHz and a bandwidth up to 3 MHz. Its lateral resolution is $\sim$0.5 km by $\sim$5 km, while its vertical resolution is selectable between 30 and 90 m (in ice). It has a penetration depth >9 km (in ice).

References

Acuna MH, Connerney JEP, Ness NF, Lin RP, Mitchell D, Carlson CW, McFadden J, Anderson KA, Reme H, Mazelle C, Vignes D, Wasilewski P, Cloutier P (1999) Global distribution of crustal magnetization discovered by the Mars Global Surveyor MAG/ER experiment. Science 284(5415):790–793. doi:10.1126/science.284.5415.790

Arvidson RE, Coradini M, Carusi A, Coradini A, Fulchignoni M, Federico C, Funiciello R, Salomone M (1976) Latitudinal variation of wind erosion of crater ejecta deposits on Mars. Icarus 27(4):503–516. doi:10.1016/0019-1035(76)90166-4

Baker VR, Strom RG, Gulick VC, Kargel JS, Komatsu G, Kale VS (1991) Ancient oceans, ice sheets and the hydrologic cycle on Mars. Nature 352:589–594. doi:10.1038/352589a0

Carr MH (1987) Water on Mars. Nature 326:30–35. doi:10.1038/326030a0

Carr MH (1996) Channels and valleys on Mars: cold climate features formed as a result of a thickening cryosphere. Planet Space Sci 44:1411–1423. doi:10.1016/S0032-0633(96)00053-0

Cartacci M, Amata E, Cicchetti A, Noschese R, Giuppi S, Langlais B, Frigeri A, Orosei R, Picardi G (2012) TEC Analysis from MARSIS subsurface data. Icarus 223(1):423–437. doi:10.1016/j.icarus.2012.12.011

Carter LM, Campbell BA, Watters TR, Phillips RJ, Putzig NE, Safaeinili A, Plaut JJ, Okubo CH, Egan AF, Seu R, Biccari D, Orosei R (2009) Shallow radar (SHARAD) sounding observations of the Medusae Fossae Formation, Mars. Icarus 199:295–302

Chapman S (1931) The absorption and dissociative or ionizing effect of monochromatic radiation of an atmosphere on a rotating Earth. Proc Phys Soc Lond 43:26–45

Di Achille G, Hynek BM (2010) Ancient ocean on Mars supported by global distribution of deltas and valleys. Nat Geosci 3:459–463. doi:10.1038/ngeo891

Frigeri A, Orosei R, Cartacci M, Cicchetti A, Mitri G, Giuppi S, Noschese R, Picardi G, Plaut JJ (2012) Radar soundings of the North Polar Cap of Mars. Paper presented at 3rd Moscow solar system symposium, IKI – Space Research Institute of Russian Academy of Sciences, Moscow, 8–12 October 2012

Gurnett DA, Bhattacharjee A (2005) Introduction to plasma physics: with space and laboratory applications. Cambridge University Press, Cambridge

Gurnett DA, Huff RL, Morgan DD, Persoon AM, Averkamp TF, Kirchner DL, Duru F, Akalin F, Kopf AJ, Nielsen E, Safaenili A, Plaut JJ, Picardi G (2008) An overview of radar soundings of the Martian ionosphere from the Mars Express spacecraft. Adv Space Res 41:1335–1346

Head JW, Roth R (1976) Mars pedestal crater escarpments: evidence for ejecta-related emplacement. Paper presented at symposium on planetary cratering mechanics. U.S. Geological Survey, Flagstaff, AZ

Ilyushin YA, Kunitsyn VE (2004) Methods for correcting ionosphere distortions of orbital ground-penetrating radar signals. J Commun Technol Electron 49:154–165

Jordan R, Picardi G, Plaut J, Wheeler K, Kirchner D, Safaeinili A, Johnson W, Seu R, Calabrese D, Zampolini E, Cicchetti A, Huff R, Gurnett D, Ivanov A, Kofman W, Orosei R, Thompson T, Edenhofer P, Bombaci O (2009) The Mars express MARSIS sounder instrument. Planet Space Sci 57:1975–1986

Leblanc F, Witasse O, Lilensten J, Frahm RA, Safaenili A, Brain DA, Mouginot J, Nilsson H, Futaana Y, Halekas J, Holmström, M, Bertaux JL, Winningham JD, Kofman W, Lundin R (2008) Observations of aurorae by SPICAM ultraviolet spectrograph on board Mars Express: simultaneous ASPERA-3 and MARSIS measurements. J Geophys Res 113(A08311). doi:10.1029/2008JA013033

Lillis RJ, Fillingim MO, Peticolas LM, Brain DA, Lin RP, Bougher SW (2009) Nightside ionosphere of Mars: modeling the effects of crustal magnetic fields and electron pitch angle distributions on electron impact ionization. J Geophys Res 114(E11009)

Lillis RJ, Brain DA, England SL, Withers P, Fillingim MO, Safaeinili A (2010) Total electron content in the Mars ionosphere: temporal studies and dependence on solar EUV flux. J Geophys Res 115(A11314)

Luo W, Stepinski TF (2009) Computer-generated global map of valley networks on Mars. J Geophys Res 114 (E11010). doi:10.1029/2009JE003357

McCauley JF (1973) Mariner 9 evidence for wind erosion in the equatorial and mid-latitude regions of Mars. J Geophys Res 78(20):4123–4137. doi:10.1029/JB078i020p04123

Mouginot J, Kofman W, Safaeinili A, Herique A (2008) Correction of the ionospheric distortion on the MARSIS surface sounding echoes. Planet Space Sci 56:917–926

Mouginot J, Pommerol A, Beck P, Kofman W, Clifford SM (2012) Dielectric map of the Martian northern hemisphere and the nature of plain filling materials. Geophys Res Lett 39(L02202). doi:10.1029/2011GL050286

Němec F, Morgan DD, Gurnett DA, Duru F (2010) Nightside ionosphere of Mars: radar soundings by the Mars Express spacecraft. J Geophys Res 115(E12009). doi:10.1029/2010JE003663

Nielsen E, Zou H, Gurnett DA, Kirchner DL, Morgan DD, Huff R, Orosei R, Safaenili A, Plaut JJ, Picardi G (2006) Observations of vertical reflections from the topside Martian ionosphere. Space Sci Rev 126:373–388

Nunes DC, Smrekar SE, Fisher B, Plaut JJ, Holt JW, Head JW, Kadish SJ, Phillips RJ (2011) Shallow Radar (SHARAD), pedestal craters, and the lost Martian layers: Initial assessments. J Geophys 116 (E04006). doi:10.1029/2010JE003690

Picardi G, Sorge S, Seu R, Plaut JJ, Johnson WTK, Jordan RL, Gurnett DA, Provvedi F, Zampolini E, Zelli C (2000) MARSIS experiment: design and operations overview. In: Proceedings of SPIE 4084, eighth international conference on ground penetrating radar. doi:10.1117/12.383568

Picardi G, Plaut JJ, Biccari D, Bombaci O, Calabrese D, Cartacci M, Cicchetti A, Clifford SM, Edenhofer P, Farrell WF, Federico C, Frigeri A, Gurnett DA, Hagfors T, Heggy E, Herique A, Huff RL, Ivanov AB, Johnson WTK, Jordan RL, Kirchner DL, Kofman W, Leuschen CJ, Nielsen E, Orosei R, Pettinelli E, Phillips RJ, Plettemeier D, Safaeinili A, Seu R, Stofan ER, Vannaroni G, Watters TR, Zampolini E (2005) Radar soundings of the subsurface of Mars. Science 310:1925–1928

Plaut JJ, Picardi G, Safaeinili A, Ivanov AB, Milkovich SM, Cicchetti A, Kofman W, Mouginot J, Farrell WM, Phillips RJ, Clifford SM, Frigeri A, Orosei R, Federico C, Williams IP, Gurnett DA, Nielsen E, Hagfors T, Heggy E, Stofan ER, Plettemeier D, Watters TR, Leuschen CJ, Edenhofer P (2007) Subsurface radar sounding of the south polar layered deposits of Mars. Science 316:92–95. doi:10.1126/science.1139672

Safaeinili A, Kofman W, Nouvel JF, Herique A, Jordan RL (2003) Impact of Mars ionosphere on orbital radar sounder operation and data processing. Planet Space Sci 51(7–8):505–515

Safaeinili A, Kofman W, Mouginot J, Gim Y, Herique A, Ivanov AB, Plaut JJ, Picardi G (2007) Estimation of the total electron content of the Martian ionosphere using radar sounder surface echoes. Geophys Res Lett 34(L23204). doi:10.1029/2007GL032154

Seu R, Biccari D, Cartacci M, Cicchetti A, Fuga O, Giuppi S, Masdea A, Noschese R, Picardi G, Federico C, Frigeri A, Melacci PT, Orosei R, Croci R, Guelfi M, Calabrese D, Zampolini E, Marinangeli L, Pettinelli E, Flamini E, Vannaroni G (2007) The SHAllow RADar (SHARAD) Experiment, a subsurface sounding radar for MRO. Mem Soc Astron Ital Suppl 11:26–36

Skolnik MI (1980) Introduction to radar systems. McGraw-Hill, New York. ISBN 0070579091

Watters TR, Leuschen CJ, Plaut JJ, Picardi G, Safaeinili A, Clifford SM, Farrell WM, Ivanov AB, Phillips RJ, Stofan ER (2006) MARSIS radar sounder evidence of buried basins in the northern lowlands of Mars. Nature 444:905–908

Watters TR, Campbell BA, Carter LM, Leuschen CJ, Plaut JJ, Picardi G, Orosei R, Safaeinili A, Clifford SM, Farrell WM, Ivanov AB, Phillips RJ, Stofan ER (2007) Radar sounding of the Medusae Fossae Formation, Mars: equatorial ice or dry, low-density deposits? Science 318:1125–1128

Zhang Z, Nielsen E, Plaut JJ, Orosei R, Picardi G (2009) Ionospheric corrections of MARSIS subsurface sounding signals with filters including collision frequency. Planet Space Sci 57: 393–403

Chapter 6
Automatic Recognition of Impact Craters on the Martian Surface from DEM and Images

Tengyu Zhang and Shuanggen Jin

Abstract Impact craters are the most outstanding and attractive geomorphological features on the surface of the planets, showing variety and complexity of the surface morphology. The accurate recognition of impact craters on Mars is very useful to analyze and understand the relative dating of Martian surface. In this chapter, four crater-detection methods have been presented and discussed with various extent of discrimination ability on Martian images or topography data. The modified ad boosting approach demonstrates the best performance in classification of craters, while the algorithms based on topography data have low efficiency in automatic detection. Comparing to previous solutions, the modified ad boosting method has greatly improved the detecting performance of the algorithm and reduced detection time.

Keywords Automatic detection • Craters • DEM • Images

6.1 Introduction

Impact craters are natural probes of target surface properties. Recognitions of impact craters have revealed lots of findings (Cheng et al. 2004; Johnson et al. 2005; Kim et al. 2005; Sawabe et al. 2006). Surface dating is one of the most common applications by the analysis of the size and distribution frequency of craters, indicating that a catalogue with the size and distribution is very useful in combination with DEM (Digital Elevation Model) data (Crater Analysis Technique Working Group 1979). The shapes of craters on Mars have lots of different types according to the interior morphology (central peaks, peak rings, central pits, wall terraces) and ejecta structures (pedestal, pancake, rampart, lobate, fluidized, radial

---

T. Zhang (✉)
Shanghai Astronomical Observatory, Chinese Academy of Sciences, Shanghai 200030, China
e-mail: zhangty@shao.ac.cn

S. Jin
Shanghai Astronomical Observatory, Chinese Academy of Sciences, Shanghai 200030, China

Bulent Ecevit University, Zonguldak, Turkey
e-mail: sgjin@shao.ac.cn

© Springer-Verlag Berlin Heidelberg 2015
S. Jin et al. (eds.), *Planetary Exploration and Science: Recent Results and Advances*,
Springer Geophysics, DOI 10.1007/978-3-662-45052-9_6

or lunar like, transitional, or diverse) (Barlow et al. 2003). Previous investigations of the history of surface modification are based on crater morphology because the impact craters are the only surface feature with initial known shape, which is one of the most direct and effective approaches for human beings to better understand the planets.

Another important and possible application of crater recognition is for a spacecraft to choose a landing site (Leroy et al. 2001). The spacecraft pose is always refined for safe landing using conspicuous landmarks, such as craters, observed by a visual sensor. It is very necessary to propose a robust method to detect craters quickly and effectively. The detected craters in the image are matched with the existing catalogue in order to avoid the uneven or hilly sites corresponding to the craters.

All the previous studies on craters are based on the specific and reliable catalogue of craters. Several teams aim to assemble as complete a global catalogue of Martian impact craters as possible. It is laborious and complicated to construct a useful catalogue (Barlow 1988; Stepinski et al. 2009; Salamunićcar et al. 2011). The MA130301GT is one of the most complete and specific Martian crater catalogues constructed by Salamunićcar et al. (2011), which contains the major currently available manually assembled catalogues and the automatically detected catalogues. This catalogue has already integrated all the attributes available in the old catalogue (Barlow, Rodionova, Boyce, Kuzmin, Stepinski). MA130301GT provides these catalogues with (1) the correlation between various morphological descriptors from used catalogues; (2) the correlation between manually assigned attributes and automated depth/diameter measurements from MA75919T and our CDA; (3) global surface dating; (4) average errors and their standard deviations for manually and automatically assigned attributes, such as position coordinates, diameters, and depth/diameter ratios; and (5) positional accuracy of features in the used datasets according to the defined coordinate system referred to as Mars Global Digital Image Mosaic MDIM 2.1, which incorporates 1,232 globally distributed ground control points, while our catalogue contains 130,301 cross-references between each used dataset. There is a considerable improvement in comparison with the completeness of the Rodionova ($\sim$10 km), Barlow ($\sim$5 km), and Stepinski ($\sim$3 km) catalogues.

It is fundamental to develop an effective crater-detection algorithm (CDA) to construct the crater catalogue. Salamunićcar and Lončarić (2012) reported that there were 82 optical-based CDAs, 39 DEMs (Digital Elevation Maps) based on CDAs, 16 CDAs that could utilize both optical and DEM images, and only 3 CDAs from other types of images. Fifty-nine percent CDAs are optically based because most of the available images are optical ones. The second largest group of CDAs is DEM based (28 %). The reason is that it is easier to develop a CDA for DEM data, wherein issues like the position of the Sun, spacecraft, and the lunar/planetary body are irrelevant. Changes in these values cause the changes in light direction and extent of shadows and therefore make a development of optical-based CDAs more challenging. A development of universal CDAs, which can process optical images as well as DEMs, is even more challenging because algorithms applicable to the first type of images are usually not applicable to the second type and vice versa.

6 Automatic Recognition of Impact Craters on the Martian Surface... 103

In this chapter, the recognition methods of global Martian impact crater catalogue are reviewed and presented in the following. At the same time, their performances and shortcomings in detecting craters are discussed. A modified adaboosting method and results of automatic recognition of impact craters on the Martian surface are presented and discussed.

6.2 Recognition Methods

6.2.1 Recognition from Image

The exploration of the solar system by automated probes has obtained large numbers of images of the surfaces on planets and satellites. Image analysis plays an important role in archiving, retrieval, processing, and interpretation of large amounts of image data as well as classification of all the resources. However, it is difficult to automatically recognize impact craters precisely from the planetary images due to the lack of distinguishing features, heterogeneous morphology in images, and huge amount of sub-kilometer craters in high-resolution planetary images (Kim et al. 2005).

There are a number of techniques in the field of image processing and pattern recognition with the purpose of the automated detection of impact craters from images of planetary surfaces. Most previous studies focused on a given technique primarily, like template matching (Michael 2003), texture analysis (Barata et al. 2004), Hough transform (Bue and Stepinski 2007), neural networks (Smirnov et al. 2002), or genetic algorithm (Brumby et al. 2003). It is difficult to develop a single methodology with the ability of detecting circular shapes in a wide size range and on an enormous diversity of terrains. Each of those previously published methodologies for automatic crater detection has its advantages and disadvantages, but so far, none of them have shown good performance to be as robust as enough to be applied as a stand-alone procedure with satisfactory final results. Naturally, we would like to try to integrate different approaches and fully utilize their strength. In the following section, two effective and fresh approaches will be briefly introduced.

6.2.1.1 Template matching

It is difficult to develop a single methodology to recognize craters' circular shape within a quite wide size range and an enormous diversity of terrains. This proposed approach for the identification of impact craters consists of three main phases in sequence (Bandeira et al. 2007): (1) candidate selection, (2) template matching, and (3) crater selection. Thus, there is a preprocessing phase in which the areas corresponding to crater rims are identified in a gray-level image, while most of the present noise is eliminated. By using the fast Fourier transform (FFT), the binary image of the candidate area is submitted to a template-matching procedure. Therefore, there will be a probability volume, 3D matrix constructed for further analysis to determine the location and dimension of the impact craters that are present in a test image.

Candidate Selection

Because of the crash of bolide on the planetary surface, the impact craters are bowl-shaped depressions due to the propagation of a shock wave. As a result, they show a roughly regular circular shape when imaged from orbit. Because the marked contours by the elevated crater rims produce shadows, the difficulty in correctly detecting craters has increased. Furthermore, the shadows are far from regular so that they cannot be used as recognition features applied in detection. Therefore, some preprocessing steps need to be done in this section.

1. A mask M with $n \times n$ elements is centered over each pixel I_{uv} of the original image. The local mean m, the minimum and maximum of the values included in the mask, will be computed, which produce a new image A defined by Eq. 6.1.

$$A_{uv} = \max\left[m\,(M) - \min(M), \ \max(M) - m\,(M)\right] \tag{6.1}$$

2. After the first step, a threshold value T can be obtained.

$$T = \alpha\left[\max(A) - \min(A)\right] + \min(A) \tag{6.2}$$

where α is a constant (0.25), which is chosen after a thorough search for the best possible results for this type of image.

A binary image will be created when the threshold is applied to compute the new pixel B_{uv} and it is of highly contrasted local edges. This image is considered to be the best candidate for the application on the following steps.

Matching

After the first step of preprocessing the image, it is necessary to locate impact craters obtained in the previous phase. A simplified binary model of the specified features is constructed to serve as a template for the matching of craters rims. It is also necessary to incorporate a thickness which can represent the irregularity of the natural shapes of impact craters. Therefore, a binary template applied in this section is constructed like a white crown that is surrounded by a background in the shape of a black square, and each radius must be analyzed to be a corresponding different template.

Many matching methods can be applied for the purpose to calculate the similarity of the template and the scene image. In this section, the correlation between the scene image and a series of templates is computed through an implementation of FFT, which makes this problem a popular question in this field, including image processing and pattern recognition. The radii of the implemented templates vary from interior minimum 5 pixels to the maximum 100 pixels. The correlation will be easily obtained and be taken as the probability that a given pixel is the center of a crater with radius r.

Crater Detection

The identification of impact craters on an image is achieved through the analysis of the probability volume created as a result of the template-matching procedure that was followed in the previous phase. Each element of the probability volume is surrounded by six neighbors considering the 3D connectivity, and only the value of above 30 % will be remained as potential candidates for the next step analysis. The probability volume is used to compute the regional maxima by dilation. They reflect the presence not only of interesting features that most like craters but also some structures of other origins which results in some false detections. The maxima can be computed using a two-step procedure.

1. H-maxima transform is applied to suppress all maxima whose heights are smaller than a certain value h.

$$\text{HMAX}_h(P) = R_P(P - h) \tag{6.3}$$

2. The extended maxima EMAX can be given on the following after the first step:

$$\text{EMAX}_h(P) = \text{RMAX}[\text{HMAX}_h(P)] \tag{6.4}$$

The signature of a crater can be found in the horizontal planes of the volume; each corresponds to a given radius. Only those that contain extended maxima must be analyzed. Two more steps will be undertaken in order to identify those that are actually markers of craters. First, use an area-opening transform to filter the small objects that cannot be craters. Second, analyze the roundness of the remaining objects, and those objects whose roundness outside a given range (0.76–1.8) are eliminated. At the end of this sequence, the point with coordinates (u, v), which corresponds to the center of mass of each surviving object, is considered to be the center of a crater with radius r and probability p.

6.2.1.2 A Modified Adaboosting Method

Adaboosting approach is a widely used method in the field of pattern recognition, which is proved to be very effective for the automatic face detection (Viola and Jones 2004). Some results have been obtained in the field of crater detection based on this

Fig. 6.1 Nine types of Haar-like features

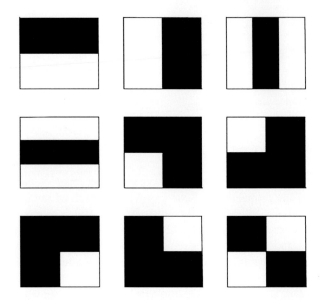

approach (Martins et al. 2009; Ding et al. 2011). The detection of impact craters on planetary surfaces can be formulated as a classical pattern recognition problem where the information extracted from different locations in the images is assigned to one of two classes: crater or noncrater. The decision of crater or noncrater is given by a series of binary classifier using a set of image features, such as Haar-like features (Fig. 6.1).

Algorithm Building the strong, texture-based classifier through Adaboosting

Input: Training set of crater candidates $(C_1, y_1), \ldots, (C_i, y_i)$, where C_i is the ith crater candidate image block and $y_i = 0, 1, \ldots$ for craters and noncraters, respectively

Output: The final strong classifier that assigns each C_i a probability of being a true crater

1. Initialize weights $w_{1,i} = (1/2m), (1/2l)$ for $y_i = 0, 1, \ldots$, where m and l are the number of craters and noncraters, respectively, in the training set.
2. For $t = 1$ to T, where T is a desired number of most discriminative weak classifiers.
3. Normalize the weights, $w_{t,i} = w_{t,i} / \sum_{j=1}^{n} w_{t,j}$
4. Select the weak classifier that minimizes the weighted error:

$$\varepsilon_t = \min_{f,p,\theta} \sum_i w_{t,i} |h(C_i, f, p, \theta) - y_i| \qquad (6.5)$$

5. Define $h_t(C) = h(C, f_t, p_t, \theta_t)$, where f_t, p_t, θ_t are the minimizers of ε_t.

6. Update the weights, $w_{t+1,i} = w_{t,i}\beta^{1-e_i}$, where $e_i = 0$ if C_i is classified correctly $e_i = 1$ if C_i is classified incorrectly $\beta_t = \varepsilon_t/(1-\varepsilon_t)$
7. End.
8. The final strong classifier is given by

$$H(C) = \begin{cases} crater & if \sum_{t=1}^{T}\alpha_t \cdot h_t(C) \Big/ \sum_{t=1}^{T}\alpha_t > \mu \\ non\text{-}crater & otherwise \end{cases} \tag{6.6}$$

where $\alpha_t = \log(1/\beta_t)$ and μ is a threshold probability.

The Modified Weak Classifier

In the traditional algorithm, a sequence of weak classifiers was generated and combined to assemble a strong classifier through a weighted boosting approach. A set of weak classifiers $h_t(C) = h(C, f_t, p_t, \theta_t)$ is defined like this:

$$h(C; f, p, \theta) = \begin{cases} 1 & if\ p \cdot f(C) \geq p \cdot \theta \\ 0 & else \end{cases} \tag{6.7}$$

where C is an image block representing a crater candidate and f is the numeric value of each feature. The discriminative power of the weak classifier is determined by the threshold θ and a polarity variable $p \in \{1, -1\}$ which means that the feature value should be greater or smaller than the threshold. In fact, each single weak classifier always identifies candidates into craters with low confidence because of its low discriminative power. However, in each round of the training process, a best weak classifier will be selected with the minimum weighted errors. If all these weak classifier has an ability of correct classification with a detection rate higher than 0.5, the final strong classifier can perform well by combining many weak classifiers.

Figure 6.2 shows the proportion of crater and noncrater candidates in all candidates when the feature value increases. Considering the distribution of the sum weight of all the candidates, we can conclude that the feature value of crater candidates distributes in a relatively narrow interval when comparing to noncrater candidates. It has been shown that only a single side threshold employed to determine craters will result in some false results, like the noncrater candidates with very high or low features considered to be craters. Therefore, the classifier with a dual threshold is constructed for the detection. It can be easily proved that smaller classification errors will be obtained when a weak classifier with a dual threshold is employed, which is introduced on the following:

$$h(C; f, \theta_1, \theta_2) = \begin{cases} 1 & if\ \theta_1 \leq f(C) < \theta_2 \\ 0 & else \end{cases} \tag{6.8}$$

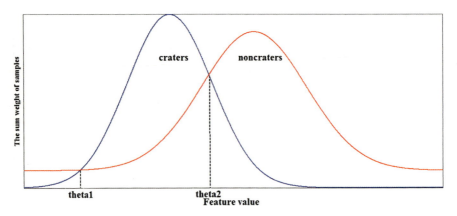

Fig. 6.2 The distribution of the sum weight of the candidates

Therefore, we should find these two thresholds in the next step, as the following:

1. According to the distribution range of the numeric value of the features, we will divide the ranges into some intervals to calculate the sum weight of craters $g(x)$ and noncraters $ng(x)$.
2. For every interval x_i, we can subtract $ng(x)$ from $g(x)$ and get $h(x) = g(x) - ng(x)$.
3. After obtaining the value of max $(h(x))$, the $x_{\max}$ corresponding to $h(x)_{\max}$ can be found. On the left and right side of $x_{\max}$, x_1 and x_2 can be got to satisfy the equation: $h(x) = 0$. In fact, the function $h(x)$ is not continuous, so the requirement cannot be satisfied strictly. Therefore, the value of x will be selected when $h(x)$ vary from negative to positive or positive to negative. The classification errors can be minimized when we use x_1 and x_2 as two thresholds.

Postprocessing

In this methodology, it is possible that we allow for multiple candidates, classified as "craters," to correspond to a single crater, and it cannot be distinguished automatically by the machine. Thus, in order to eliminate the repeat count from the approach, we need to do further work and analysis in this step. Based on the research by Bandeira et al. (2010), the craters labeled candidates fulfilling the following criteria are taken to correspond to a single crater and grouped together:

$$\frac{|d_i - d_j|}{\max(d_i, d_j)} \leq \alpha \quad \text{and} \quad \frac{dis(c_i, c_j)}{\max(d_i, d_j)} \leq \gamma \quad (6.9)$$

Here d_i and d_j are the diameters of craters i, j, where c_i and c_j are their centers. The values of α, γ are determined experimentally with a best choice for free parameters $\alpha = \gamma = 0.5$. The criteria for the selection is that the candidate

6   Automatic Recognition of Impact Craters on the Martian Surface...   109

of the highest value of H(C) within the group is considered the best one for the positive identification of the craters, while the remaining ones are discarded without becoming false negatives.

6.2.2   Recognition from DEM

All the image-based crater-detection approaches discussed above involve complicated, multistep algorithms to combat inherent limitations of imagery data. On the other hand, there are many factors, such as illumination effects, that increase the possibility of the biased estimate of detection results. Furthermore, it has some fundamental limits because of the 2D imagery data. Imagery data is well suited for human visual interpretation but ill-suited for automated processing because images are skewed representations of the landscapes they portray. However, we can easily obtain DEM data which are more direct and well-suited descriptors of planetary surface than images. Interestingly, no complete DEM-based crater-detection algorithms have been developed in the past. The less interest in DEM-based algorithm probably results from the scarcity and limited resolution of planetary topography data.

Currently, the available Martian DEMs are constructed from the Mars Orbiter Laser Altimeter Mission Experiment Gridded Data Record (MEGDR) with a resolution of $1/128°$ or ~ 500 m at the equator. It has limited the size of crater that can be detected to 3–5 km at most in diameter because of the coarse resolution of the DEM. This is on par with the lowest size of craters in the Barlow catalogue and small enough to extract a catalogue of scientific significance, and it has not met the increasing requirement in navigation or research of Martian surface history. In the near future, significantly higher-resolution DEMs of selected portions of Martian surface will be compiled from the Mars Express high-resolution stereo camera data which will greatly improve the ability of detecting the small size craters and contribute to our further research.

Detecting craters from DEM is, in principle, more convenient than from visual images because craters are landforms which can be calculated as terrain attributes to be distinguished from other morphological features. As the topographic data can be more easily available, more scientific teams pay attention in this field and have some basic tests to find an effective algorithm. In this section, we will introduce two new approaches in details.

6.2.2.1   Detecting Craters by Hough Transform

Preprocessing the DEM

A DEM is a raster dataset, where every pixel is assigned an elevation value and labeled by a set of coordinates with x and y. Profile curvature, which is the curvature in the gradient direction reflecting the change of slope angle, has been proved to show good performance as crater rim indicator.

$$\text{cur}(x, y) = \frac{\frac{\partial^2 h}{\partial x^2}\left(\frac{\partial h}{\partial x}\right)^2 + 2\frac{\partial^2 h}{\partial x \partial y}\frac{\partial h}{\partial x}\frac{\partial h}{\partial y} + \frac{\partial^2 h}{\partial y^2}\left(\frac{\partial h}{\partial y}\right)^2}{m\sqrt{n^3}} \quad (6.10)$$

where $m = (\partial h/\partial x)^2 + (\partial h/\partial y)^2$ and $n = m + 1$, the $\partial h/\partial x$ and $\partial h/\partial y$ are separately the first derivative of the elevation in the x and y direction, and $\partial^2 h/\partial x^2$ is the second derivative. Then, we use the value of curvature to produce a binary image of the site according to the following transformation:

$$I_k(x, y) = \begin{cases} 1 \text{ (black)}, & \text{for } k(x, y) \leq k_{th} \\ 0 \text{ (white)}, & \text{for } k(x, y) > k_{th} \end{cases} \quad (6.11)$$

Here, k_{th} is a threshold value for concave areas. The chosen value of k_{th} represents a tradeoff between selectivity and the presence of noise. Choosing k_{th} close to Min[$k(x, y)$] selects only areas with the highest concavity, eliminating noise but also misses the rims of smaller or degraded craters. In this test, we have chosen to use relatively large value of $k_{th} = -0.001$ in order to increase detection chances for small craters (Fig. 6.3).

Site Segmentation

Craters are enclosed topographic basins, which mean that the "flooding" algorithm can be used to determine crater areas in an idealized situation. This is because a real Martian landscape includes enclosed basins that are not craters, some craters are not basins due to degradation of their rims, and there are superimposed craters that form only a single fragment. All these realities prevent flooding algorithm from becoming

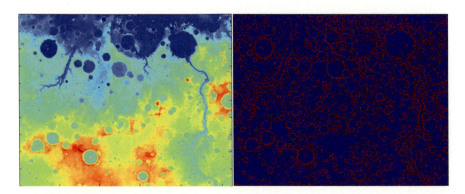

Fig. 6.3 The DEM and curvature map of the test site

6 Automatic Recognition of Impact Craters on the Martian Surface...

Fig. 6.4 The *left* is the binary image generated by the flooded algorithm, and the *right* one is the segmentation image by bounding box

as a stand-alone crater-detection algorithm. However, it can be used as assistant approach to reduce the computational cost by segmentation of the site image.

It will produce numerous fragments after the segmentation of the image into connected components, which are labeled by $i = 1, 2, \ldots, k$. Each fragment contains some degree of topographic depression, but not all of them are craters and some are even not craters. For each fragment, we calculate a bounding box. To allow for situations where craters can be slightly larger than fragments due to low top off point caused by rim degradation, each of them is slightly larger than the extent of its enclosed fragment. Because I_f and I_k are co-registered images, applying bounding boxes B_i to I_k divides a binary image of threshold profile curvature into small images, without cutting through the craters. We detect craters from each small image separately. This procedure guarantees that all detectable craters in the entire site are detected (possibly more than once). The benefit of site segmentation is the much-reduced computational cost of crater detection (Fig. 6.4).

Selecting Crater Candidate

After the first step in processing DEM data, we can easily identify the crater rims by using the binary profile curvature map. There are a few more steps before we can detect craters in the binary images. First, the morphological closing operation followed by a thinning operation needs to be done with a 3×3 structuring element to smooth the edges of structures and to eliminate small holes in them. Then, in order to reduce all lines to a single pixel thickness, we utilize the thinning operation for skeletonization. In the next step, we can use the circular Hough transform to detect crater candidates (Bue and Stepinski 2007). In the third step, we will examine each candidate using our confirmation algorithm which either accepts a candidate as a crater and adds it to the catalogue or rejects it. Finally, an elimination algorithm will be run on the entire catalogue to remove duplicate detections.

6.2.2.2 Crater Detection by Using Morphological Characteristics

Detection of Crater Centers

Two different strategies are developed in this process to detect the centers of simple and complex crater candidates according to their different morphological characteristics (Xie et al. 2013). In the first strategy, we can identify and extract the crater centers by local minima filter. Ordinarily, a pixel will be recognized as a local minimum if its value remains the same after being filtered. Here, we use a multi-scale square window to search the test image. However, during the experiments, there are usually several connected minima located at the bottom of a single crater. This will reduce the method's efficiency in the following steps. Thus, during filtering, if another minimum has already been detected within the extent of the filter, the new one will be marked and finally removed from the results. The radius of the filter determines the minimal size of detectable craters. To examine the method's ability to detect small craters, the radius of the filter is set to three pixels in this letter as a crater's morphology cannot be effectively presented with fewer pixels.

For the complex craters with central peaks or other structure instead of local minima, it is necessary to develop another strategy to handle them. In this section, we will detect the crater centers by using the features they have in common in the slope matrices of DEM. No matter what structures a complex crater possesses, there is always a flat and symmetrical area between its wall and center. Even in complex situations, when craters are superposed and intersected, the flat areas of different ones will be split by the walls and thus can be easily distinguished as well.

Detection of Crater Rims

After the detection of crater centers in the first step, we will detect the rims of simple craters and complex craters in two different datasets, respectively. The original DEMs are used as the basic data to detect the rims of simple craters. In the radial direction of a detected crater center, a local maximum can represent the rim of the crater. Thus, a complete revolution of scan lines from a crater center is used to detect and mark the rim points. However, numerous pixels will be repeatedly scanned on a frequent basis, and this will greatly reduce the efficiency of the CDA. Thus, a square scan frame is used to replace scan lines and the length of the square will increase gradually. Here the original side length of the frame is three pixels. The following points in the same direction will be marked and no longer be checked after a rim pixel is detected.

When it comes to complex craters, there are much more complex local variations in DEMs, particularly the central peaks near the centers, disturbing the process of detection of the radial local maxima. Thus, the slope matrices are used instead.

Circle Fitting and Crater Checking

The circle fitting method used in this CDA is least square circle fitting. Based on the coordinates of the detected rim pixels, we can get the best fit circles representative of craters with the criteria by finding the minimal value of errors, $E(a, b, r)$.

$$E(a, b, r) = \sum \left[(x - a)^2 + (y - b)^2 - r^2 \right]^2 \qquad (6.12)$$

where a and b are the coordinates and r is the radius of a fitted circle and x, y are the coordinates of rim pixel. In order to remove some false candidates in the detection, three criteria, including a restriction on rim completeness, the slope filter, and the depression filter, are used. It is normally checked by comparing the sum of discrete pixels (SDPs) on the rim and the perimeter of the corresponding circle. There is some kind of linear relationship between the SDP and the ratio of the radius R of the circle and the resolution g of the DEM. The coefficients of the relationship are presented as follows:

$$SDP = 5.658 \ R/g + 11.95 \qquad (6.13)$$

The threshold can then be set according to the applied DEM data. Secondly, the slope filter will be aimed at checking the average slope near the rims of craters. Finally, the depression filter will be used to check the ratio of the area of depression to that of non-depression inside a crater. After all these steps have been done, the final detection results can be obtained.

6.3 Validation of Methods

It is a fact that different geologists or even the same geologist at different times would assign slightly different coordinates and radius for the same crater. It has great possibility that two different craters have similar coordinates or radius in the catalogue. Also, it has to be expected that different CDAs would not assign identical coordinates and radius to the same crater, which may have already been labeled in a ground truth (GT) catalogue. A ground truth catalogue, which contains the locations and sizes of known craters, is an important element in the evaluation of CDAs developed for a wide range of applications.

The importance of crater registration lies in the fact that, in general, every identification (manual or machine) of a given crater results in a different set of assigned parameters. Manual crater mappers assign coordinates of crater center and its diameter based on their criteria. Since this is done by human beings, the resulting catalogues are considered to be ground truth. However, manual assignment of the crater center and diameter may vary even between successive determinations by the same human operator, so in fact we really have a fuzzy GT. Different automated methods use different algorithms to assign those parameters and in principle can be even less correct. The issue is how to reconcile those measurements. In an attempt to solve this issue, there is a framework proposed by Salamunićcar (Salamunićcar and Lončarić 2008) for the solution of objective CDA evaluation which is an effective approach for this problem.

The proposed framework consists of the following elements that must be strictly defined for objective evaluation of CDAs:

1. Measurement of differences between craters
2. Specification of test field data
3. Specification of a ground truth catalogue
4. Selection of methodology for training and testing
5. Measurement of CDA performance

In order to produce subjective results, it is necessary to develop a solution to automate the matching process. According to craters' radiuses, the relative values represented the differences in position and size between two craters can be obtained. A recommended definition is introduced in this paper for the measurement of crater difference f_m. Let r_1 and r_2 be radiuses of two craters, where $r_1 \geq r_2$; let d be the distance between the crater centers, and let $f_c \geq 0$ be a constant value called craters' difference factor. If and only if the following two equations are satisfied, the two craters are considered to belong to the same crater:

$$f_m = \max\left(\frac{r_1}{r_2} - 1, \ \frac{d}{r_2}\right) \tag{6.14}$$

$$f_m < f_c \tag{6.15}$$

If the threshold is defined for correct matching f_c, it can be determined by checking all possible combinations within the GT catalogue.

6.4 Results and Discussion

In this chapter, we have already presented four comprehensive methods in automatically detecting craters from images and from digital topography separately. Each of them has shown their abilities in detection of craters. In order to evaluate the performance of different methods, the quality factors true detection rate (TDR) and false detection rate (FDR) are introduced, and they are computed as follows:

$$\text{TDR} = \frac{\text{TD}}{\text{GT}} \cdot 100 \ \% \tag{6.16}$$

$$\text{FDR} = \frac{\text{FD}}{\text{TD} + \text{FD}} \cdot 100 \ \% \tag{6.17}$$

6 Automatic Recognition of Impact Craters on the Martian Surface... 115

where TD is the number of true detections, FD is the number of false detections, and GT is the number of ground truth. The ground truth served as the basis for a quantitative assessment of the methodology.

6.4.1 Performance Evaluation

6.4.1.1 Recognition Results from Images

Four regions are chosen for the test in Bandeira et al. (2007). The average value of the TDR is 86.57 %, which is an amazing result comparing to the ever published results with automatic detection method. In region C, the method shows the best performance in detecting craters maybe because of the relatively low number of craters. However, four regions all show good consistency between each other even though the number of GT is substantially different. The difference between them may result from the slightly different geomorphological settings in regions.

Because of the good performances in face recognition field, the Adaboosting approach has been introduced to detecting craters. Adaboosting is a machine learning method, which is different from the unsupervised method, like template matching. By using the series of weak classifiers from the training set, we can finally build a strong classifier for the classification. The detections will be automatically evaluated as true or false craters by comparison with the manually built truth. It is clearly that the performance varies with the change of the threshold. When a lower threshold is determined in the classification, higher true detection rate (TDR) and false detection rate (FDR) are obtained at the same time.

In Tables 6.1 and 6.2, we have shown the performances of two automatic crater-detection methods within the planetary images. When the threshold is set to be 0.75, the modified Adaboosting method performs better than the template-matching method.

6.4.1.2 Recognition Results from DEM

Since the resolution of DEM data on Mars cannot be obtained as high as the planetary images, we can only detect some relatively large craters based on DEM

Table 6.1 Detection rates

	Original method		Improved method	
Threshold	TDR (%)	FDR (%)	TDR (%)	FDR (%)
0.55	92.3	40.1	97.3	40.3
0.65	88.2	25.6	94.2	28.3
0.75	83.9	17.6	91.2	18.6
0.85	75.5	8.8	85.2	10.1

Table 6.2 Detection rates

Region	GT	TD	TDR	FD	FDR
<5 km	344	256	74.4	125	32.8
5–10 km	190	123	64.7	72	36.9
10–15 km	65	52	80	15	22.4
15–20 km	26	21	80.8	3	12.5
20–25 km	15	13	86.6	4	23.5
>25 km	32	30	93.8	4	11.8
Total	672	495	73.4	223	31.6

Table 6.3 Detection rates

Region	TDR	FDR
A	55.2	7.8
B	73.3	6.8

data. In this work by Bue and Stepinski (2007), the test site is chosen in a heavily cratered terrain covering almost 1.0×10^6 km^2. The craters listed in Barlow catalogue serve as the ground truth to evaluate the performance of this method.

A method based on the geomorphological features of Martian craters is introduced by Xie et al. (2013). In Table 6.3, we list the performance of this method.

6.4.2 Discussion

There are two categories of crater-detection algorithm, which are image based and topography based separately. We have presented these methods and shown their performances. Generally speaking, better detection rate can be obtained when the approach is based on images rather than topography. It is because the resolution of planetary images is always higher than topography contributing to detecting more small craters. The heavily degraded craters and overlapped craters are difficult to be detected in DEM because of the limited resolutions. As more and more collections of images and topography data at higher spatial resolution can be obtained, more efficient detection methods are very necessary, even combining images with topography data.

6.5 Conclusion

As some algorithms have shown their good performances, the successful detection of fresh well-formed craters is not hard to achieve. However, the detection of eroded craters as well as the craters which are partially erased is challenging because, with the increase of sensitivity, the number of false detections increases as well. In this chapter, four crater-detection methods have been presented and discussed with various extent of discrimination ability on planetary images or topography data. The modified Adaboosting approach demonstrates the best performance in classification of craters. The algorithms which are based on topography data are of low efficiency in automatic detection. However, considering that topography data can provide 3D structure of craters, it is necessary to develop new algorithms to improve efficiency. In the future, more work should be done to investigate the possibility of using a hybrid method by combining optical images with topography data.

References

Bandeira L, Saraiva J, Pina P (2007) Impact crater recognition on Mars based on a probability volume created by template matching. IEEE Trans Geosci Remote Sens 45(12):4008–4015. doi:10.1109/TGRS.2007.904948

Bandeira L, Ding W, Stepinski TF (2010) Automatic detection of sub-km craters using shape and texture information. Lunar and Planetary Science XLI (CD-ROM), Abs. 1144

Barata T, Alves EI, Saraiva J, Pina P (2004) Automatic recognition of impact craters on the surface of Mars. In: Campilho A, Kamel M (eds.) Image analysis and recognition. Lecture notes in computer science, vol 3212. Springer, New York, pp 489–496

Barlow NG (1988) Crater size-frequency distributions and a revised Martian relative chronology. Icarus 75(2):285–305

Barlow NG, Barnes CW, Barnouin-Jha OS, Boyce JM, Chapman CR, Costard FM, Craddock RA, Garvin JB, Greeley R, Hare TM, Kuzmin RO, Mouginis-Mark PJ, Newsom HE, Sakimoto SEH, Stewart ST, Soderblom LA (2003) Utilizing GIS in Martian impact crater studies. In: Proceedings of the ISPRS WG IV/9 extraterrestrial mapping workshop: advances in planetary mapping (Abstract)

Brumby S, Plesko C, Asphaug E (2003) Evolving automated feature extraction algorithms for planetary science. In: Proceedings of ISPRS WGIV/9: extraterrestrial mapping workshop—advances planetary mapping, Houston, TX

Bue BD, Stepinski TF (2007) Machine detection of Martian impact craters from digital topography data. IEEE Trans Geosci Remote Sens 45(1):265–274

Cheng Y, Goguen J, Johnson A, Leger C, Matthies L, San Martin M, Willson R (2004) The Mars exploration Rovers descent image motion estimation system. IEEE Intell Syst 19(3):13–21

Crater Analysis Techniques Working Group (1979) Standard techniques for presentation and analysis of crater size-frequency data. Icarus 37:467–474

Ding W, Stepinski T, Mu Y, Bandeira L, Vilalta R, Wu Y, Lu Z, Cao T, Wu X (2011) Sub-kilometer crater discovery with boosting and transfer learning. ACM Trans Intell Syst Technol 2(4):39

Johnson AR, Willson J, Goguen J, Alexander, Meller D (2005) Field testing of the Mars exploration rovers descent image motion estimation system. In: Proceedings of the IEEE international conference on Robotics and Automation, Barcelona, Spain, April 2005, pp 4463–4469

Kim J, Muller JP, Van Gasselt S, Morley J, Neukum G (2005) Automated crater detection: a new tool for Mars cartography and chronology. Photogramm Eng Remote Sens 71(10):1205–1217

Leroy B, Medioni GG, Johnson E, Matthies L (2001) Crater detection for autonomous landing on asteroids. Image Vis Comput 19(11):787–792

Martins R, Pina P, Marques J, Silveira M (2009) Crater detection by a boosting approach. IEEE Geosci Remote Sens Lett 6:127–131

Michael G (2003) Coordinate registration by automated crater recognition. Planet Space Sci 51(9):563–568

Salamunićcar G, Lončarić S (2008) GT-57633 catalogue of Martian impact craters developed for evaluation of crater detection algorithms. Planet Space Sci 56(15):1992–2008. doi:10.1016/j.pss.2008.09.010

Salamunićcar G, Lončarić S (2012) Crater detection algorithms: a survey of the first decade of intensive research. In: Veress B, Szigethy J (eds) Horizons in earth science research, vol 8. Nova Science, New York, pp 93–123

Salamunićcar G, Lončarić S, Pina P, Bandeira L, Saraiva J (2011) MA130301GT catalogue of Martian impact craters and advanced evaluation of crater detection algorithms using diverse topography and image datasets. Planet Space Sci 59(1):111–131

Sawabe Y, Matsunaga T, Rokugawa S (2006) Automated detection and classification of lunar craters using multiple approaches. Adv Space Res 37(1):21–27

Smirnov AA (2002) Exploratory study of automated crater detection algorithm. Technical report, Boulder, Colorado, USA. http://www.cs.colorado.edu/~rossnd/fcdmf/CraterPaper.pdf

Stepinski TF, Mendenhall MP, Bue BD (2009) Machine cataloging of impact craters on Mars. Icarus 203(1):77–87. doi:10.1016/j.icarus.2009.04.026

Viola P, Jones M (2004) Robust real-time face detection. Int J Comput Vis 57(2):137–154

Xie Y, Tang G, Yan S, Lin H (2013) Crater detection using the morphological characteristics of Chang'E-1 digital elevation models. IEEE Geosci Remote Sens Lett 10(4):885–889

Chapter 7
Upper Ionosphere of Mars During Solar Quiet and Disturbed Conditions

S.A. Haider

Abstract In this chapter, we have described upper and lower ionospheric measurements, which have been obtained from radio occultation experiment onboard Mariners 6, 7 and 9; Mars 4 and 5; Viking 1 and 2; Mars Global Surveyor; and Mars Express. The ionisation sources like solar EUV, X-ray and particle radiations have been discussed. Observations on the upper ionosphere of Mars during disturbances like aurorae, solar flares, solar energetic particles and coronal mass ejections are also described. The understanding of complex behaviour of Martian ionosphere requires a balanced effort in the area of theoretical modelling. Therefore, we have also reported modelling of the upper ionosphere of Mars during quiet and disturbed conditions. At present measurements on the ionosphere of Mars are limited to middle- and high-latitude region. The low-latitude ionosphere of Mars is not observed. The physics of the low-latitude ionosphere could be very different from middle- and high-latitude ionosphere. Therefore, it is necessary to look for opportunities for obtaining observations at low-latitude region.

Keywords Ionosphere of Mars

7.1 Introduction

The upper ionosphere of Mars has been explored mostly with the radio occultation experiment onboard Mariner 6, 7 and 9 (Fjeldbo et al. 1970; Kliore et al. 1973); Mars 2, 3, 4, 6 and 7 (Kolosov et al. 1975; Vasiliev et al. 1975; Savich and Samovol 1976); Viking 1 and 2 (Fjeldbo et al. 1977); and more recently by Mars Global Surveyor (MGS) (Hinson et al. 1999; Tyler et al. 2001) and Mars Express (MEX) (Pätzold et al. 2005). In addition to radio occultation experiment, MEX also carried Mars Advanced Radar for Subsurface and Ionosphere Sounding (MARSIS) experiment, which provided electron density profiles well above the main ionospheric peak (Gurnett et al. 2005). The daytime ionosphere models, which

---

S.A. Haider (✉)

Space and Atmospheric Science Division, Physical Research Laboratory, Navrangpura, Ahmedabad, Gujarat, India

e-mail: haider@prl.res.in

© Springer-Verlag Berlin Heidelberg 2015

S. Jin et al. (eds.), *Planetary Exploration and Science: Recent Results and Advances*, Springer Geophysics, DOI 10.1007/978-3-662-45052-9_7 consider ionisation by solar radiation (Winchester and Rees 1995), explain the total electron concentration obtained by the radio occultation experiments. The nightside ionospheric models considering the precipitation of solar wind electrons agree with the Viking radio observations rather than other models (Haider et al. 2002). In spite of the relatively large number of measurements and models, the effect of coronal mass ejection (CME), magnetic storms, aurorae and solar flares are not understood in detail in the upper ionosphere of Mars. This chapter describes observations on the upper ionosphere of Mars during quiet and disturbed conditions. The modelling of the upper ionosphere is also described in this chapter.

7.2 Martian Ionosphere and Source

UV and X-ray radiations are the major ionising sources in the upper atmosphere of Mars. The photons of these frequencies contain sufficient energy to dislodge an electron from a neutral gas upon absorption. The reverse process to ionisation is recombination in which a free electron is captured by a positive ion. At Mars the main reaction in the ionosphere is the dissociative recombination of O_2^+ and the energy is carried away in the form of kinetic energy by two resulting O atoms. The ionisation depends primarily on the sun and its activity. The amount of ionisation in the Martian ionosphere varies greatly with the amount of radiation received from the sun. Thus, there is a diurnal and seasonal effect on Mars. During the northern winter/northern summer, Mars is away/close to the sun. Thus, it will receive less solar UV radiation during winter than summer. The activity of the sun is also associated with the sunspot cycle, with more radiation occurring with more sunspots. Although Mars has a strong magnetic field originating in the crust, these fields are not global and therefore do not deflect solar wind particles which can ionise the upper atmosphere of Mars. However, the interaction of solar wind with inhomogeneous crustal fields is a major cause of spatial and temporal variations and thus in the ionospheric chemistry, dynamics and energetic (Nagy et al. 2004; Brain 2006; Withers 2009; Mendillo et al. 2011). The upper ionosphere of Mars is divided into E and F region. These two regions are described in the following sections.

7.2.1 E Region Ionosphere

Mars ionospheric E layer, which has its peak density at about 115 km, is produced by X-ray radiation in the daytime ionosphere at wavelength range 10–90 Å (Haider et al. 2002; Rishbeth and Mendillo 2004; Haider et al. 2009a, b). At night the E layer disappears because the primary source of ionisation is no longer present. The vertical structure of the E layer is primarily determined by the competing effects of ionisation and recombination. The ion composition calculation shows that O_2^+ and NO^+ are the major ions in the E region with N_2^+ and O^+ as minor ions (Haider et al. 2012). Figure 7.1 shows production rate (photoionisation + photoelectron impact

7 Upper Ionosphere of Mars During Solar Quiet and Disturbed Conditions

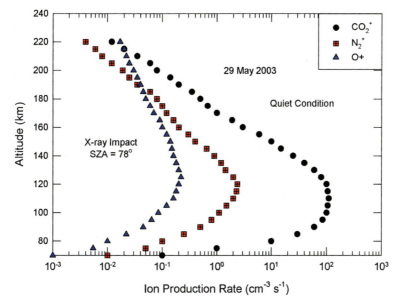

Fig. 7.1 Altitude profile of predicted ion production rates of CO_2^+, N_2^+ and O^+ in the E region of Martian ionosphere due to absorption of X-ray radiation (1–8 Å) during quiet conditions (Reprinted from Haider et al. (2012) with permission from Springer)

ionisation rates) on 29 May 2003 for major ions CO_2^+, N_2^+ and O^+ in the E region of Martian ionosphere due to absorption of solar X-ray radiation during quiet condition. X-rays produce major ion CO_2^+ in the E region of the Martian ionosphere and the maximum ion production occurs in the altitude range ~90–110 km (Haider et al. 2012).

7.2.2 F Region Ionosphere

The F region is formed mainly owing to photoionisation of neutral atoms/molecules by EUV radiation from 90 to 1,026 Å. Precipitating energetic charged particles also contribute to ionisation of neutral species. The major gases CO_2, N_2, O_2, O, Ar and CO are ionised by EUV radiations. The maximum ionisation in the F region occurs at altitude ~125–135 km in the dayside ionosphere of Mars for subsolar conditions (cf. Hanson et al. 1977; Fox and Dalgarno 1979; Bougher et al. 1990; Fox 2009). At higher solar zenith angles of the dayside, the peak ionisation occurs at higher altitudes. Recently, Haider et al. (2009a, 2010) have compared the ionospheric F regions of Earth and Mars. They have found that the thickness of the F layer and location of its peak density decreases by a factor of 1.6–1.8 in the Martian ionosphere as compared to that observed in the Earth's ionosphere because of smaller neutral scale height; the solar EUV energy is

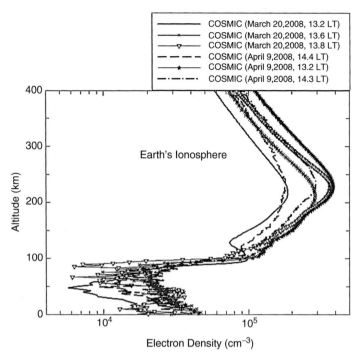

Fig. 7.2 Six sample profiles of electron density representing D, E and F layers as observed in Earth's ionosphere by radio occultation experiment onboard COSMIC satellites at different locations at nearly the same local time in the afternoon (Reproduced from Haider et al. (2009a) by permission of John Wiley & Sons Ltd.)

deposited within smaller altitude range in the upper ionosphere of Mars as compared to the corresponding altitude range in the upper ionosphere of Earth. Figures 7.2 and 7.3 show sample profiles of electron density observed at nearly the same time, location and solar conditions in the ionospheres of Earth and Mars, respectively (Haider et al. 2009a). Six profiles have been selected in the Earth's ionosphere from COSMIC measurements carried out on 20 March 2008 at coordinates (72.7°N, 153.4°W), (51.8°N, 49.5°W), (49.8°N, 144.5°W), (67.3°N, 146°W) and (54°N, 126.3°W) and on 9 April 2008 at coordinates (58.4°N, 93.6°W). These observations were carried out in the afternoon at nearly the same local time (13.2, 13.6, 13.8, 14.4, 13.2 and 14.3) during a low solar activity period $F_{10.7} = 68$. Six profiles of electron density have been selected from MGS observations carried out on 20 March 2005 at coordinates (74.4°N, 106.5°W), (74.3°N, 49.2°W), (74.2°N, 209°W), (70°N, 156°W) and (70°N, 127°W) and on 9 April 2005 at coordinates (70°N, 99°W). (The data are not available in 2008 because MGS failed to work after 9 June 2006.) These observations were again performed in the afternoon at nearly the same local times (12.95, 12.96, 12.97, 13.75, 13.76 and 13.74) with low solar activity (equivalent $F_{10.7} = 39$ at Mars).

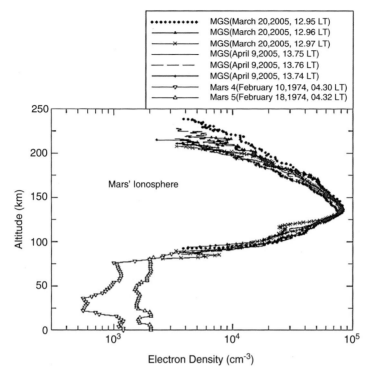

Fig. 7.3 Six sample profile of electron density representing E and F layers in the Martian ionosphere as observed by radio occultation experiment onboard MGS at nearly the same local time in the afternoon. Density profiles representing D layer between altitude range ∼25–35 km is observed by radio occultation experiment onboard Mars 4 and Mars 5 at 04:30 LST (Reproduced from Haider et al. (2009a) by permission of John Wiley & Sons Ltd.)

7.2.3 Solar Wind Impact Ionisation

Outside the crustal magnetic field region, Mars has an induced magnetosphere. As a result solar wind dynamic pressure compresses the interplanetary magnetic field into the Martian ionosphere. Shinagawa and Cravens (1989) developed a one-dimensional magnetohydrodynamic (MHD) model to study the role of electromagnetic forces in the Martian ionosphere. They found a good agreement between model and Viking observations by adding an extra heat source caused by the solar wind in the upper ionosphere. Later Shinagawa and Bougher (1999) developed a two-dimensional MHD model and studied two cases of solar wind dynamic pressure on Mars. In the first case, solar wind velocity was high ∼450 km/s, which exceeded the maximum ionospheric pressure of Mars. In the second case, solar wind velocity was below ∼300 km/s. This study showed that the upper ionosphere of Mars is significantly influenced by solar wind. Using a three-dimensional model, Ma et al. (2004) have reported that solar wind plays an important role above 250 km. Similar conclusions were also drawn by Haider et al. (2010a) using a one-dimensional model with non-zero upward flux boundary condition. Below this height, these models are in agreement with the earlier modelling results.

The nightside ionosphere of Mars was first measured by the radio occultation experiment onboard Mars 4 and Mars 5 at solar zenith angles of 127° and 106° in February 1974 under solar minimum condition (Savich and Samovol 1976). Later Zhang et al. (1990) reported that about 60 % of the radio occultation profiles obtained in the night-time from Viking 1/2 do not show a well-defined peak during low solar activity. For the remaining 40 % profiles, the average nightside peak value was about 5×10^3 cm^{-3}, at an altitude of about 150 km. This peak is produced by solar wind electron transportation from dayside to nightside atmosphere across the terminator (cf. Verigin et al. 1991; Haider et al. 1992; Fox et al. 1993). Kallio and Janhunen (2001) have calculated ion production rates in the nightside Martian ionosphere due to H^+-H impact process. This source was found to be an important ionisation process for the nightside ionosphere of Mars. Using this source of ionisation, Haider et al. (2002) predicted peak electron densities of 3.5×10^3 cm^{-3} and 2.0×10^3 cm^{-3} at solar zenith angles of 105° and 127°, respectively. They found that fast hydrogen atoms penetrate deeper into Martian atmosphere and lose their energy at lower altitudes as compared to solar wind electron impact ionisations. Above 200 km, the photoelectrons produced during the dayside that travel to the nightside are found to be an important process that contributes about 30–40 % photoelectron flux in the night-time ionosphere for SZA $\sim$127° (Fox et al. 1993).

7.3 Chemistry and Effects of the Upper Ionosphere

The chemistry of ions O_2^+, NO^+, CO_2^+, O^+, N_2^+ and CO^+ in the upper ionosphere of Mars has been studied by several investigators (cf. Chen et al. 1978; Hanson et al. 1977; Fox et al. 1993; Haider 1997; Ma et al. 2004; Duru et al. 2008; Haider et al. 2010, 2012). Solar EUV is the major ionisation source for the production of these ions in the dayside, while it is electron impact ionisation for the nightside ionosphere of Mars. Major ions are O_2^+, NO^+ and CO_2^+ below about 200 km above which O^+ dominates. The ion NO^+ is mainly produced due to reaction of O_2^+ with N and NO, and it is entirely destroyed by dissociative recombination. The density of NO^+ is directly proportional to the densities of N and NO (Haider et al. 2009a). The other major source of this ion is the reaction of CO^+ and N_2^+ with CO_2. The ion CO_2^+ is lost by collision with atomic oxygen and is the one of the main sources of the dominant ion O_2^+. Dissociative recombination is an important loss process for O_2^+ at all altitudes. The dominant ion O_2^+ is produced in the dayside and nightside ionosphere mainly due to the reaction of CO_2^+ with O (Barth 1985). Among the minor ions, CO^+ is lost in the charge exchange reaction with CO_2. This process destroys almost all CO^+ ions in the Martian ionosphere. Solar wind electron impact ionisation is the major source of CO^+ in the nightside ionosphere (Haider et al. 2007). Fox (2009) found that the charge exchange reaction

7 Upper Ionosphere of Mars During Solar Quiet and Disturbed Conditions

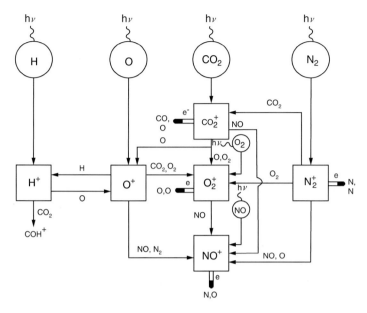

Fig. 7.4 Ionospheric chemical reaction scheme of Mars upper ionosphere (Reproduced from Chen et al. (1978) by permission of John Wiley & Sons Ltd.)

between CO_2^+ and O is the dominant source of O^+ at heights below 190 km. Above this altitude solar radiation is an important production mechanism for O^+. The ion N_2^+ is produced by photoionisation and is lost by reaction with CO_2. The chemical reaction scheme for the upper ionosphere of Mars is shown in Fig. 7.4.

7.3.1 Effects of Solar Flares on the Upper Ionosphere

Several solar flares of classes M and X were registered on the sun during the maximum phase of solar cycle 23. Responses to these flares in the Martian ionosphere have been reported by several investigators. Using MARSIS data, Nielsen et al. (2006) reported that the maximum electron density in the Martian ionosphere suddenly increased from 1.8×10^5 to 2.4×10^5 cm^{-3} at 08:39 UT on 15 September 2005. This coincided closely in time with the increase in solar X-ray fluxes measured from GOES 12 spacecraft at the Earth, as shown in Fig. 7.5. Mendillo et al. (2006) examined ionospheric data of MGS and reported two elevated electron density profiles during the solar X-ray flares which occurred on 15 April and 26 April 2001 at 13:50 UT and 13:10 UT, respectively. They found 200 % enhancement in the electron density profiles during these flares, as shown in Fig. 7.6. Haider et al. (2009b) studied solar flare of 13 May 2005 and the following CMEs and reported their effects in the E region ionosphere. Mahajan et al. (2009) surveyed all

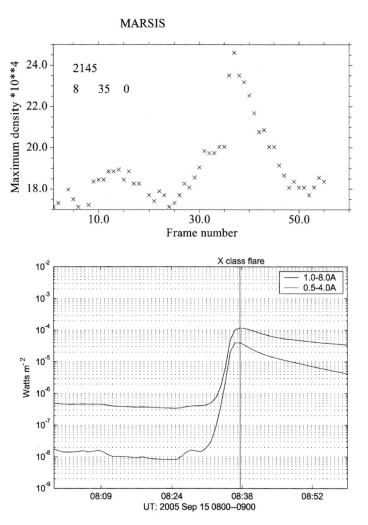

Fig. 7.5 Sudden increase of electron density maximum (*top panel*) simultaneous to an increase in solar X-ray flux (*bottom panel*) of wavelength 0.5–4 Å (*lower curve*) and 1–8 Å (*top curve*). The X-ray data were observed on 15 September 2005 by GOES spacecraft when MARSIS observed the sudden increase in electron density (Reprinted from Nielsen et al. (2006) with kind permission from Springer)

the electron density profiles measured by MGS and reported effects of seven X-ray flares in the Martian ionosphere. They found elevated electron densities in the E region during all the flares but in the F region during some flares only.

Recently, Haider et al. (2012) have modelled solar X-ray fluxes measured by GOES 12 during the periods 29 May to 3 June 2003, 15–20 January 2005 and 12–18 May 2005 and investigated effects on electron densities produced by individual X-ray flares that occurred within these intervals. They have reproduced

7 Upper Ionosphere of Mars During Solar Quiet and Disturbed Conditions 127

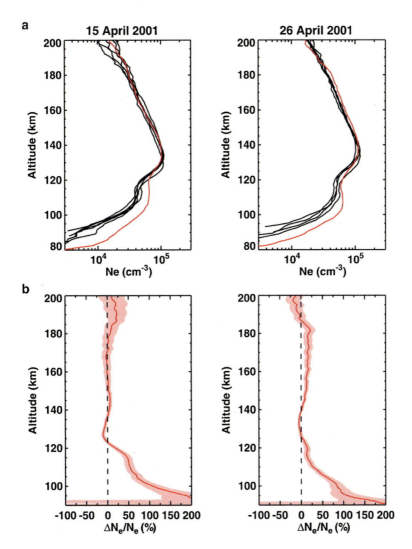

Fig. 7.6 (**a**) Electron density profiles on Mars obtained for 15 April and 26 April 2001. Two profiles in *red* at 14:15 and 13:16 UT show significant enhancement at low altitude because of solar flares which peaked in X-ray fluxes at Earth at 13:50 and 13:10 UT, respectively. On 15 April there were five MGS profiles before the flare at 2:28, 6:23, 8:21, 10:19 and 12:17 UT and none after the flare. On 26 April pre-flare profiles were available at 9:20 and 11:18 UT and post-flare at 17:11 and 19:09 UT. (**b**) % differences between the flare affected profiles and the averages of the other profiles on each day. The shading gives 1-σ standard error in relative change in electron density (N_e) (Reprinted from Haider and Mahajan (2014) with kind permission from Springer)

the major characteristics and magnitudes of the measured solar X-ray spectra by model calculation. Figure 7.7a, c, e present the calculated and measured solar X-ray flux distribution from 29 May to 3 June 2003, 15 to 20 January 2005 and 12 to 18 May 2005, respectively, in the wavelength range 1–8 Å. Figure 7.7b, d, f show

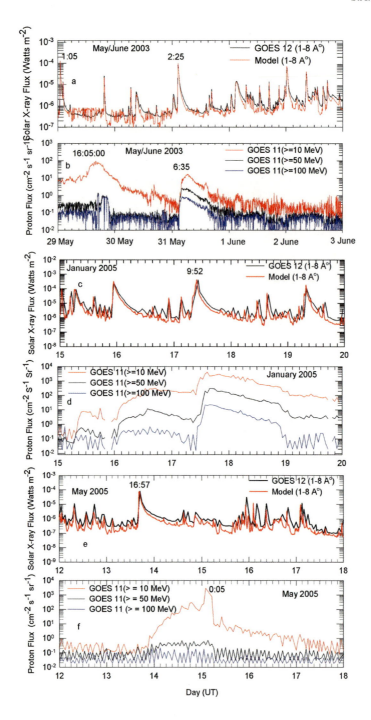

the distribution of proton fluxes with time as observed by GOES 11 at three energies ≥ 10, ≥ 50 and ≥ 100 MeV thus indicating that CMEs erupted from the active regions on the sun for several hours after each flare. The flare of 17 January was very strong, and protons of all three energies were accelerated from the heliosphere of the sun (Fig. 7.7d). The X-ray and proton fluxes for the flares of 29 May and 31 May 2003 were not as large as those for the flares of 17 January and 13 May 2005. Figure 7.7a–c represent time series of the measured and calculated total electron content (TEC) of E region ionosphere for each respective day between 29 May and 3 June 2003, 15 and 20 January 2005 and 12 and 18 May 2005, respectively. MGS observed the electron density profiles a few hours before and immediately after these solar flares. Before flaring, the ionosphere of Mars was calm. Soon after the solar flares, an increase by a factor of ∼4–5 in the TEC was estimated. The effect of these flares endured each day for about 1–2 h in the E region ionosphere of Mars. During the quiet period the value of predicted TEC is higher by a factor of 1.5–2 than the observed value. This is due to the fact that electron-ambient-electron collisions were neglected in this model. This process will reduce the modelled TEC.

7.3.2 Effect of CMEs on the Upper Ionosphere

Studying and understanding the effect of CME is a key area of research in planetary aeronomy. Haider et al. (2012) were the first to detect the effect of CMEs in the E region ionosphere of Mars. They examined MGS data between 30 and 31 May 2003, 2 and 3 June 2003 and 16 and 17 May 2005 following the flares of 29 May, 31 May 2003 and 13 May 2005, respectively (see Fig. 7.8a–c). They found that the physical processes of magnetic storms (the after effects of CMEs) are different on Earth and Mars. During a magnetic storm, shock waves driven by CME compress the Earth's magnetosphere leading to increased energetic particle precipitation into the ionosphere. This leads to sudden increase in the electron density. The magnetic storm effects on Mars, on the other hand, have quiet different characteristics. Mars has no dipolar magnetic field. Therefore, solar wind interacts directly with the Martian ionosphere, which acts as an obstacle and diverts the solar wind around

---

Fig. 7.7 (**a**) Solar X-ray flux measured by GOES 12 (*black line*) and calculated by model (*red line*) between 29 May and 3 June 2003 (**b**) Proton flux distributions between 29 May and 3 June 2003 measured by GOES 11 at three energies: ≥ 10 MeV (*red colour*), ≥ 50 MeV (*black colour*) and ≥ 100 MeV (*blue colour*). (**c**) Solar X-ray flux measured by GOES 12 (*black line*) and calculated by model (*red line*) between 15 and 20 January 2005. (**d**) Proton flux distribution between 15 and 20 January 2005 measured by GOES 11 at three energies: ≥ 10 MeV (*red colour*), ≥ 50 MeV (*black colour*) and ≥ 100 MeV (*blue colour*). (**e**) Solar X-ray flux measured by GOES 12 (*black line*) and calculated by model (*red line*) between 12 and 18 May 2005. (**f**) Proton flux distribution between 12 and 18 May 2005 measured by GOES 11 at three energies: ≥ 10 MeV (*red colour*), ≥ 50 MeV (*black colour*) and ≥ 100 MeV (*blue colour*) (Reprinted from Haider et al. (2012) with kind permission from Springer)

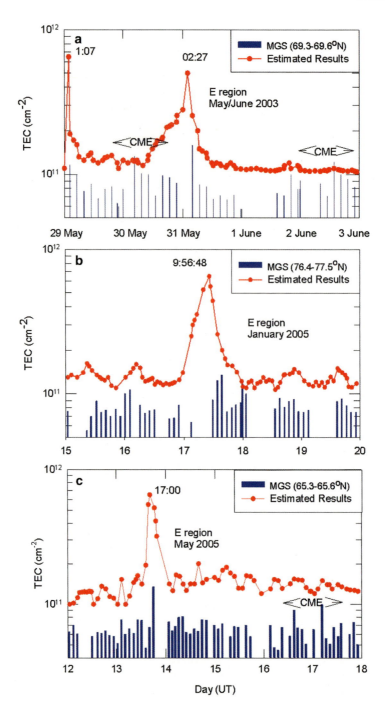

7 Upper Ionosphere of Mars During Solar Quiet and Disturbed Conditions 131

it (Acuña et al. 1998). Actually, it is a magnetic barrier, which diverts the bulk of solar wind shock around the planet. There exists a sharp boundary (ionopause) between the magnetised magnetosheath and the ionosphere, as shown in Fig. 7.9. The location of the magnetosheath and ionopause can change with the solar wind dynamic pressure. Thermal pressure balances magnetic pressure at the ionopause boundary. Mass loading of the magnetosheath does effect the flow behind the shock such that the pickup of neutrals ionised outside the ionopause (i.e. pick up ions) contribute to further stagnation of this flow and to the growth of the magnetic barrier. Consequently, this contributes to the development of the magnetotail.

MGS has observed a magnetosheath at about 435 km on the sunlit hemisphere of Mars during quiet conditions (Mitchell et al. 2000). In the magnetosheath the planetary neutrals are mainly H atoms of the hydrogen corona. Fast hydrogen atoms are produced by charge exchange between solar wind protons and hydrogen corona in this region. These energetic proton-hydrogen atoms have the same energies as the solar wind protons and move in the same direction as that of the fast protons just before the collisions (Haider et al. 2002). In this way the magnetosheath of Mars can be compressed similar to that observed in the Earth's magnetosphere (Dandouras et al. 2007) and the accelerated solar wind protons get turned into fast hydrogen atoms at lower altitudes. To verify that the flare of 13 May 2005 has an effect on the magnetosheath of Mars, Haider et al. (2012) analysed the magnetic field data obtained from MGS at altitudes $\sim$420 and $\sim$430 km from 12 to 18 May 2005. The variation of magnetic field measured in the magnetosheath region of Mars is shown in Fig. 7.10. There are two broad peaks in the magnetic field at an altitude of $\sim$420 km on 15 and 17 May with the values $\sim$50 nT and 40 nT at 21:50 UT and 02:52 UT, respectively. These values are larger than that the magnetic field normally observed at altitude $\sim$430 km by a factor of $\sim$2.5. Before and after these times, the magnetic field does not change significantly between these two altitudes. This suggests that the CME arrived at Mars on 15 May at about 21:50 UT and compressed its magnetosheath by about 10–15 km.

Haider et al. (2012) also ran a three-dimensional kinetic solar wind model (Hakamada-Akasofu-Fry version 2/HAFv.2) to confirm the arrival of CME at Mars following the solar flare of 13 May 2005. This model does not provide any way to distinguish between the effects on the ionosphere of magnetic storm from the CME shocks and the energetic particles from that shock. Figure 7.11a–h shows simulated ecliptic plane profile of IMF (about to 2 AU) from 15 to 18 May 2005. The simulation confirmed that the CME reached Mars on 15 May 2005 after its

Fig. 7.8 (**a**) Measured TEC (*blue colour*) and predicted TEC (*red colour*) in the E region ionosphere of Mars for the period 29 May to 3 June 2003. (**b**) Measured TEC (*blue colour*) and predicted TEC (*red colour*) in the E region ionosphere of Mars for the period 15–20 January 2005. (**c**) Measured TEC (*blue colour*) and predicted TEC (*red colour*) in the E region ionosphere of Mars for the period 12–18 May 2005. CME arrival and its effect on 30–31 May 2003, 2–3 June 2003 and 16–17 May 2005 are marked by an *arrow line* (Reproduced from Haider et al. (2012) by permission of John Wiley & Sons Ltd.)

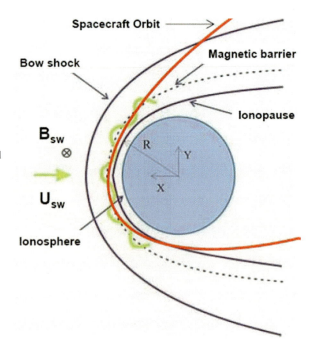

Fig. 7.9 Schematic diagram of the ionopause between magnetised magnetosheath and the ionosphere of Mars. U_{sw} and B_{sw} represent the velocity and magnetic field directions of solar wind on Mars, respectively. Magnetic barrier, bow shock and ionopause locations are also shown orbit. Mangalyaan will be launched in November 2013 and has elliptical orbit, which can pass from low-latitude region of Mars (Details are given in the website: http://en.wikipedia.org/wiki/Mangalyaan)

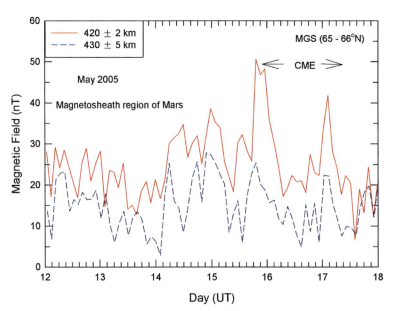

Fig. 7.10 Magnetic field in the magnetosheath region of Mars at altitude 420 ± 2 km (*red line*) and 430 ± 5 km (*blue line*) passing through MGS at latitude $65°$–$66°$N between 12 and 18 May 2005. The major peaks at 21:50 UT and 02:52 confirms the arrival of CME at Mars on 15 and 16 May, respectively. *Dashed blue line* shows the magnetic field magnetosheath of Mars under quiet condition (Reprinted from Haider et al. (2012) with permission from Elsevier)

time of arrival at Earth. A second CME-related shock was predicted to reach Mars on 17 May 2005 at 16:00 UT (Fig. 7.11g). The arrival of the resultant shock was predicted to reach Mars between 16 and 17 May 2005 (Fig. 7.11d–g). The direction of the CME was away from Mars on 18 May (Fig. 7.11h). The overall effect of CME arrival at Mars lasted for about 2 days. As a result TEC increased suddenly (see Fig. 7.8c) by factors of $\sim$2–3 between 16 and 17 May 2005.

7.3.3 Effect of SEPs on the Upper Ionosphere

Solar energetic particle (SEP) events are a part of major disturbances in the heliosphere (Schnjver and Siscoe 2010). These events are mostly composed of protons with about 10 % He^+ and <1 % heavier elements. There are two types of SEP events: impulsive and gradual (Cane et al. 1986). Impulsive events are relatively of short duration (<1 day) with a high proton content. Gradual events are of longer duration (days), have higher fluxes, display a wider spread in longitude and are associated with fast CMEs. Mckenna-Lawlor et al. (2012) reported three major factors in connection with effect of SEP radiations at the Martian surface: (1) shadowing by the planet, which cuts off $\sim$50 % of SEP primary particle flux, (2) atmospheric attenuation, which shields out SEP primaries characterised by relatively low energies and (3) backscattering particles, mostly neutrons, due to the interaction of high energy particles with soil material. The Mars Energetic Radiation Environment models (MEREM) were developed by ESA and NASA to study the SEP radiation close to the Martian environment. The output of these models gives (1) particle influence, (2) effective dose and (3) ambient dose equivalent with the Martian atmosphere and in the planetary orbit about Mars (Mckenna-Lawlor et al. 2012).

It is known that the energetic particle densities in the solar wind are significantly enhanced during SEP events. The observations made by MARSIS have demonstrated that SEP events modify the ionosphere of Mars (Morgan et al. 2006). It needs to be mentioned that this instrument did not detect its usual reflections from the surface of Mars during an SEP event, indicating that radio waves which usually pass smoothly through the ionosphere were fully absorbed (Withers 2011). Sheel et al. (2012) investigated the effect of SEP event of 29 September 1989 on the ionosphere of Mars. This event was observed by IMP and GOES satellites during the disturbed condition of the sun (Lovell et al. 1998). Figure 7.12 shows electron density profiles predicted for the 29 September 1989 SEP event, with and without photoionisation process. It can be noted that the SEP event caused electron density to exceed 2×10^5 cm^{-3} at $\sim$120–140 km, a value much larger than typically observed by MGS in the dayside ionosphere of Mars (cf. Haider et al. 2011). Sheel et al. had carried out two model calculations – photochemical and generalised models. Solid lines represent the calculation of photochemical equilibrium model. The black dashed line shows the electron density profile obtained from a generalised model. The generalised model neglects the energy deposition at high energies.

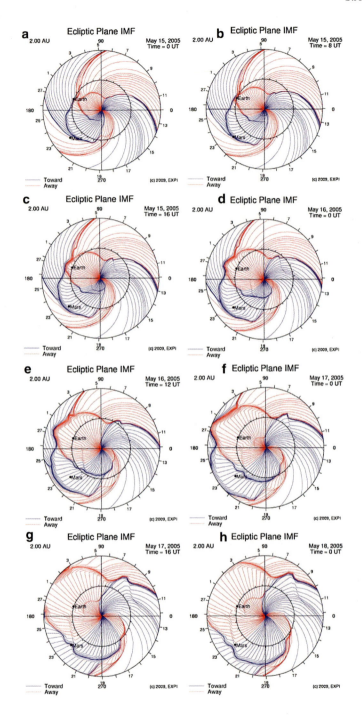

7 Upper Ionosphere of Mars During Solar Quiet and Disturbed Conditions

Therefore, electron density decreases exponentially with height in this model. The photochemical equilibrium model includes photoionisation process, chemical reactions and their production and loss processes under steady state condition. There is a reasonable agreement between two model calculations in the absence of photoionisation. Before MARSIS observations, Leblanc et al. (2002) had simulated the vertical profile of energy deposition rate in the Martian ionosphere for the SEP event of 20 October 1995. However, they did not study the ionospheric effects of this SEP event at Mars. Their results are comparable with Sheel et al. (2012) in the vicinity of ionisation peaks produced by this SEP event.

7.4 Auroral Ionosphere

Aurorae are often seen in the upper atmosphere of Earth at high latitudes following the occurrence of solar flares and CMEs. Since Mars has no strong remnant of an ancient intrinsic magnetic field in the northern hemisphere (Acuña et al. 1998), Fox (1992) argued that Mars should have Venus-like diffuse aurora in this region. However, auroral events are not observed in the northern hemisphere, but effects of magnetic storms have been detected in the ionosphere of Mars (Haider et al. 2009b, 2012). Bertaux et al. (2005) have discovered southern aurora in the night-time atmosphere (solar zenith angle 117.5°) of Mars using the 'Spectroscopy for the Investigation of the Characteristic of the Atmosphere of Mars' (SPICAM) experiment onboard MEX. Figure 7.13a shows the limb observations carried for 450–750 s at wavelength range of 100–350 Å in the night-time atmosphere. H Lyman-α (121.6 nm) and NO bands (181–298 nm) are clearly visible in this figure. Figure 7.13b represents an auroral spectrum integrated over the wavelength range of the NO bands (181–298 nm) as a function of time for the five spatial bins. There is a strong peak in all the bins. These spectra were observed at a tangent altitude of 19 km. Local time was ~21:00 h and longitude and latitude were 198.4° and −46.3°, respectively. These measurements were carried out using nadir looking direction in the strong crustal magnetic field region where the field lines are nearly open (Lundin et al. 2006; Mitchell et al. 2007; Haider et al. 2010). Leblanc et al. (2008) found a good correlation among the measured auroral emission by SPICAM, the measured downward/or upward flux of electrons by ASPERA-3 and the TEC recorded by MARSIS. They found that TEC increased when there was a precipitation of high flux auroral electrons into the atmosphere of Mars. During the

---

Fig. 7.11 Ecliptic plane simulations out to 2 AU of IMF and solar wind disturbances predicted by the HAFv.2 model during 15–18 May 2005 at selected time: (**a**) 15 May at 0.00 UT, (**b**) 15 May at 8:00 UT, (**c**) 15 May at 16:00 UT, (**d**) 16 May at 0:00 UT, (**e**) 16 May at 12:00 UT, (**f**) 17 May at 0:00 UT, (**g**) 17 May at 16:00 UT and (**h**) 18 May at 0:00 UT. IMF lines are shown in *red* (away sectors) and in *blue* (towards sectors). The locations of the Earth and Mars are indicated by *black dots* (Reproduced from Haider et al. (2012) by permission of John Wiley & Sons Ltd.)

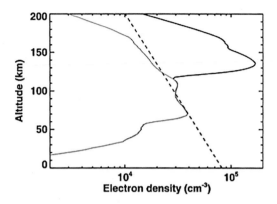

Fig. 7.12 Electron density profile predicted for 29 September 1989 SEP event, neglecting photoionisation (*grey solid line*) and including photoionisation (*black solid line*). These two profiles are identical below 100 km. The *black dashed line* shows the electron density profile from generalised model (Reproduced from Sheel et al. (2012) by permission of John Wiley & Sons Ltd.)

auroral events ASPERA-3 measured an increase in the electron flux by an order of magnitude. Before and after this event, the electron flux decreased by two orders of magnitude. Leblanc et al. too found a broader peak in TEC during the SPICAM measurements of an auroral event. This suggests that electron density increased significantly in the auroral ionosphere of Mars during this event. Figure 7.14a–g displays the measurements by SPICAM (Fig. 7.14g), MGS/Electron Reflectometer (ER) (Fig. 7.14f), ASPERA-3 (Figs. 7.14d, e), MARSIS (Fig. 7.14c) and the MEX altitude (Fig. 7.14a) and MEX latitude (Fig. 7.14b). MEX observations were carried out on 26 January 2006 during orbit # 2621. Track of MEX during this orbit was at local time of 20:30 h at a longitude of 180° that is above the most intense crustal magnetic field recorded by MGS (Mitchell et al. 2007; Haider et al. 2010). SPICAM observed a significant increase of auroral light between 14:04:01 UT and 14:04:15 UT.

It is found that the energy distribution of the downward electron flux measured by ASPERA-3 exhibited non-Maxwellian features similar to those observed in the V-shaped potential structure of Earth's auroral zone (cf. Fang et al. 2010; Lei and Zhang 2009; Lillis et al. 2009; Fillingim et al. 2010; Lundin et al. 2011). Ip (2012) proposed a mechanism for the acceleration of ions and electrons in the Martian auroral zone. Figure 7.15 represents a schematic view of inverted V-shaped potential structures, where electrons are precipitating downward and conic ions are escaping upward in the presence of strong crustal magnetic fields. Ip suggested that only those ions, which are created in the crustal fields connected to the interplanetary magnetic fields, would be able to escape. The ions which are created in the magnetic flux tubes of closed field lines will be trapped in a bouncing motion. The southern aurora may not be visible to the human eyes because they were observed in the ultraviolet wavelength region.

7 Upper Ionosphere of Mars During Solar Quiet and Disturbed Conditions 137

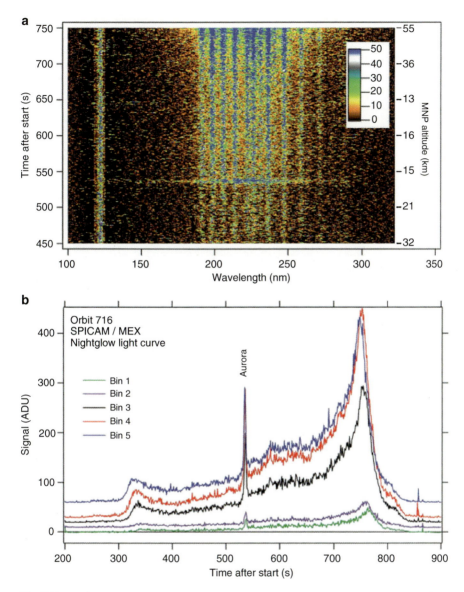

Fig. 7.13 (**a**) Spectra recorded during the grazing limb observation between 450 and 750 s. Altitudes of the Mars nearest point (MNP) of the line of sight are indicated at the right. It contains H Lyman-α emission at 121.6 nm and well-structured band (190–270 nm) of NO. The intensity in ADU (analogue-to-digital units) per pixel is colour-coded. (**b**) Auroral peak is sharp and different from NO spectrum. Signal intensity in ADU for five bins (each averaged from 181 to 298 nm) as a function of time between 200 and 900 s are plotted in Fig. 7.13 (Reprinted from Haider and Mahajan (2014) with kind permission from Springer)

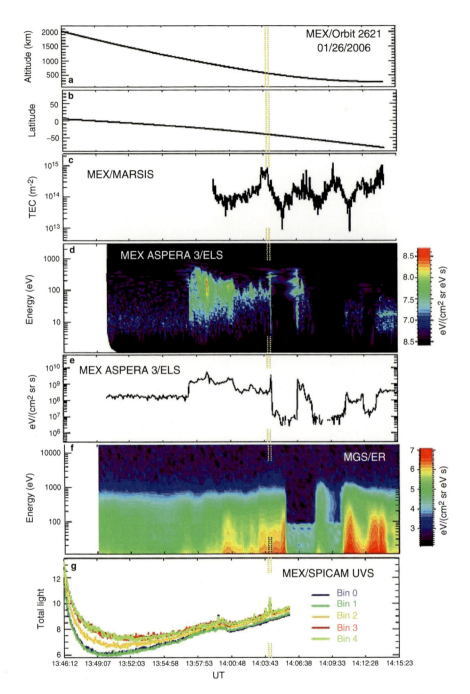

Fig. 7.14 Time series of MEX measurements during orbit 2621. Shown is the MEX (**a**) altitude, (**b**) latitude, (**c**) MEX/MARSIS TEC, (**d**) MEX/ASPERA-3/ELS electron measurement, (**e**) MEX/ASPERA-3/ELS energy flux, (**f**) MGS/MAG/ER electron energy flux and (**g**) MEX/SPICAM measurements. Indicated on each by *vertical dotted lines* are the periods during which an aurora event has been identified in SPICAM/UVS observations (Reproduced from Leblanc et al. (2008) by permission of John Wiley & Sons Ltd.)

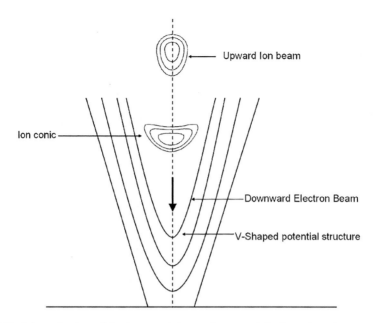

Fig. 7.15 Schematic view of the downward electron beams and the formation of upward flux of oxygen ions at the foot points of the Martian crustal magnetic fields. Transverse heating of the O+ ions and other ion species will lead to the generation of ion conics, which are to be converted to ion beams at higher altitude. Parallel electric fields will be maintained by the V-shaped potential structure (Reprinted from Ip (2012) with permission from Elsevier)

7.5 Ionospheric Models

Fox et al. (1996) carried out a model calculation of the thermosphere and ionosphere of Mars and calculated neutral densities, temperature, ion production rates and densities of ions and electrons during solar minimum and maximum conditions. They found that the electron density increased by a factor of ∼3 during solar maximum condition in comparison to that estimated for solar minimum condition. These model results were found to be in good agreement with the radio occultation measurements made by Viking 1 and Mariners 6 and 7 during solar minimum and solar maximum conditions, respectively. Fox et al. (1996) used Eq. 7.1 to calculate the ion and electron densities in the Martian ionosphere. In this calculation, the model atmospheres for low and high solar activities were used from the MTGCM model of Bougher et al. 1990. Using the photochemical model, Mendillo et al. (2004) estimated noontime TEC at subsolar latitude of Mars for aphelion and perihelion positions during solar minimum and solar maximum periods. In this calculation neutral atmosphere and solar flux were taken from Bougher et al. (1990) and Tobiska et al. (2000), respectively, for solar maximum/minimum conditions. The results of these model calculations are shown in Fig. 7.16. It can be seen that TEC increased in the Mars' ionosphere by a factor of ∼2 during solar maximum

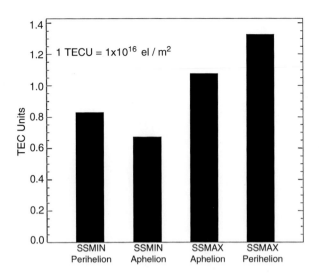

Fig. 7.16 Model results for noontime ionospheric TEC in the altitude range 100–200 km at the subsolar latitude on Mars for aphelion/perihelion positions during solar maximum/minimum conditions (Reproduced from Mendillo et al. (2004) by permission of John Wiley & Sons Ltd.)

conditions. It should be noted that TEC in the Earth's ionosphere is larger during solar maximum conditions by one to two orders of magnitude (Rishbeth and Mendillo 2004) not because of Earth's closer distance to the sun, but due to the non-photochemical layer F2, providing major contribution to the TEC integral. Thus, photochemical models in Mars' ionosphere exhibit minimum solar cycle variations.

Using MARSIS data obtained from MEX, Lillis et al. (2010) have reported that disturbed solar and space weather conditions can produce prolonged higher TEC values, while an individual SEP event causes a short-lived absolute increase in TEC. They also found a relationship between TEC and both He-II line irradiance and $F_{10.7}$ solar radio flux as power laws with exponents of 0.54 and 0.44, respectively. As mentioned before, Haider et al. (2012) used radio occultation data of MGS at high latitudes (65.3–65.6°N, 69.3–69.6°N and 74.6–77.5°N) and studied the effects of solar X-ray flares on TEC in the E region of the Martian ionosphere. Modelling of flare-induced solar X-ray fluxes, ion production rates, electron densities and TEC were carried out for solar flare events that occurred on 29 and 31 May 2003 and 17 January and 13 May 2005. They found that solar X-ray flares caused enhancements in the electron density by a factor of 5–6 during disturbed condition. The observed and estimated electron density profiles for quiet and disturbed conditions are represented in Fig. 7.17 for solar X-ray flare of 31 May 2003. It should be noted that the measured electron density profile cannot be reproduced completely by this model. This is due to the fact that while E and F layers in the Martian ionosphere at ~90–110 km and ~130–140 km are produced due to absorption of both solar X-ray

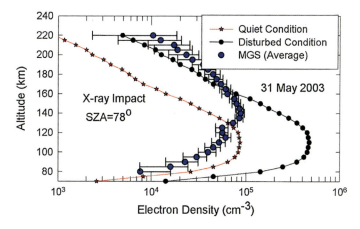

Fig. 7.17 Model calculation of electron densities at quiet (*red line* with *star*) and disturbed conditions (*black line* with *circle*) for SZA 78° on 31 May 2003. Eight profiles were observed by MGS on 31 May 2003. These profiles are averaged and plotted with error bars (*blue colour circle*) in this figure for comparison with model calculations (Reproduced from Haider et al. (2012) by permission of John Wiley & Sons Ltd.)

and EUV radiations, respectively, in this calculation only X-ray fluxes were used as input, which produce the E region of the Martian ionosphere. Haider et al. (2012) have reported that the modelled E layer peak height compares well with that of the MGS observations during quiet condition. MGS did not measure electron density during the maximum phase of the flare.

The Boltzmann equation is represented by continuity, momentum and energy equations. The continuity equation is written as

$$\frac{\partial n_s}{\partial t} + \nabla \cdot (n_s \boldsymbol{u}_s) = P_s - L_s \quad (7.1)$$

where P_s is the production rate of species s including primary production by photoionisation and photoelectron impact ionisation (or by collisional ionisation by energetic electrons or ions) as well as the production due to chemical reactions and L_s is the loss rate of species s due to chemistry of various reactions. This equation has been used by many investigators to study the ionosphere of Mars (e.g. Fox et al. 1993; Fox 2009; Fox and Yeager 2006, 2009; Haider et al. 2010 and references therein). The momentum equation for species s can be written as

$$n_s m_s \left[\frac{\partial \boldsymbol{u}_s}{\partial t} + \boldsymbol{u}_s \cdot \nabla \boldsymbol{u}_s\right] = -\nabla p_s + n_s e_s (\boldsymbol{E} + \boldsymbol{u}_s \times \boldsymbol{B}) + n_s m_s \boldsymbol{g}$$
$$- n_s m_s \sum_j v_{sj} (\boldsymbol{u}_s - \boldsymbol{u}_j) + p_s m_s (\boldsymbol{u}_s - \boldsymbol{u}_n) \quad (7.2)$$

where g is the acceleration due to gravity and ν_{sj} is the momentum transfer collision frequency between species s and j. The charge on species s is equal to $\pm$ e depending on whether s denotes electrons or ions; m_s is the mass of charged particles; E and B are the electric and magnetic fields, respectively; p_s is the pressure ($=n_s k_B T_s$ where k_B and T_s are Boltzmann constant and temperature, respectively); and u_n is the speed of the neutral particle. The temperature for species s can be calculated by the energy equation

$$\frac{3}{2} k_B n_s \frac{\partial T_s}{\partial t} - \frac{\partial}{\partial z} \left(K_s \frac{\partial T_s}{\partial z} \right) = Q_s - L_s \tag{7.3}$$

where Q_s and L_s are heating and cooling rates, respectively, and K_s is coefficient of thermal conductivity. This energy equation leaves out dynamical terms such as heat advection, which is less important for electrons than heat conduction and local heating and cooling. Chen et al. (1978) and Rohrbaugh et al. (1979) have solved energy equations to study ion and electron temperatures in the dayside ionosphere of Mars. In the magnetised plasma, the momentum equation includes the magnetic field (cf., Eq. 7.2), which can be derived from the so-called magnetic induction equation. The induction equation can be written as

$$\frac{\partial B}{\partial t} = \nabla \times (u_s \times B) - \nabla \times \left(\frac{\eta}{\mu_o} \nabla \times B \right) \tag{7.4}$$

This equation is obtained by combining Faraday's law, the generalised Ohm's law and Ampere's law and is known as the magnetic convection-diffusion equation. The first term on the right-hand side of Eq. 7.4 is magnetic convection term, and the second term is the magnetic diffusion term.

Apart from continuity, energy and momentum equations, there are MHD and hybrid models, which are self-consistent Mars-solar wind plasma approaches (cf. Nagy et al. 2004; Ma et al. 2008; Kallio et al. 2010). MHD model provides a high-resolution, three-dimensional simulation of the Martian ionosphere, which contains both a solar wind and a self-consistent ionosphere. In this model, all ions are assumed to have the same bulk velocity. Furthermore, it assumes a Maxwellian velocity distribution function, while pickup ions O^+ are highly non-Maxwellian on Mars. The hybrid model is also a three-dimensional model and has been used to study the global modelling of the Mars-solar wind interaction (Modolo et al. 2006; Brecht and Ledvina 2006; Ledvina et al. 2008; Brain et al. 2010; Kallio et al. 2010). This model represents electrons as a massless fluid. The advantage of this model is that it includes kinetic effects such as finite gyro radius of ions, wave particle interaction and instabilities associated with the non-Maxwellian velocity distribution function. Hybrid model has a low resolution. It does not include Martian crustal magnetic anomalies because the limited resolution makes it impossible to include a realistic crustal magnetic field model.

7.6 Summary

The understanding of complex behaviour of upper ionosphere of Mars required a balanced effort in theoretical modelling, experiments and analysis of the observations. The ability to combine observations with numerical models is critical in predicting ionospheric phenomena. Thus, the models based on fundamental principles became important tool to understand physical, chemical and dynamical processes related to upper ionosphere of Mars. In this chapter we have described ionosphere of Mars during quiet and disturbed conditions of the sun. We have also described various ionospheric models. The mid- and high-latitude ionospheres have been observed from past radio occultation experiments. The ionosphere of Mars has not been observed at low latitude ($<30°$). The plasma density at low latitude can be observed from a spacecraft orbiting at low inclination angle $\leq 30°$ covering the low latitude $\sim 30°N–30°S$. Future missions like Mars-Next, ExoMars and Mars Exploration with Lander and Orbiter (MELOS) are in pipeline to explore Mars. The objectives of these missions are to address key science questions on Mars 'upper ionosphere, solar wind interaction and escape to outer space'. It is expected that new Mars missions will provide missing measurements.

Acknowledgement We acknowledge Shuanggen Jin for his encouragements to write this chapter. Author is grateful to Planetary Data System for providing us radio science data for modelling and analysis. Some part of this chapter is previously published in Space Science Reviews by S.A. Haider and K.K. Mahajan, Lower and Upper Ionosphere of Mars, 2014 doi: 10.1007/s11214-014-0058-2.

References

Acuña MH et al (1998) Magnetic field and plasma observations at Mars: initial results of the Mars Global Surveyor Mission. Science 279:1676–1680

Barth CA (1985) Photochemistry of the atmosphere of Mars. In: Levine JS (ed) The photochemistry of the atmospheres. Academic, New York, pp 337–392

Bertaux J-L, Leblanc F, Witasse O, Quemerais E, Lilensten J, Stern SA, Sandel B, Korablev O (2005) Discovery of an aurora on Mars. Nature 435:790–794

Bougher SW, Roble RG, Ridley EC, Dickinson RE (1990) The Mars thermosphere II. General circulation with coupled dynamical and composition. J Geophys Res 95:14811–14827

Brain DA (2006) Mars Global Surveyor measurements of the Martian solar wind interaction. Space Sci Rev 126:77–112

Brain D et al (2010) A comparison of global models for the solar wind interaction with Mars. Icarus 206:149–151

Brecht SH, Ledvina SA (2006) The solar wind interaction with the Martian ionosphere/atmosphere. Space Sci Rev 126:15–38

Cane HV, McGuire RE, von Rosenvinge TT (1986) Two classes of solar energetic particle events associated with impulsive and long-duration soft X-ray flares. Astrophys J 301:448–459

Chen RH, Cravens TE, Nagy AF (1978) The Martian ionosphere in light of the Viking observations. J Geophys Res 83:3871–3876

Dandouras I, Reme H, Cao JB, Escoubet P, Brandt PC (2007) Abstract on magnetosphere response to the 2005 and 2006 extreme solar events as observed by the cluster and double star spacecraft: solar extreme events. In: Symposium held at Athens in September

Duru F, Gurnett DA, Morgan DD, Modolo R, Nagy AF, Najib D (2008) Electron densities in the upper ionosphere of Mars from the excitation of electron plasma oscillations. J Geophys Res 113:A07302

Fang X, Liemohn MW, Nagy AF, Luhmann J, Ma Y (2010) On the effect of the Martian crustal magnetic field on atmospheric erosion. Icarus 206:130–138

Fillingim MO, Peticolas LM, Lillis RJ, Brain DA, Halekas JS, Lummerzheim D, Bougher SW (2010) Localized ionization patches in the nighttime ionosphere of Mars and their electrodynamic consequences. Icarus 206:112–119

Fjeldbo G, Kliore A, Seidel B (1970) The Martian 1969 occultation measurements of the upper atmosphere of Mars. Radio Sci 5:381–386

Fjeldbo G, Sweetnam D, Brenkle J, Christensen E, Farless D, Mehta J, Seidel B, Michael W Jr, Wallio A, Grossi M (1977) Viking radio occultation measurements of Martian atmosphere and topography: primary mission covering age. J Geophys Res 82:4317–4324

Fox JL (1992) Airglow and Aurora in the atmosphere of Venus and Mars. In: Luhmann JG, Tatrallyay M, Pepin RO (eds) Venus and Mars: atmosphere, ionosphere, and solar wind interactions, vol 66, Geophysical monograph series. AGU, Washington, DC, pp 191–222

Fox JL (2009) Morphology of the dayside ionosphere of Mars: implications for ion outflows. J Geophys Res 114:E12005. doi:10.1029/2009JE003432

Fox JL, Dalgarno A (1979) Ionization, luminosity, and heating of the upper atmosphere of Mars. J Geophys Res 84:7315–7331

Fox JL, Yeager KE (2006) Morphology of the near termination Martian ionosphere: a comparison of models and data. J Geophys Res 111:A10309

Fox JL, Yeager KE (2009) MGS electron density profiles: analysis of the peak magnitudes. Icarus 200:468–479

Fox JL, Brannon JF, Porter HS (1993) Upper limits to the nightside ionosphere of Mars. Geophys Res Lett 20:1339–1342

Fox JL, Zhon P, Bougher SW (1996) The Martian thermosphere/ionosphere at high and low solar activities. Adv Space Res 17(11):203

Gurnett DA et al (2005) Radar soundings of the ionosphere of Mars. Science 310:1929–1933

Haider SA (1997) Chemistry on the nightside ionosphere of Mars. J Geophys Res 102:407–416. doi:10.1029/96JA02353

Haider SA, Mahajan KK (2014) Lower and upper Ionosphere of Mars. Space Sci Rev 182:19–84. doi:10.1007/s11214-014-0058-2

Haider SA, Kim J, Nagy AF, Keller CN, Verigin MI, Gringauz KI, Shutte NM, Szego K, Kiraly P (1992) Calculated ionization rates, ion densities, and airglow emission rates due to precipitating electrons in the nightside ionosphere of Mars. J Geophys Res 97(A7):10637–10641. doi:10.1029/92JA00317

Haider SA, Seth SP, Kallio E, Oyama KI (2002) Solar EUV and electron-proton-hydrogen atom produced ionosphere on Mars: comparative studies of particle fluxes and ion production rates due to different processes. Icarus 159:18–30. doi:10.1006/icar.2002.6919

Haider SA, Singh V, Choksi VR, Maguire WC, Verigin MI (2007) Calculated densities of $H_3O^+(H_2O)_n$, $NO_2^-(H_2O)_n$, $CO_3^-(H_2O)_n$ and electron in the nighttime ionosphere of Mars: impact of solar wind electron and galactic cosmic rays. J Geophys Res 112:A12309. doi:10.1029/2007JA012530

Haider SA, Abdu MA, Batista IS, Sobral JH, Luan X, Kallio E, Maguire WC, Verigin MI, Singh V (2009a) D, E, and F layers in the daytime at high-latitude terminator ionosphere of Mars: comparison with Earth's ionosphere using COSMIC data. J Geophys Res 114:A03311. doi:10.1029/2008JA13709

Haider SA, Abdu MA, Batista IS, Sobral JH, Kallio E, Kallio E, Maguire WC, Verigin MI (2009b) On the responses to solar X-ray flare and coronal mass ejection in the ionosphere of Mars and Earth. Geophys Res Lett 36:L13104. doi:10.1029/2009GL038694

7 Upper Ionosphere of Mars During Solar Quiet and Disturbed Conditions 145

Haider SA, Seth SP, Brain DA, Mitchell DL, Majeed T, Bougher SW (2010) Modeling photoelectron transport in the Martian ionosphere at Olympus Mons and Syrtis Major: MGS observations. J Geophys Res 115:A08310. doi:10.1029/2009JA014968

Haider SA, Mahajan KK, Kallio E (2011) Mars ionosphere: a review of experimental results and modeling studies. Rev Geophys 49:RG4001

Haider SA, McKenna-Lawlor SMP, Fry CD, Jain R, Joshipura KN (2012) Effects of solar X-ray flares in the E region ionosphere of Mars: first model results. J Geophys Res 117:A05326

Hanson WB, Sanatani S, Zuccaro R (1977) The Martian ionosphere as observed by the Viking retarding potential analyzers. J Geophys Res 82:4351–4363

Hinson DP, Simpson RA, Twicken JD, Tyler GL, Flassar FM (1999) Initial results from radio occultation measurements with Mars Global Surveyor. J Geophys Res 104:26997–27012

Ip WH (2012) ENA diagnostic of auroral activity at Mars. Planet Space Sci 63/64:83–86

Kallio E, Janhunen P (2001) Atmospheric effects of proton precipitation in the Martian atmosphere and its connection to the Mars-solar wind interaction. J Geophys Res 106:5617–5634

Kallio E, Liu K, Javinen R, Pohjola V, Janhunen P (2010) Oxygen ion escape at Mars in a hybrid model: High energy and low energy ions. Icarus 206:152–163. doi:10.1016/j.icarus.2009.05.015

Kliore AJ, Fjeldbo G, Seidel BL, Sykes MJ, Woiceshyn PM (1973) S band radio occultation measurements of the atmosphere and topography of Mars with Mariner 9: extended mission coverage of polar and intermediate latitudes. J Geophys Res 78:4331–4351

Kolosov MA et al (1975) Results of investigating the Martian atmosphere by radio occultation using Mars 2, Mars 4 and Mars 6 spacecraft. Kosmich Issled 13:54–59

Leblanc F, Luhmann JG, Johnson RE, Chassefiere E (2002) Some expected impacts of a solar energetic particle event at Mars. J Geophys Res 107:1058

Leblanc F et al (2008) Observations of aurorae by SPICAM ultraviolet spectrograph on board Mars Express: simultaneous ASPERA-3 and MARSIS measurements. J Geophys Res 113:A08311

Ledvina SA, Ma Y-J, Kallio E (2008) Modeling and simulating flowing plasmas and related phenomena. Space Sci Rev 139:143–189

Lei L, Zhang Y (2009) Model investigation of the influence of the crustal magnetic field on the oxygen ion distribution in the near Martian tail. J Geophys Res 114:A06215. doi:10.1029/2008JA013850

Lillis RJ, Fillingim MO, Peticolas LM, Brain DA, Lin RP, Bougher SW (2009) The nightside ionosphere of Mars: modeling the effects of crustal magnetic fields and electron pitch angle distributions on electron impact ionization. J Geophys Res 114:E11009

Lillis RJ, Brain DA, England SL, Withers P, Fillingim MO, Safaeinili A (2010) Total electron content in the Mars ionosphere: temporal studies and dependence on solar EUV flux. J Geophys Res 115:A11314

Lovell JL, Dulding ML, Humble JE (1998) An extended analysis of the September 1989 Cosmic ray ground level enhancement. J Geophys Res 103:23733–23742. doi:10.1029/98JA02100

Lundin R et al (2006) Plasma acceleration above Martian magnetic anomalies. Science 311:980. doi:10.1126/science.1122071

Lundin R, Barabash S, Dubinin E, Winingham D, Yamauchi M (2011) Low latitude acceleration of ionospheric ions at Mars. Geophys Res Lett 38:L08108. doi:10.1029/2011GL047064

Ma Y, Nagy AF, Sokolov IV, Hanse KC (2004) Three dimensional, multispecies, high spatial resolution MHD studies of the solar wind interaction with Mars. J Geophys Res 109:A07211

Ma Y et al (2008) Plasma flow and related phenomena in planetary aeronomy. Space Sci Rev 139:311–353. doi:10.1007/s11214-008-9389-1

Mahajan KK, Lodhi NK, Singh S (2009) Ionospheric effects of solar flares at Mars. Geophys Res Lett 36:L15207. doi:10.1029/2009GL039454

McKenna-Lawlor S, Goncalves P, Keating A, Reitz G, Matthia D (2012) Overview of energetic particle hazards during prospective manned mission to Mars. Planet Space Sci 63/64:12–132

Mendillo M, Pi X, Smith S, Martinis C, Wilson J, Hinson D (2004) Ionospheric effects upon a satellite navigation system at Mars. Radio Sci 39:RS2028

Mendillo M, Withers P, Hinson D, Rishbeth H, Reinisch B (2006) Effects of solar flares on the ionosphere of Mars. Science 311:1135–1138

Mendillo M, Lollo A, Withers P, Matta M, Pätzold M, Tellmann S (2011) Modeling Mars' ionosphere with constraints from same-day observations by Mars Global Surveyor and Mars Express. J Geophys Res 116:A11303

Mitchell DL, Lin RP, Rème H, Crider DH, Cloutier PA, Connerney JEP, Acuña MH, Ness NF (2000) Oxygen auger electrons observed in Mars ionosphere. Geophys Res Lett 27:1871–1874

Mitchell DL, Lillis RJ, Lin RP, Connerney JEP, Acuña MH (2007) A global map of Mars' crustal magnetic field based on electron reflectometer. J Geophys Res 112:E01002

Modolo R, Chanteur GM, Dubinin E, Matthews AP (2006) Simulated solar wind plasma interaction with the Martian exosphere: influence of the solar EUV flux on the bow shock and the magnetic pile-up boundary. Ann Geophys 24:3403–3410

Morgan DD, Gurnett DA, Kirchner DL, Huff RL, Brain DA, Boynton WV, Acuña MH, Plaut JJ, Picardi G (2006) Solar control of radar wave absorption by the Martian ionosphere. Geophys Res Lett 33:L13202

Nagy AF et al (2004) The plasma environment of Mars. Space Sci Rev 111:33–114

Nielsen E, Zou H, Gurnett DA, Kirchner DL, Morgan DD, Huff R, Orosei R, Safaeinili A, Plaut JJ, Picardi G (2006) Observations of vertical reflections from the topside Martian ionosphere. Space Sci Rev 126:373–388

Pätzold M, Tellmann S, Haüsler B, Hinson D, Schaa R, Tyler GL (2005) A sporadic third layer in the ionosphere of Mars. Science 310:837–839

Rishbeth H, Mendillo M (2004) Ionospheric layers of Mars and Earth. Planet Space Sci 52: 849–852

Rohrbaugh RP, Nisbet JS, Bleuler E, Herman JR (1979) The effects of energetically produced O^+_2 on the ion temperature of the Martian thermosphere. J Geophys Res 84:3327–3336

Savich NA, Samovol VA (1976) The night time ionosphere of Mars from Mars 4 and Mars 5 dual frequency radio occultation measurements. Space Res XVI:1009–1010

Schnjver CJ, Siscoe GL (2010) Heliophysics. Cambridge University Press, Cambridge

Sheel V, Haider SA, Withers P, Kozarev K, Jun I, Kang S, Gronoff G, Simon Wedlund C (2012) Numerical simulation of the effects of a solar energetic particle event on the ionosphere of Mars. J Geophys Res 117:A05312

Shinagawa H, Bougher SW (1999) A two-dimensional MHD model of the solar wind interaction with Mars. Earth Planets Space 51:55–62

Shinagawa H, Cravens TE (1989) A one-dimensional multispecies magnetohydrodynamic model of the day side ionosphere of Mars. J Geophys Res 94:6506–6516

Tobiska WK, Woods T, Eparvier F, Viereck R, Floyd L, Bouwer D, Rottman G, White OR (2000) The solar 2000 empirical solar irradiance model and forecast tool. J Atmos Sol Terr Phys 62:1233–1250. doi:10.1016/S1364-6826(00)00070-5

Tyler GL et al (2001) Radio science observations with Mars Global Surveyor: orbit insertion through one Mars year in mapping orbit. J Geophys Res 106:23327–23348. doi:10.1029/2000JE001348

Vasiliev MB et al (1975) Preliminary results of dual frequency radio occultation of the Martian ionosphere with the aid of Mars 5 spacecraft. Kosm Issled 13:48–51

Verigin MI, Gringauz KI, Shutte NM, Haider SA, Szego K, Kiraly P, Nagy AF, Gombosi TI (1991) On the possible source of the ionization in the nighttime Martian ionosphere 1. Phobos 2 HARP electron spectrometer measurements. J Geophys Res 96:19307–19313

Winchester C, Rees D (1995) Numerical models of the Martian coupled thermosphere and ionosphere. Adv Space Res 15(4):51

Withers P (2009) A review of observed variability in the dayside ionosphere of Mars. Adv Space Res 44:277–307

Withers P (2011) Attenuation of radio signals by the ionosphere of Mars: theoretical development and application to MARSIS observations. Radio Sci 46:RS2004

Chapter 8
Mars Astrobiology: Recent Status and Progress

Antonio de Morais M. Teles

Abstract In this chapter, we begin making a brief review on the history of the studies of planet Mars. Then, we review the modern research (with recent status, data, results, and progress) on the search for possible extinct or extant life on Mars, with special emphasis on the search for the presence (in the past and/or presently) of liquid water within Mars' surface and subsurface – a prerequisite for the evolution from geochemical state to biogeochemical state, as we are aware of here on planet Earth. Through the chapter, I present some proposals of mine about the astrobiology of Mars. We also analyze recent astrobiological experiments on board the International Space Station (ISS) for the future exploration of Mars. And we make a brief review on the evolution of equipment for its exploration and of future manned presence on the beautiful planet Mars – the "Red Planet."

Keywords Mars • Astrobiology • Biogeochemistry • Space • Planets • Exploration • Terraforming

8.1 Introduction: Past Search for Life on Mars

Well, for centuries, people have speculated about the possibility of life on Mars due to the planet's proximity and similarity to Earth. Serious searches for evidence of life began in the nineteenth century. Mars' polar ice caps were observed as early as the mid-seventeenth century, and they were first proven to grow and shrink alternately, in the summer and winter of each hemisphere, by William Herschel in the latter part of the eighteenth century.

By the mid-nineteenth century, astronomers knew that Mars had certain other similarities to Earth, for example, that the length of a day on Mars was almost the same as a day on Earth. They also knew that its axial tilt was similar to Earth's,

---

A. de Morais M. Teles (✉)
Brazilian Center of Physics Research, Rua Dr. Xavier Sigaud, 150, 3° andar, LAFEX, Urca, Rio de Janeiro, RJ 22290-180, Brazil
e-mail: antonioamore@yahoo.com

© Springer-Verlag Berlin Heidelberg 2015
S. Jin et al. (eds.), *Planetary Exploration and Science: Recent Results and Advances*,
Springer Geophysics, DOI 10.1007/978-3-662-45052-9_8

which meant it experienced seasons just as Earth does – but of nearly double the length owing to its much longer year. These observations led to the increase in speculation that the darker albedo features were water and the brighter ones were land. It was therefore natural to suppose that Mars may be inhabited by some form of life.

Spectroscopic analysis of Mars' atmosphere began in earnest in 1894, when US astronomer William Wallace Campbell showed that neither water nor oxygen was present in the Martian atmosphere (Chambers 1999). By 1909, better telescopes and the best perihelic opposition of Mars since 1877 showed a Martian surface without extended traces of life forms, as vegetation.

8.2 Modern Mars Astrobiology Research

Why Mars is so special?

Let us try to answer this simple but profound question.

First of all, a definition of astrobiology. Astrobiology is the study of the origin, evolution, distribution, and future of life in the universe: extraterrestrial life and life on Earth. This interdisciplinary field encompasses the search for habitable environments in the solar system and habitable planets outside the solar system, the search for evidence of prebiotic chemistry, laboratory and field research into the origins and early evolution of life on Earth, and studies of the potential for life to adapt to challenges on Earth and in outer space. Astrobiology addresses the question of whether life exists beyond Earth and how humans can detect it if it does. The term exobiology is similar but more specific – it covers the search for life beyond Earth and the effects of extraterrestrial environments on living things.

Serious searches for evidence of life continue today via telescopic investigations and landed missions. Modern scientific inquiry has emphasized the search for water, chemical biosignatures in the soil and rocks at the planet's surface, and biomarker gases in the atmosphere (Fig. 8.1).

Mars is of particular interest for the study of the origins of life because of its similarity to the early Earth. This is especially so since Mars has a cold climate and lacks plate tectonics or continental drift, so it has remained almost unchanged since the end of the Hesperian period. At least two-thirds of Mars' surface is more than 3.5 billion years (Gyrs) old, and Mars may thus hold the best record of the prebiotic conditions leading to abiogenesis, even if life does not or has never existed there (McKay and Stoker 1989; Gaidos and Selsis 2007). It remains an open question whether life currently exists on Mars or has existed there in the past (Sagan 1980).

On January 24, 2014, NASA reported that current studies on the planet Mars by the Curiosity and Opportunity rovers will now be searching for evidence of ancient life, including a biosphere based on autotrophic, chemotrophic, and/or

8 Mars Astrobiology: Recent Status and Progress

Fig. 8.1 Planet Mars as photographed by the NASA's robotic Hubble Space Telescope spacecraft (http://hubblesite.org/newscenter/archive/releases/2001/24/image/a/ – NASA/STScI/AURA, June 26, 2001)

chemolithoautotrophic microorganisms, as well as ancient water, including fluvial–lacustrine environments (plains related to ancient rivers or lakes) that may have been habitable (Grotzinger 2014; Various 2014). The search for evidence of habitability, taphonomy (related to fossils), and organic carbon on planet Mars is now a primary NASA objective (Grotzinger and MSL Science Team 2014) (Fig. 8.2).

Chemical, physical, geological, and geographic attributes shape the environments on Mars. Isolated measurements of these factors may be insufficient to deem an environment habitable, but the sum of measurements can help predict locations with greater or lesser habitability potential.

The two current ecological approaches for predicting the potential habitability of the Martian surface use 19 or 20 environmental factors, with emphasis on water availability, temperature, presence of nutrients, an energy source, and protection from solar ultraviolet and galactic cosmic radiation (Schuerger et al. 2012).

Fig. 8.2 Picture of Mars, taken by the Spirit rover on Sol 454. One sol is one Martian day. This image shows a region in the "Columbia Hills" inside Gusev Crater. The view features two interesting outcrops in the middle distance and "Clark Hill" in the left background. The outcrop on the right, with rover tracks leading from it, is "Larry's Lookout." On the left is the Methuselah outcrop, with apparent layering (http://photojournal.jpl.nasa.gov/catalog/PIA07855, OrigCaption – NASA/JPL-Caltech/Cornell University, April 20, 2005)

8.2.1 Past Mars

Recent models have shown that even with a dense CO_2 atmosphere, early Mars was, in fact, colder than Earth. However, transiently warm conditions related to impacts or volcanism could have produced conditions favoring the formation of the late Noachian valley networks, even though the mid–late Noachian global conditions were probably icy. Local warming of the environment by volcanism and impacts would have been sporadic, but there should have been many events of water flowing at the surface of Mars. Both the mineralogical and the morphological evidence indicate a degradation of habitability from the mid-Hesperian onward. The exact causes are not well understood but may be related to a combination of processes including loss of early atmosphere, or impact erosion, or both (Westall et al. 2013).

The loss of the Martian magnetic field strongly affected surface environments through atmospheric loss and increased radiation; this change significantly degraded surface habitability (Summons et al. 2011). When there was a magnetic field, the atmosphere would have been protected from erosion by solar wind, which would ensure the maintenance of a dense atmosphere, necessary for liquid water to exist on the surface of Mars (Dehant et al. 2007).

The loss of the atmosphere was accompanied by decreasing temperatures. A part of the liquid water inventory sublimed and was transported to the poles, while the rest became trapped in a subsurface ice layer.

Observations on Earth and numerical modeling have shown that a crater-forming impact can result in the creation of a long-lasting hydrothermal system when ice is present in the crust. For example, a 130-km large crater could sustain an active hydrothermal system for up to 2 Ma, that is, long enough for microscopic life to emerge (Westall et al. 2013).

Soil and rock samples studied in 2013 by NASA's Curiosity rover's on board instruments brought about additional information on several habitability factors. The rover team identified some of the key chemical ingredients for life in this soil, including sulfur, nitrogen, hydrogen, oxygen, phosphorus and possibly carbon, as well as clay minerals, suggesting a long-ago aqueous environment – perhaps a lake or an ancient streambed – that was neutral and not too salty. On December 9, 2013, NASA reported that, based on evidence from Curiosity studying Aeolis Palus, Gale Crater contained an ancient freshwater lake which could have been a hospitable environment for microbial life.

The confirmation that liquid water once flowed on Mars, of the existence of nutrients, and of the previous discovery of a past magnetic field that protected the planet from cosmic and solar radiation together strongly suggests that Mars could have had the environmental factors to support life. However, the assessment of past habitability is not in itself evidence that Martian life has ever actually existed. If it did, it was probably microbial, existing communally in fluids or on sediments, either free-living or as biofilms, respectively (Dehant et al. 2007; Summons et al. 2011; Various 2014).

8.2.2 Present Mars

No definitive evidence for biosignatures or organics of Martian origin has been identified, and assessment will continue not only through the Martian seasons but also back in time as the Curiosity rover studies what is recorded in the depositional history of the rocks in Gale Crater. While scientists have not identified the minimum number of parameters for the determination of habitability potential, some teams have proposed hypotheses based on simulations.

Scientists do not know the minimum number of parameters for the determination of habitability potential, but they are certain it is greater than one or two of the factors in the table below. Similarly, for each group of parameters, the habitability threshold for each is to be determined. Laboratory simulations show that whenever multiple lethal factors are combined, the survival rates plummet quickly. There are no full-Mars simulations published yet that include all of the biocide factors combined.

Habitability factors

Water	Liquid water activity (a_w)
	Past/future liquid (ice) inventories
	Salinity, pH, and Eh of available water
Chemical environment	*Nutrients*
	C, H, N, O, P, S, essential metals, essential micronutrients
	Fixed nitrogen
	Availability/mineralogy
	Toxin abundances and lethality
	Heavy metals (e.g., Zn, Ni, Cu, Cr, As, Cd, etc., some essential but toxic at high levels)
	Globally distributed oxidizing soils
Energy for metabolism	*Solar* (surface and near-surface only)
	Geochemical (subsurface)
	Oxidants
	Reductants
	Redox gradients
Conducive physical conditions	Temperature
	Extreme diurnal temperature fluctuations
	Low pressure (is there a low-pressure threshold for terrestrial anaerobes?)
	Strong ultraviolet germicidal irradiation
	Galactic cosmic radiation and solar particle events (long-term accumulated effects)
	Solar UV-induced volatile oxidants, e.g., O^{2-}, O^-, H_2O_2, O_3
	Climate/variability (geography, seasons, diurnal, and eventually obliquity variations)
	Substrate (soil processes, rock microenvironments, dust composition, shielding)
	High CO_2 concentrations in the global atmosphere

8.2.2.1 Subsurface

Although Mars soils are likely not to be overtly toxic to terrestrial microorganisms, life on the surface of Mars is extremely unlikely because it is bathed in radiation and it is completely frozen. But extremophiles could hypothetically be able to thrive through such harsh conditions. Therefore, the best potential locations for discovering life on Mars may be at subsurface environments that have not been studied yet. The extensive volcanism in the past possibly created subsurface cracks and caves within different strata where liquid water could have been stored, forming large aquifers with deposits of saline liquid water, minerals, organic molecules, and geothermal heat – potentially providing a habitable environment away from the harsh surface conditions.

8.2.2.2 Surface Brines

Although liquid water does not appear at the surface of Mars, several modeling studies suggest that potential locations on Mars could include regions where thin films of salty liquid brine or perchlorate may form near the surface that may provide a potential location for terrestrial salt- and cold-loving microorganisms (halophilic psychrophile). Various salts present in the Martian soil may act as antifreeze and could keep water liquid well below its normal freezing point, if water was present at certain favorable locations. Astrobiologists are keen to find out more, as not much is known about these brines at the moment. The briny water may or may not be habitable to microbes from Earth or Mars. Another researcher argues that although chemically important, thin films of transient liquid water are not likely to provide suitable sites for life. In addition, an astrobiology team asserted that the activity of water on salty films, the temperature, or both are less than the biological thresholds across the entire Martian surface and shallow subsurface.

The damaging effect of ionizing radiation on cellular structure is one of the prime limiting factors on the survival of life in potential astrobiological habitats. Even at a depth of 2 m beneath the surface, any microbes would probably be dormant, cryopreserved by the current freezing conditions, and so metabolically inactive and unable to repair cellular degradation as it occurs. Also, solar ultraviolet (UV) radiation proved particularly devastating for the survival of cold-resistant microbes under simulated surface conditions on Mars, as UV radiation was readily and easily able to penetrate the salt–organic matrix that the bacterial cells were embedded in. In addition, NASA's Mars Exploration Program states that life on the surface of Mars is unlikely, given the presence of superoxides that break down organic (carbon-based) molecules on which life is based.

8.2.2.3 Cosmic Radiation

In 1965, the Mariner 4 probe discovered that Mars had no global magnetic field that would protect the planet from potentially life-threatening cosmic radiation and solar radiation; observations made in the late 1990s by the Mars Global Surveyor confirmed this discovery. Scientists speculate that the lack of magnetic shielding helped the solar wind blow away much of Mars' atmosphere over the course of several billion years.

As a result, the planet has been vulnerable to radiation from space for about 4 billion years. Currently, ionizing radiation on Mars is typically two orders of magnitude (or 100 times) higher than on Earth. Even the hardiest cells known could not possibly survive the cosmic radiation near the surface of Mars for that long. After mapping cosmic radiation levels at various depths on Mars, researchers have concluded that any life within the first several meters of the planet's surface would be killed by lethal doses of cosmic radiation. The team calculated that the cumulative damage to DNA and RNA by cosmic radiation would limit retrieving viable dormant cells on Mars to depths greater than 7.5 m below the planet's surface.

Even the most radiation-tolerant Earthly bacteria would survive in dormant spore state only 18,000 years at the surface; at 2 m – the greatest depth at which the ExoMars rover will be capable of reaching – survival time would be 90,000 to half million years, depending on the type of rock.

The Radiation Assessment Detector (RAD) on board the Curiosity rover is currently quantifying the flux of biologically hazardous radiation at the surface of Mars today and will help determine how these fluxes vary on diurnal, seasonal, solar cycle, and episodic (flare, storm) timescales. These measurements will allow calculations of the depth in rock or soil to which this flux, when integrated over long timescales, provides a lethal dose for known terrestrial organisms.

Research published in January 2014 of data collected by the RAD instrument revealed that the actual absorbed dose measured is 76 mGy/year at the surface and that ionizing radiation strongly influences chemical compositions and structures, especially for water, salts, and redox-sensitive components such as organic matter (Hassler and MSL Science Team 2014).

Regardless of the source of Martian organic matter (meteoritic, geological, or biological), its carbon bonds are susceptible to breaking and reconfiguration with the surrounding elements by ionizing charged particle radiation. These improved subsurface radiation estimates give insight into the potential for the preservation of possible organic biosignatures as a function of depth as well as survival times of possible microbial or bacterial life forms left dormant beneath the surface. The report concludes that the in situ surface measurements – and subsurface estimates – constrain the preservation window for Martian organic matter following exhumation and exposure to ionizing radiation in the top few meters of the Martian surface (Hassler and MSL Science Team 2014).

8.2.2.4 Nitrogen Fixation

After carbon, nitrogen is arguably the most important element needed for life. Thus, measurements of nitrate over the range of 0.1–5 % are required to address the question of its occurrence and distribution. There is nitrogen (as N_2) in the atmosphere at low levels, but this is not adequate to support nitrogen fixation for biological incorporation. Nitrogen in the form of nitrate, if present, could be a resource for human exploration both as a nutrient for plant growth and for use in chemical processes. On Earth, nitrates correlate with perchlorates in desert environments, and this may also be true on Mars. Nitrate is expected to be stable on Mars and to have formed in shock and electrical processes. Currently, there is no data on its availability.

8.2.2.5 Low Pressure

Further complicating the estimates of the habitability of the Martian surface is the fact that very little is known on the growth of microorganisms at pressures close to the conditions found on the surface of Mars. Some teams determined that some bacteria may be capable of cellular replication down to 25 mbar, but that is still above the atmospheric pressures found on Mars (range 1–14 mbar). In another study, 26 strains of bacteria were chosen based on their recovery from spacecraft assembly facilities, and only *Serratia liquefaciens* strain ATCC 27592 exhibited growth at 7 m bar, 0 °C, and CO_2-enriched anoxic atmospheres.

8.3 Water on Mars

Water on Mars exists today almost exclusively as ice, with a small amount present in the atmosphere as vapor. The only place where water ice is visible at the surface is at the north polar ice cap (Carr 1996). Abundant water ice is also present beneath the permanent carbon dioxide ice cap at the Martian south pole and in the shallow subsurface at more temperate latitudes (Feldman et al. 2004). More than five million cubic kilometers of ice have been identified at or near the surface of modern Mars, enough to cover the whole planet to a depth of 35 m (Christensen 2006). Even more ice is likely to be locked away in the deep subsurface.

Some liquid water may occur transiently on the Martian surface today but only under certain conditions (Hecht 2002). No large standing bodies of liquid water exist because the atmospheric pressure at the surface averages just 600 Pa (0.087 psi) – about 0.6 % of Earth's mean sea level pressure – and because the global average temperature is far too low (210 K (−63 °C)), leading to either rapid evaporation or freezing. Before about 3.8 billion years ago, Mars may have had a denser atmosphere and higher surface temperatures (Pollack 1987), allowing vast amounts of liquid water on the surface, possibly including a large ocean that may have covered one-third of the planet. Water has also apparently flowed across the surface for short periods at various intervals more recently in Mars' history. On December 9, 2013, NASA reported that, based on evidence from the Curiosity rover studying Aeolis Palus, Gale Crater contained an ancient freshwater lake which could have been a hospitable environment for microbial life.

Many lines of evidence indicate that water is abundant on Mars and has played a significant role in the planet's geologic history. The present-day inventory of water on Mars can be estimated from spacecraft imagery, remote sensing techniques (spectroscopic measurements, radar, etc.), and surface investigations from landers and rovers. Geological evidence of past water includes enormous outflow channels carved by floods; ancient river valley networks, deltas, and lake beds; and the detection of rocks and minerals on the surface that could only have formed in liquid water. Numerous geomorphic features suggest the presence of ground ice (permafrost) and the movement of ice in glaciers, both in the recent past and present. Gullies and slope lineae along cliffs and crater walls suggest that flowing water continues to shape the surface of Mars, although to a far lesser degree than in the ancient past.

Although the surface of Mars was periodically wet and could have been hospitable to microbial life billions of years ago, the current environment at the surface is dry and subfreezing, probably presenting an insurmountable obstacle for living organisms. In addition, Mars lacks a thick atmosphere, ozone layer, and magnetic field, allowing solar and cosmic radiation to strike the surface unimpeded. The damaging effects of ionizing radiation on cellular structure are another one of the prime limiting factors on the survival of life on the surface. Therefore, the best potential locations for discovering life on Mars may be in subsurface environments.

Understanding water on Mars is vital to assess the planet's potential for harboring life and for providing usable resources for future human exploration. For this reason, "Follow the Water" was the science theme of NASA's Mars Exploration Program (MEP) in the first decade of the twenty-first century. Discoveries by the 2001 Mars Odyssey, Mars Exploration Rovers (MERs), Mars Reconnaissance Orbiter (MRO), and Mars Phoenix Lander have been instrumental in answering key questions about water's abundance and distribution on Mars. The ESA's Mars Express Orbiter has also provided essential data in this quest. The Mars Odyssey, Mars Express, MER Opportunity rover, MRO, and Mars Science Lander Curiosity rover are still sending back data from Mars, and discoveries continue to be made.

On January 24, 2014, NASA reported that current studies on Mars by the Curiosity and Opportunity rovers will now be searching for evidence of ancient life, including a biosphere based on autotrophic, chemotrophic, and/or chemolithoautotrophic microorganisms, as well as ancient water, including fluvial–lacustrine environments (plains related to ancient rivers or lakes) that may have been habitable. The search for evidence of habitability, taphonomy (related to fossils), and organic carbon on planet Mars is now a primary NASA objective (Various 2014).

8.3.1 Evidence from Rocks and Minerals

Today, it is widely accepted that Mars had abundant water very early in its history, but all large areas of liquid water have since disappeared. A fraction of this water is retained on modern Mars as both ice and locked into the structure of abundant water-rich materials, including clay minerals (phyllosilicates) and sulfates. Studies of hydrogen isotopic ratios indicate that asteroids and comets from beyond 2.5 astronomical units (AU) provide the source of Mars' water, which currently totals 6–27 % of the Earth's present ocean.

8.3.1.1 Water in Weathering Products (Aqueous Minerals)

The primary rock type on the surface of Mars is basalt, a fine-grained igneous rock made up mostly of the mafic silicate minerals olivine, pyroxene, and plagioclase feldspar. When exposed to water and atmospheric gases, these minerals chemically weather into new (secondary) minerals, some of which may incorporate water into their crystalline structures, either as H_2O or as hydroxyl (OH). Examples of hydrated (or hydroxylated) minerals include the iron hydroxide goethite (a common component of terrestrial soils), the evaporate minerals gypsum and kieserite, opaline silica, and phyllosilicates (also called clay minerals), such as kaolinite and montmorillonite. All of these minerals have been detected on Mars.

One direct effect of chemical weathering is to consume water and other reactive chemical species, taking them from mobile reservoirs like the atmosphere and hydrosphere and sequestering them in rocks and minerals. The amount of water in the Martian crust stored in hydrated minerals is currently unknown but may be quite large. For example, mineralogical models of the rock outcroppings examined by instruments on the Opportunity rover at Meridiani Planum suggest that the sulfate deposits there could contain up to 22 % water by weight.

On Earth, all chemical weathering reactions involve water to some degree. Thus, many secondary minerals do not actually incorporate water but still require water to form. Some examples of anhydrous secondary minerals include many carbonates, some sulfates (e.g., anhydrite), and metallic oxides such as the iron oxide mineral hematite. On Mars, a few of these weathering products may theoretically form without water or with scant amounts present as ice or in thin molecular-scale films (monolayers). The extent to which such exotic weathering processes operate on Mars is still uncertain. Minerals that incorporate water or form in the presence of water are generally termed "aqueous minerals."

Aqueous minerals are sensitive indicators of the type of environment that existed when the minerals formed. The ease at which aqueous reactions occur (Gibbs free energy) depends on the pressure, temperature, and concentrations of the gaseous and soluble species involved.

Two important properties are pH and oxidation–reduction potential (Eh). For example, the sulfate mineral jarosite forms only in low pH (highly acidic) water. Phyllosilicates usually form in water of neutral to high pH (alkaline). Eh is a measure of the oxidation state of an aqueous system. Together, Eh and pH indicate the types of minerals that are thermodynamically most likely to form from a given set of aqueous components. Thus, past environmental conditions on Mars, including those conducive to life, can be inferred from the types of minerals present in the rocks.

8.3.1.2 Hydrothermal Alteration

Aqueous minerals can also form in the subsurface by hydrothermal fluids migrating through pores and fissures. The heat source driving a hydrothermal system may be nearby magma bodies or residual heat from large impacts. One important type of hydrothermal alteration in the Earth's oceanic crust is serpentinization, which occurs when seawater migrates through ultramafic and basaltic rocks. The water–rock reactions result in the oxidation of ferrous iron in olivine and pyroxene to produce ferric iron (as the mineral magnetite) yielding molecular hydrogen (H_2) as a by-product. The process creates a highly alkaline and reduces (low) Eh environment favoring the formation of certain phyllosilicates (serpentine minerals) and various carbonate minerals, which together form a rock called serpentinite. The hydrogen gas produced can be an important energy source for chemosynthetic organisms or it can react with CO_2 to produce methane gas, a process that has been considered as a non-biological source for the trace amounts of methane reported in the Martian atmosphere. Serpentine minerals can also store a lot of water (as hydroxyl) in their crystal structure. A recent study has argued that hypothetical serpentinites in the ancient highland crust of Mars could hold as much as a 500-m-thick global equivalent layer (GEL) of water. Some serpentine minerals have been detected on Mars, and widespread outcroppings are evident from remote sensing data from NASA's Mars Reconnaissance Orbiter spacecraft. McLaughlin Crater, one of the deepest craters on Mars, contains evidence for Mg–Fe-bearing clays and carbonates that probably formed in an alkaline, groundwater-fed lacustrine setting (Michalski et al. 2013). This fact could hypothetically indicate the presence of large amounts of serpentinite hidden at depth in the Martian crust. Deposits formed as a result of groundwater upwelling on Mars, such as those in McLaughlin Crater, could preserve critical evidence of a deep biosphere on Mars (Michalski et al. 2013) and may be a good place to look for signs of life that pooled there from underground (Fig. 8.3).

8.3.1.3 Weathering Rates

The rates at which primary minerals convert to secondary aqueous minerals vary. Primary silicate minerals crystallize from magma under pressures and temperatures vastly higher than conditions at the surface of a planet. When exposed to a surface environment, these minerals are out of equilibrium and will tend to interact with available chemical components to form more stable mineral phases. In general, the silicate minerals that crystallize at the highest temperatures (solidify first in a cooling magma) weather the most rapidly. On Earth and Mars, the most common mineral to meet this criterion is olivine, which readily weathers to clay minerals in the presence of water. Olivine is widespread on Mars, suggesting that Mars' surface has not been pervasively altered by water; abundant geological evidence suggests otherwise.

8.3.2 Evidence of Water in Martian Meteorites

Over 60 meteorites have been found that came from Mars. Some of them contain evidence that they were exposed to water when on Mars. Some Martian meteorites called basaltic shergottites appear (from the presence of hydrated carbonates and sulfates) to have been exposed to liquid water prior to ejection into space. It has been shown that another class of meteorites, the nakhlites, were suffused with liquid water around 620 Ma ago and that they were ejected from Mars around 10.75 Ma ago by an asteroid impact. They fell to Earth within the last 10,000 years.

In 1996, a group of scientists reported the possible presence of microfossils in the Allan Hills 84001, a meteorite from Mars. Many studies disputed the validity

Fig. 8.3 Layers with carbonate content inside McLaughlin Crater on Mars. This view of layered rocks on the floor of McLaughlin Crater shows sedimentary rocks that contain spectroscopic evidence for minerals formed through interaction with water. The High Resolution Imaging Science Experiment (HiRISE) camera on NASA's Mars Reconnaissance Orbiter recorded the image. A combination of clues suggests this 2.2-km-deep (1.4-mile-deep) crater once held a lake fed by groundwater. Part of the evidence is the identification of clay and carbonate minerals within layers visible near the center of this image. The mineral identifications come from the Compact Reconnaissance Imaging Spectrometer for Mars (CRISM), also on the Mars Reconnaissance Orbiter. The scene covers an area about 550 m (one-third of a mile) across, at 337.6° east longitude, 21.9° north latitude. It indicates the location of layers bearing clay and carbonate minerals and includes a scale bar of 100 m (328 ft). North is up (http://www.nasa.gov/mission_pages/MRO/multimedia/pia16710.html and http://www.nature.com/ngeo/journal/v6/n2/abs/ngeo1706.html – NASA/JPL-Caltech/University of Arizona, 2012)

of the fossils. It was found that most of the organic matter in the meteorite was of terrestrial origin. More about Martian meteorites will be treated with detail later in this chapter.

8.3.3 Geomorphic Evidence

8.3.3.1 Lakes and River Valleys

The 1971 Mariner 9 spacecraft caused a revolution in our ideas about water on Mars. Huge river valleys were found in many areas. Images showed that floods

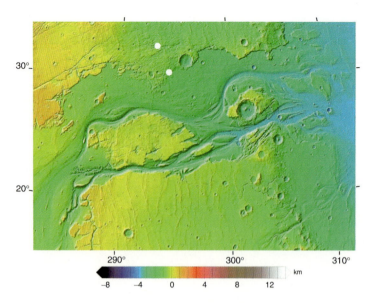

Fig. 8.4 Kasei Vallis – a major outflow channel – seen in MOLA elevation data. Flow was from the bottom left to right. Image is approximately 1,600 km across. The channel system extends another 1,200 km south of this image to Echus Chasma – Mars Global Surveyor (http://themis.mars.asu.edu/feature/10 – NASA/JPL-Caltech/Arizona State University, 2006)

of water broke through dams, carved deep valleys, eroded grooves into bedrock, and traveled thousands of kilometers. Areas of branched streams, in the southern hemisphere, suggested that rain once fell. The numbers of recognized valleys have increased through time. Research published in June 2010 mapped 40,000 river valleys on Mars, roughly quadrupling the number of river valleys that had previously been identified. Martian waterworn features can be classified into two distinct classes: 1) dendritic (branched), terrestrial-scale, widely distributed, Noachian-age valley networks and 2) exceptionally large, long, single-thread, isolated, Hesperian-age outflow channels. Recent work suggests that there may also be a class of currently enigmatic, smaller, younger (Hesperian to Amazonian) channels in the mid-latitudes, perhaps associated with the occasional local melting of ice deposits (Fig. 8.4).

Some parts of Mars show inverted relief. This occurs when sediments are deposited on the floor of a stream and then become resistant to erosion, perhaps by cementation. Later, the area may be buried. Eventually, erosion removes the covering layer and the former streams become visible since they are resistant to erosion. Mars Global Surveyor found several examples of this process. Many inverted streams have been discovered in various regions of Mars, especially in the Medusae Fossae Formation, Miyamoto Crater, Saheki Crater, and Juventae Plateau.

A variety of lake basins have been discovered on Mars. Some are comparable in size to the largest lakes on Earth, such as the Caspian Sea, Black Sea, and Lake

Baikal. Lakes that were fed by valley networks are found in the southern highlands. There are places that are closed depressions with river valleys leading into them. These areas are thought to have once contained lakes; one is in Terra Sirenum which had its overflow move through Ma'adim Vallis into Gusev Crater, explored by the Mars Exploration Rover Spirit. Another is near Parana Vallis and Loire Vallis. Some lakes are thought to have formed by precipitation, while others were formed from groundwater. Lakes are estimated to have existed in the Argyre basin, the Hellas basin, and maybe in Valles Marineris. It is likely that at times in the Noachian, very many craters hosted lakes. These lakes are consistent with a cold, dry (by Earth standards) hydrological environment somewhat like that of the Great Basin of the western USA during the Last Glacial Maximum.

Research from 2010 suggests that Mars also had lakes along parts of the equator. Although earlier research had showed that Mars had a warm and wet early history that has long since dried up, these lakes existed in the Hesperian Epoch, a much later period.

Using detailed images from NASA's Mars Reconnaissance Orbiter, the researchers speculate that there may have been increased volcanic activity, meteorite impacts, or shifts in Mars' orbit during this period to warm Mars' atmosphere enough to melt the abundant ice present in the ground.

Volcanoes would have released gases that thickened the atmosphere for a temporary period, trapping more sunlight and making it warm enough for liquid water to exist. In this study, channels were discovered that connected lake basins near Ares Vallis. When one lake filled up, its waters overflowed the banks and carved the channels to a lower area where another lake would form. These dry lakes would be targets to look for evidence (biosignatures) of past life.

On September 27, 2012, NASA scientists announced that the Curiosity rover found direct evidence for an ancient streambed in Gale Crater, suggesting an ancient "vigorous flow" of water on Mars. In particular, analysis of the now dry streambed indicated that the water ran at 3.3 km/h (0.92 m/s), possibly at hip depth. Proof of running water came in the form of rounded pebbles and gravel fragments that could have only been weathered by strong liquid currents. Their shape and orientation suggests long-distance transport from above the rim of the crater, where a channel named Peace Vallis feeds into the alluvial fan.

8.3.3.2 Lake Deltas

Researchers have found a number of examples of deltas that formed in Martian lakes. Finding deltas is a major sign that Mars once had a lot of liquid water. Deltas usually require deep water over a long period of time to form. Also, the water level needs to be stable to keep sediment from washing away. Deltas have been found over a wide geographical range, though there is some indication that deltas may be concentrated around the edges of the putative former northern ocean of Mars (Fig. 8.5).

Fig. 8.5 Delta in Eberswalde Crater – Mars Global Surveyor (http://www.msss.com/mars_images/moc/2003/11/13/ – NASA/JPL-Caltech/MSSS, November 13th, 2003)

8.3.3.3 Groundwater

By 1979, it was thought that outflow channels formed in single, catastrophic ruptures of subsurface water reservoirs, possibly sealed by ice, discharging colossal quantities of water across an otherwise arid Mars surface. In addition, evidence in favor of heavy or even catastrophic flooding is found in the giant ripples in the Athabasca Vallis. Many outflow channels begin at chaos or chasma features, providing evidence for the rupture that could have breached a subsurface ice seal.

The branching valley networks of Mars are not consistent with their formation by sudden catastrophic release of groundwater, both in terms of their dendritic shapes which do not come from a single outflow point and in terms of the discharges which apparently flowed along them. Instead, some authors have argued that they were formed by slow seepage of groundwater from the subsurface essentially as springs. In support of this interpretation, the upstream ends of many valleys in such networks begin with box canyon or "amphitheater" heads, which on Earth are typically associated with groundwater seepage. There is also little evidence of finer-scale channels or valleys at the tips of the channels, which some authors have interpreted as showing the flow appearing suddenly from the subsurface with appreciable discharge rather than accumulating gradually across the surface.

8 Mars Astrobiology: Recent Status and Progress

Fig. 8.6 The preservation and cementation of aeolian dune stratigraphy in Burns Cliff slope in Endurance Crater are thought to have been controlled by flow of shallow groundwater. The layers show different types of deposition of sulfate-rich sediments, with ancient dunes on the bottom, ancient groundwater on the middle, and ancient water streams on the top of the layers. Image by the Opportunity rover (http://photojournal.jpl.nasa.gov/catalog/PIA07110 – NASA/JPL-Caltech/Cornell University, December 17, 2007)

Others have disputed the strong link between amphitheater heads of valleys and formation by groundwater for terrestrial examples and have argued that the lack of fine-scale heads to valley networks is due to their removal by weathering or impact gardening. Most authors accept that most valley networks are at least partly influenced and shaped by groundwater seep processes (Fig. 8.6).

Groundwater also plays a vital role in controlling broad-scale sedimentation patterns and processes on Mars. According to this hypothesis, groundwater with dissolved minerals came to the surface, in and around craters, and helped to form layers by adding minerals – especially sulfate – and cementing sediments. In other words, some layers may be formed by groundwater rising up depositing minerals and cementing existing, loose, aeolian sediments.

The hardened layers are consequently more protected from erosion. This process may occur instead of layers forming under lakes. A study published in 2011 using data from the Mars Reconnaissance Orbiter show that the same kinds of sediments exist in a large area that includes Arabia Terra. It has been argued that areas which we know from satellite remote sensing are rich in sedimentary rocks are also those areas which are most likely to experience groundwater upwelling on a regional scale.

8.3.4 Mars Ocean Hypothesis

The Mars ocean hypothesis proposes that the Vastitas Borealis basin was the site of an ocean of liquid water at least once and presents evidence that nearly a third of the surface of Mars was covered by a liquid ocean early in the planet's geologic history. This ocean, dubbed Oceanus Borealis, would have filled the Vastitas Borealis basin in the northern hemisphere, a region which lies 4–5 km (2.5–3 miles) below the mean planetary elevation. Two major putative shorelines have been suggested: a higher one, dating to a time period of approximately 3.8 billion years ago and concurrent with the formation of the valley networks in the Highlands, and a lower one, perhaps correlated with the younger outflow channels. The higher one, the "Arabia shoreline," can be traced all around Mars except through the Tharsis volcanic region. The lower, the "Deuteronilus," follows the Vastitas Borealis formation.

A study in June 2010 concluded that the more ancient ocean would have covered 36 % of Mars. Data from Mars Orbiter Laser Altimeter (MOLA), which measures the altitude of all terrain on Mars, was used in 1999 to determine that the watershed for such an ocean would have covered about 75 % of the planet. Early Mars would have required a warmer climate and denser atmosphere to allow liquid water to exist at the surface. In addition, the large number of valley networks strongly supports the possibility of a hydrological cycle on the planet in the past.

The existence of a primordial Martian ocean remains controversial among scientists, and the interpretations of some features as "ancient shorelines" has been challenged. One problem with the conjectured 2-billion-year-old (2 Gyrs) shoreline is that it is not flat – i.e., does not follow a line of constant gravitational potential. This could be due to a change in distribution in Mars' mass, perhaps due to volcanic eruption or meteor impact; the Elysium volcanic province or the massive Utopia basin that is buried beneath the northern plains has been put forward as the most likely cause.

8.3.5 Present Water Ice

A significant amount of surface hydrogen has been observed globally by the Mars Odyssey's neutron spectrometer and gamma ray spectrometer. This hydrogen is thought to be incorporated into the molecular structure of ice, and through stoichiometric calculations, the observed fluxes have been converted into concentrations of water ice in the upper meter of the Martian surface. This process has revealed that ice is both widespread and abundant on the modern surface. Below 60 degrees of latitude, ice is concentrated in several regional patches, particularly around the Elysium volcanoes, Terra Sabaea, and northwest of Terra Sirenum, and exists in concentrations up to 18 % ice in the subsurface.

Above 60 degrees latitude, ice is highly abundant. Pole wards on 70 degrees of latitude; ice concentrations exceed 25 % almost everywhere and approach 100 % at the poles. More recently, the SHARAD and MARSIS radar sounding instruments have begun to be able to confirm whether individual surface features are ice rich. Due to the known instability of ice at current Martian surface conditions, it is thought that almost all of this ice must be covered by a veneer of rocky or dusty material.

The Mars Odyssey's neutron spectrometer observations indicate that if all the ice in the top meter of the Martian surface were spread evenly, it would give a water equivalent global layer (WEG) of at least ~14 cm – in other words, the globally averaged Martian surface is approximately 14 % water. The water ice currently locked in both Martian poles corresponds to a WEG of 30 m, and geomorphic evidence favors significantly larger quantities of surface water over geologic history, with WEG as deep as 500 m. It is believed that part of this past water has been lost to the deep subsurface and part to space, although the detailed mass balance of these processes remains poorly understood. The current atmospheric reservoir of water is important as a conduit allowing gradual migration of ice from one part of the surface to another on both seasonal and longer timescales. It is insignificant in volume, with a WEG of no more than 10 μm (Fig. 8.7).

8.3.5.1 Ice Patches

On July 28, 2005, the European Space Agency announced the existence of a crater partially filled with frozen water; some then interpreted the discovery as an "ice lake." Images of the crater, taken by the high-resolution stereo camera on board the European Space Agency's Mars Express Orbiter, clearly show a broad sheet of ice in the bottom of an unnamed crater located on Vastitas Borealis, a broad plain that covers much of Mars' far northern latitudes, at approximately 70.5° North and 103° East. The crater is 35 km wide and about 2 km deep. The height difference between the crater floor and the surface of the water ice is about 200 m. ESA scientists have attributed most of this height difference to sand dunes beneath the water ice, which are partially visible. While scientists do not refer to the patch as a "lake," the water ice patch is remarkable for its size and for being present throughout the year. Deposits of water ice and layers of frost have been found in many different locations on the planet.

As more and more of the surface of Mars has been imaged by the modern generation of orbiters, it has become gradually more apparent that there are probably many more patches of ice scattered across the Martian surface. Many of these putative patches of ice are concentrated in the Martian mid-latitudes (~30 – 60° N/S of the equator). For example, many scientists believe that the widespread features in those latitude bands variously described as "latitude-dependent mantle" or "pasted-on terrain" consist of dust- or debris-covered ice patches, which are

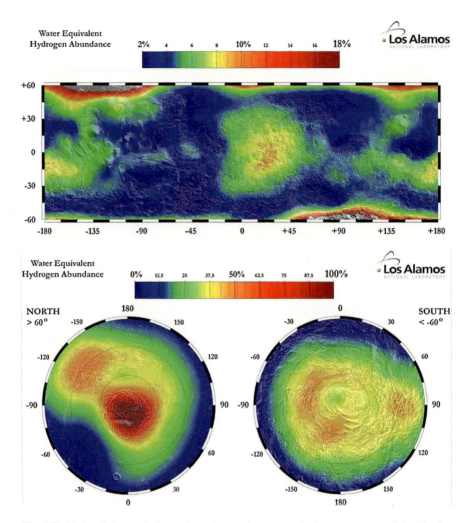

Fig. 8.7 (*Top*) and (*bottom*): Proportion of water ice present in the upper meter of the Martian surface for lower (*top*) and higher (*bottom*) latitudes. The percentages are derived through stoichiometric calculations based on epithermal neutron fluxes. These fluxes were detected by the neutron spectrometer aboard the Mars Odyssey spacecraft (http://mars.jpl.nasa.gov/odyssey/gallery/latestimages/20030724a.html – NASA/JPL-Caltech/LANL, 2003)

slowly degrading. A cover of debris is required both to explain the dull surfaces seen in the images that do not reflect like ice and also to allow the patches to exist for an extended period of time without subliming away completely. These patches have been suggested as possible water sources for some of the enigmatic channelized flow features like gullies also seen in those latitudes.

8.3.5.2 Equatorial Frozen Sea

Surface features consistent with existing pack ice have been discovered in the southern Elysium Planitia. What appear to be plates of broken ice, ranging in size from 30 to 30 km, are found in channels leading to a flooded area of approximately the same depth and width as the North Sea. The plates show signs of breakup and rotation that clearly distinguish them from lava plates elsewhere on the surface of Mars. The source for the flood is thought to be the nearby geological fault Cerberus Fossae which spewed water as well as lava aged some 2–10 Ma. It was suggested that the water exited the Cerberus Fossae then pooled and froze in the low, level plains and that such lakes may still exist. Not all scientists agree with these conclusions.

8.3.5.3 Polar Ice Caps

Both the northern polar cap (Planum Boreum) and the southern polar cap (Planum Australe) are thought to grow in thickness during winter and partially sublime during summer. In 2004, the MARSIS radar sounder on the Mars Express satellite targeted the southern polar cap and was able to confirm that ice there extends to a depth of 3.7 km (2.3 mi) below the surface.

In the same year, the OMEGA instrument on the same orbiter revealed that the cap is divided into three distinct parts, with varying contents of frozen water depending on latitude (Bibring et al. 2004).

The first part is the bright part of the polar cap seen in images, centered on the pole, which is a mixture of 85 % CO_2 ice and 15 % water ice. The second part comprises steep slopes known as scarps, made almost entirely of water ice, that ring and fall away from the polar cap to the surrounding plains. The third part encompasses the vast permafrost fields that stretch for tens of kilometers away from the scarps and is not obviously part of the cap until the surface composition is analyzed. NASA scientists calculate that the volume of water ice in the south polar ice cap, if melted, would be sufficient to cover the entire planetary surface to a depth of 11 m (36 ft). Observations over both poles and more widely over the planet suggest melting all the surface ice would produce a water equivalent global layer 35 m deep (Fig. 8.8).

On July 2008, NASA announced that the Phoenix lander had confirmed the presence of water ice at its landing site near the northern polar ice cap (at 68.2° latitude). This was the first ever direct observation of ice from the surface. Two years later, the shallow radar on board the Mars Reconnaissance Orbiter took measurements of the north polar ice cap and determined that the total volume of water ice in the cap is 821,000 cubic kilometers (197,000 cubic miles). That is equal to 30 % of the Earth's Greenland ice sheet or enough to cover the surface of Mars to a depth of 5.6 m. Both polar caps reveal abundant fine internal layers when examined in HiRISE (High Resolution Imaging Science Experiment is a camera on board the Mars Reconnaissance Orbiter) and Mars Global Surveyor imagery.

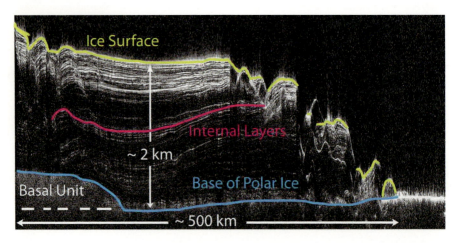

Fig. 8.8 Cross section of a portion of the north polar ice cap of Mars, derived from satellite radar sounding – Mars Reconnaissance Orbiter (http://photojournal.jpl.nasa.gov/catalog/PIA13164 – NASA/JPL-Caltech/Italian Space Agency, 2010)

Many researchers have attempted to use this layering to attempt to understand the structure, history, and flow properties of the caps, although their interpretation is not straightforward.

Lake Vostok in Antarctica may have implications for liquid water still existing on Mars because if water existed before the polar ice caps on Mars, it is possible that there is still liquid water below the ice caps.

8.3.5.4 Ground Ice

For many years, various scientists have suggested that some Martian surfaces look like periglacial regions on Earth. By analogy with these terrestrial features, it has been argued for many years that these are regions of permafrost. This would suggest that frozen water lies right beneath the surface. A common feature in the higher latitudes, patterned ground, can occur in a number of shapes, including stripes and polygons. On Earth, these shapes are caused by the freezing and thawing of soil.

There are other types of evidence for large amounts of frozen water under the surface of Mars, such as terrain softening, which rounds sharp topographical features. Theoretical calculations and analysis have tended to bear out the possibility that these features are formed by the effects of ground ice. Evidence from Mars Odyssey's gamma ray spectrometer and direct measurements with the Phoenix lander have corroborated that many of these features are intimately associated with the presence of ground ice.

Some areas of Mars are covered with cones that resemble those on Earth where lava has flowed on top of frozen ground. The heat of the lava melts the ice and then changes it into steam. The powerful force of the steam works its way through the

Fig. 8.9 The European Space Agency's (ESA) Mars Express Orbiter robotic spacecraft obtained this view of an unnamed impact crater located on Vastitas Borealis, a broad plain that covers much of Mars' far northern latitudes. The circular patch of bright material located at the center of the crater is residual water ice. The colors are very close to natural, but the vertical relief of the topography is exaggerated three times. This patch of ice is present all year round, remaining after frozen carbon dioxide overlaying; it disappears during the Martian summer (http://www.dlr.de/mars/en/DesktopDefault.aspx/tabid-4677/7747_read-11947/gallery-1/gallery_read-Image.8.5131/ – Image credit: ESA/DLR/FU Berlin, July 28, 2005)

lava and produces such rootless cones. These features can be found, for example, in Athabasca Vallis, associated with lava flowing along this outflow channel. Larger cones may be made when the steam passes through thicker layers of lava (Fig. 8.9).

8.3.5.5 Scalloped Topography

Certain regions of Mars display scallop-shaped depressions. The depressions are suspected to be the remains of a degrading ice-rich mantle deposit. Scallops are caused by ice sublimating from frozen soil. This mantle material was probably deposited from the atmosphere as ice formed on dust when the climate was different due to changes in the tilt of the Mars pole (please see the subitem "Ice Ages," below). The scallops are typically tens of meters deep and from a few hundred to a few thousand meters across. They can be almost circular or elongated.

Some appear to have coalesced causing a large heavily pitted terrain to form. The process of forming the terrain may begin with sublimation from a crack. There are often polygonal cracks where scallops form, and the presence of scalloped topography seems to be an indication of frozen ground.

170 A. de Morais M. Teles

These scalloped features are superficially similar to "Swiss cheese" features, found around the south polar cap. "Swiss cheese" features are thought to be due to cavities forming in a surface layer of solid carbon dioxide, rather than water ice – although the floors of these holes are probably H_2O rich.

8.3.5.6 Glaciers

Many large areas of Mars either appear to host glaciers or carry evidence that they used to be present. Much of the areas in high latitudes, especially the Ismenius Lacus quadrangle, are suspected to still contain enormous amounts of water ice. Recent evidence has led many planetary scientists to believe that water ice still exists as glaciers across much of the Martian mid- and high latitudes, protected from sublimation by thin coverings of insulating rock and/or dust.

In January 2009, scientists released the results of a radar study of the glacier-like features called lobate debris aprons in an area called Deuteronilus Mensae, which found widespread evidence of ice lying beneath a few meters of rock debris. Glaciers are associated with fretted terrain and many volcanoes. Researchers have described glacial deposits on Hecates Tholus, Arsia Mons, Pavonis Mons, and Olympus Mons. Glaciers have also been reported in a number of larger Martian craters in the mid-latitudes and above.

Glacier-like features on Mars are known variously as viscous flow features, Martian flow features, lobate debris aprons, or lineated valley fill, depending on the form of the feature, its location, the landforms it is associated with, and the author describing it. Many, but not all, small glaciers seem to be associated with gullies on the walls of craters and mantling material. The lineated deposits known as lineated valley fill are probably rock-covered glaciers which are found on the floors of most channels within the fretted terrain found around Arabia Terra in the northern hemisphere. Their surfaces have ridged and grooved materials that deflect around obstacles. Lineated floor deposits may be related to lobate debris aprons, which have been proven to contain large amounts of ice by orbiting radar. For many years, researchers interpreted that features called "lobate debris aprons" were glacial flows, and it was thought that ice existed under a layer of insulating rocks. With new instrument readings, it has been confirmed that lobate debris aprons contain almost pure ice that is covered with a layer of rocks.

Moving ice carries rock material and then drops it as the ice disappears. This typically happens at the snout or edges of the glacier. On Earth, such features would be called moraines, but on Mars, they are typically known as moraine-like ridges, concentric ridges, or arcuate ridges. Because ice tends to sublime rather than melt on Mars and because Mars' low temperatures tend to make glaciers "cold based" (frozen down to their beds and unable to slide), the remains of these glaciers and the ridges they leave do not appear to be exactly same as normal glaciers on Earth. In particular, Martian moraines tend to be deposited without being deflected by the underlying topography, which is thought to reflect the fact that the ice in Martian glaciers is normally frozen down and cannot slide. Ridges of debris on the surface of the glaciers indicate the direction of ice movement. The surface of some glaciers has rough textures due to sublimation of buried ice. The ice evaporates without melting and leaves behind an empty space. Overlying material then collapses into the void. Sometimes chunks of ice fall from the glacier and get buried in the land surface. When they melt, a more or less round hole remains. Many of these "kettle holes" have been identified on Mars.

Despite strong evidence for glacial flow on Mars, there is little convincing evidence for landforms carved by glacial erosion, e.g., U-shaped valleys, crag and tail hills, arêtes, and drumlins. Such features are abundant in glaciated regions on Earth, so their absence on Mars has proven puzzling. The lack of these landforms is thought to be related to the cold-based nature of the ice in most recent glaciers on Mars. Because the solar insolation reaching the planet, the temperature and density of the atmosphere, and the geothermal heat flux are all lower on Mars than they are on Earth, modeling suggests that the temperature of the interface between a glacier and its bed stays below freezing and the ice is literally frozen down to the ground. This prevents it from sliding across the bed, which is thought to inhibit the ice's ability to erode the surface.

8.3.6 Ice Ages

Mars has experienced large-scale changes in the amount and distribution of ice on its surface in its relatively recent geological past, and as on Earth, these are known as ice ages. Ice ages on Mars are very different from the ones that the Earth experiences. During a Martian ice age, the poles get warmer, and water ice then leaves the ice caps and is deposited in mid-latitudes. The moisture from the ice caps travels to lower latitudes in the form of deposits of frost or snow mixed with dust. The atmosphere of Mars contains a great deal of fine dust particles; the water vapor condenses on these particles which then fall down to the ground due to the additional weight of the water coating.

When ice at the top of the mantling layer returns to the atmosphere, it leaves behind dust which serves to insulate the remaining ice. The total volume of water removed is a few percent of the ice caps or enough to cover the entire surface of the planet under 1 m of water. Much of this moisture from the ice caps results in a thick smooth mantle with a mixture of ice and dust. This ice-rich mantle, a few meters thick, smoothen the land at lower latitudes, but in places it displays a bumpy texture. Multiple stages of glaciations probably occurred. Because there are few craters on the current mantle, it is thought to be relatively young. It is thought that this mantle was laid in place during a relatively recent ice age. Ice ages are driven by changes in Mars' orbit and tilt, which can be compared to terrestrial Milankovich cycles. Orbital calculations show that Mars wobbles on its axis far more than Earth does.

The Earth is stabilized by its proportionally large moon, so it only wobbles a few degrees. Mars may change its tilt – also known as its obliquity – by many tens of degrees. When this obliquity is high, its poles get much more direct sunlight and heat; this causes the ice caps to warm and become smaller as ice sublimes. Adding to the variability of the climate, the eccentricity of the orbit of Mars changes twice as much as Earth's eccentricity. As the poles sublime, the ice is deposited closer to the equator, which receives somewhat less solar insolation at these high obliquities. Computer simulations have shown that a 45° tilt of the Martian axis would result in ice accumulation in areas that display glacial landforms. A 2008 study provided evidence for multiple glacial phases during Late Amazonian glaciation at the dichotomy boundary on Mars.

8.3.7 Evidence for Recent Flows

8.3.7.1 Branched Gullies

Liquid water cannot exist in a stable form on the surface of Mars with its present low atmospheric pressure and low temperature, except at the lowest elevations for a few hours. So, a geological mystery commenced when observations from NASA's Mars Reconnaissance Orbiter revealed gully deposits that were not there 10 years ago, possibly caused by flowing salty water (brine) during the warmest months on Mars. The images (below) were of two craters called Terra Sirenum and Centauri Montes which appear to show the presence of liquid water flows on Mars at some point between 1999 and 2001.

There is disagreement in the scientific community as to whether or not gullies are formed by liquid water. It is also possible that the flows that carve gullies are dry or perhaps lubricated by carbon dioxide. Even if gullies are carved by flowing water at the surface, the exact source of the water and the mechanisms behind its motion are not well understood (Fig. 8.10).

In August 2011, NASA announced the discovery by Nepalese student Lujendra Ojha of current seasonal changes on steep slopes below rocky outcrops near crater rims in the southern hemisphere. Dark streaks were seen to grow downslope during the warmest part of the Martian summer, then to gradually fade through the rest of the year, recurring cyclically between years. The researchers suggested these marks were consistent with salty water (brines) flowing downslope and then evaporating, possibly leaving some sort of residue. Because these flows have been the flows form and fade in sync with heat flux into the surface, many scientists feel these recurrent slope lineae are probably the best candidates for features formed by flowing water on Mars today. The rate of growth of these features has been shown to be consistent with shallow groundwater flow downslope through a sandy substrate (Figs. 8.11 and 8.12).

8.3.8 Habitability Assessment

Life is understood to require liquid water, but it is not the only essential requirement for life. These requirements include water, an energy source, and materials necessary for cellular growth, while all under appropriate environmental conditions.

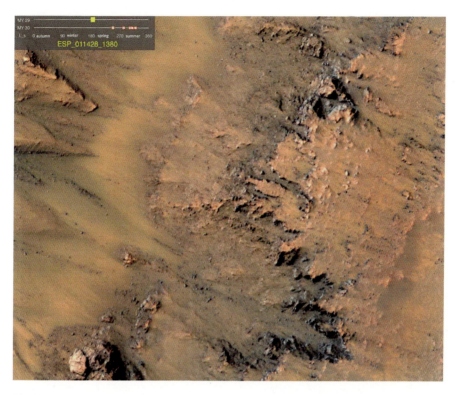

Fig. 8.10 Warm-season flows on the slope of Newton Crater show features that might be evidence of salty liquid water active on Mars today – Mars Reconnaissance Orbiter (http://www.nasa.gov/mission_pages/MRO/multimedia/pia14472.html – NASA/JPL-Caltech/University of Arizona, 2010–2011)

The confirmation that liquid water once flowed on Mars, of the existence of nutrients, and of the previous discovery of a past magnetic field that protected the planet from cosmic and solar radiation together strongly suggests that Mars could have had the environmental factors to support life. To be clear, the finding of past habitability is not evidence that Martian life has ever actually existed.

When there is a magnetic field, the atmosphere is protected from erosion by solar wind and it ensures the maintenance of a dense atmosphere, necessary for liquid water to exist on the surface of Mars.

The two current ecological approaches for predicting the potential habitability of the Martian surface use 19 or 20 environmental factors, with emphasis on water availability, temperature, presence of nutrients, an energy source, and protection from solar ultraviolet and galactic cosmic radiation (MEPAG Special Regions – Science Analysis et al. 2006). In particular, the damaging effect of ionizing radiation on cellular structure is one of the prime limiting factors on the survival of life in potential astrobiological habitats. Even at a depth of 2 m beneath the surface, any microbes would likely be dormant, cryopreserved by the current freezing conditions, and so metabolically inactive and unable to repair cellular degradation as it occurs.

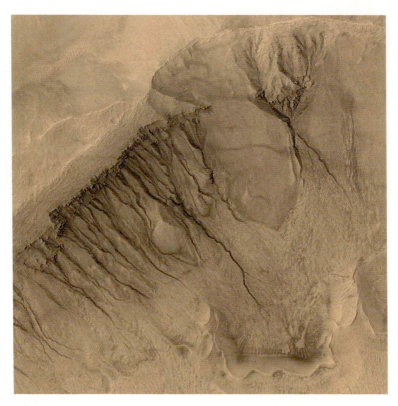

Fig. 8.11 Gullies, similar to those formed on Earth, are visible on this image from Mars Global Surveyor (http://www.msss.com/mars_images/moc/news2002/gullies_streaks/ and http://photojournal.jpl.nasa.gov/catalog/PIA03872 – NASA/JPL-Caltech/MSSS, 2002)

Therefore, the best potential locations for discovering life on Mars may be at subsurface environments that have not been studied yet. The extensive volcanism in the past possibly created subsurface cracks and caves within different strata, and liquid water could have been stored in these subterraneous places, forming large aquifers with deposits of saline liquid water, minerals, organic molecules, and geothermal heat – potentially providing a current habitable environment away from the harsh surface conditions.

8.3.9 Findings by Robotic Probes

8.3.9.1 Mars 1

Mars 1 was the first spacecraft launched successfully to Mars in 1962. Some data was collected during flyby at approximately 193,000 km, but communication

8 Mars Astrobiology: Recent Status and Progress

Fig. 8.12 Millimetric-scale gullies within sand ridges at a beach on Earth, resembling meter-to-kilometric-scale gullies on Mars. This shows interesting geochemical-physical similarities between the two planets and, also, the scale-invariance property of fractal geometry embedded within such geological processes (Photograph by the author: de Morais 2009)

was lost before reaching closer to Mars. With Mars 2 and Mars 3 in 1971–1972, information was obtained on the nature of the surface rocks and altitude profiles of the surface density of the soil, its thermal conductivity, and thermal anomalies detected on the surface of Mars. The missions found that its northern polar cap has a temperature below $-110\,°C$ and that the water vapor content in the atmosphere of Mars is 5,000 times less than on Earth. No signs of life were found.

8.3.9.2 Mariner 4

Mariner 4 probe performed the first successful flyby of planet Mars, returning the first pictures of the Martian surface in 1965. The photographs showed an arid Mars without rivers, oceans, or any signs of life. Further, it revealed that the surface (at least the parts that it photographed) was covered with craters, indicating a lack of plate tectonics and weathering of any kind for the last 4 billion years. The probe also found that Mars has no global magnetic field that would protect the planet from potentially life-threatening cosmic rays. The probe was able to calculate the atmospheric pressure on the planet to be about 0.6 kPa (compared to Earth's 101.3 kPa), meaning that liquid water could not exist on the planet's surface. After Mariner 4, the search for life on Mars changed to a search for bacteria-like living organisms rather than for multicellular organisms, as the environment was clearly too harsh for these.

8.3.9.3   Mariner 9

The images acquired by the Mariner 9 Mars orbiter, launched in 1971, revealed the first direct evidence of past water in the form of dry riverbeds, canyons (including the Valles Marineris, a system of canyons over about 4,020 km (2,500 mi) long), evidence of water erosion and deposition, weather fronts, fogs, and more. The findings from the Mariner 9 missions underpinned the later Viking program. The enormous Valles Marineris canyon system is named after Mariner 9 in honor of its achievements.

8.3.9.4   Viking Program

Viking Orbiters

Liquid water is necessary for known life and metabolism, so if water was present on Mars, the chances of it having supported life may have been determinant. The Viking orbiters found evidence of possible river valleys in many areas, erosion, and, in the southern hemisphere, branched streams.

Viking Landers: Biological Experiments

The primary mission of the Viking landers by mid-1970s was to carry out experiments designed to detect any moisture and microorganisms in Martian soil because the favorable conditions for the evolution of multicellular organisms ceased some 4 billion years ago on Mars. The tests were formulated to look for microbial life similar to that found on Earth. Of the four experiments, only the Labeled Release (LR) experiment returned a positive result, showing increased $^{14}CO_2$ production on the first exposure of soil to water and nutrients. All scientists agree on two points from the Viking missions: that radiolabeled $^{14}CO_2$ evolved in the Labeled Release experiment and that the GCMS detected no organic molecules. However, there are vastly different interpretations of what those results imply.

By discovering many geological forms that are typically formed from large amounts of water, the two Viking orbiters and the two landers caused a revolution in our knowledge about water on Mars. Huge outflow channels were found in many areas. They showed that floods of water broke through dams, carved deep valleys, eroded grooves into bedrock, and traveled thousands of kilometers. Large areas in the southern hemisphere contained branched valley networks, suggesting that rain once fell. Many craters look as if the impactor fell into mud. When they were formed, ice in the soil may have melted, turned the ground into mud, and then the mud flowed across the surface. Regions, called "chaotic terrain," seemed to have quickly lost great volumes of water which caused large channels to form downstream. Estimates for some channel flows run to 10,000 times the flow of the Mississippi River. Underground volcanism may have melted frozen ice; the water then flowed away and the ground collapsed to leave chaotic terrain. Also, a general chemical analysis by the two Viking landers suggested that the surface has been either exposed to or submerged in water in the past.

8.3.9.5 Mars Global Surveyor

The Mars Global Surveyor's thermal emission spectrometer (TES) is an instrument able to determine the mineral composition on the surface of Mars. Mineral composition gives information on the presence or absence of water in ancient times. TES identified a large ($30,000 \text{ km}^2$) area in the Nili Fossae formation that contains the mineral olivine. It is thought that the ancient asteroid impact that created the Isidis basin resulted in faults that exposed the olivine. The discovery of olivine is a strong evidence that parts of Mars have been extremely dry for a long time. Olivine was also discovered in many other small outcrops within 60° north and south of the equator. The probe has imaged several channels that suggest past sustained liquid flows, two of them are found in Nanedi Vallis and in Nirgal Vallis.

8.3.9.6 Mars Pathfinder

The Pathfinder lander recorded the variation of diurnal temperature cycle. It was coldest just before sunrise, about $-78\,°C$, and warmest just after Mars noon, about $-8\,°C$. These extremes occurred near the ground which both warmed up and cooled down fast. At this location, the highest temperature never reached the freezing point of water ($0\,°C$), too cold for pure liquid water to exist on the surface.

Surface pressures varied diurnally over a 0.2 millibar range but showed two daily minima and two daily maxima. The average daily pressure decreased from about 6.75 millibars to a low of just under 6.7 millibars, corresponding to when the maximum amount of carbon dioxide had condensed on the South Pole. The atmospheric pressure measured by the Pathfinder on Mars is very low – about 0.6 % of Earth's – and it would not permit liquid water to exist on the surface.

Other observations were consistent with water being present in the past. Some of the rocks at the Mars Pathfinder site leaned against each other in a manner geologists term imbricated. It is suspected that strong flood waters in the past pushed the rocks around until they faced away from the flow. Some pebbles were rounded, perhaps from being tumbled in a stream. Parts of the ground are crusty, maybe due to cementing by a fluid containing minerals. There was evidence of clouds and maybe fog. Sojourner rover studied rocks using an Alpha Proton X-ray Spectrometer.

8.3.9.7 Mars Odyssey

The 2001 Mars Odyssey found much evidence for water on Mars in the form of images, and with its spectrometer, it proved that much of the ground is loaded with water ice. Mars has enough ice just beneath the surface to fill Lake Michigan twice. In both hemispheres, from 55° latitude to the poles, Mars has a high density of ice just under the surface; one kilogram of soil contains about 500 g of water ice. But close to the equator, there is only 2–10 % of water in the soil.

Scientists think that much of this water is also locked up in the chemical structure of minerals, such as clay and sulfates. Although the upper surface contains a few percent of chemically bound water, ice lies just a few meters deeper, as it has been shown in Arabia Terra, Amazonis quadrangle, and Elysium quadrangle that contain large amounts of water ice. Analysis of the data suggests that the southern hemisphere may have a layered structure, suggestive of stratified deposits beneath, now extinct large water mass.

The instruments aboard the Mars Odyssey are only able to study the top meter of soil, while the radar aboard the Mars Reconnaissance Orbiter can measure a few kilometers deep. In 2002, available data were used to calculate that if all soil surfaces were covered by an even layer of water, this would correspond to a global layer of water (GLW) 0.5–1.5 km deep.

Thousands of images returned from Odyssey orbiter also support the idea that Mars once had great amounts of water flowing across its surface. Some images show patterns of branching valleys; others show layers that may have been formed under lakes; even river and lake deltas have been identified.

For many years, researchers thought that glaciers existed under a layer of insulating rocks. Lineated valley fill is one example of these rock-covered glaciers. They are found on the floors of some channels. Their surfaces have ridged and grooved materials that deflect around obstacles. Lineated floor deposits may be related to lobate debris aprons, which have been shown by orbiting radar to contain large amounts of ice.

8.3.9.8 Phoenix Lander

The Phoenix mission landed a robotic spacecraft in the polar region of Mars on May 25, 2008, and it operated until November 10, 2008. One of the mission's two primary objectives was to search for a "habitable zone" in the Martian regolith where microbial life could exist, the other main goal being to study the geologic history of water on Mars. The lander has a 2.5-m robotic arm that was capable of digging shallow trenches in the regolith. There was an electrochemistry experiment which analyzed the ions in the regolith and the amount and type of antioxidants on Mars.

The Viking program data indicate that oxidants on Mars may vary with latitude, noting that Viking 2 saw fewer oxidants than Viking 1 in its more northerly position. Phoenix landed further north still. Phoenix's preliminary data revealed that Mars soil contains perchlorate and thus may not be as life-friendly as thought earlier. The pH and salinity level were viewed as benign from the standpoint of biology. The analyzers also indicated the presence of bound water and CO_2.

The Phoenix lander also confirmed the existence of large amounts of water ice in the northern region of Mars. This finding was predicted by previous orbital data and theory and was measured from orbit by the Mars Odyssey instruments. Phoenix lander's camera showed subsurface water ice exposed by the landing retrorockets, in an unexpected technique called "diffusive gas explosive erosion," which uncovered 18 cm of Martian soil. On June 19, 2008, NASA announced that dice-sized clumps of bright material in the "Dodo-Goldilocks" trench, dug by the robotic arm, had vaporized over the course of four days, strongly implying that the bright clumps were composed of water ice which sublimes following exposure. Even though CO_2 (dry ice) also sublimes under the conditions present, it would do so at a rate much faster than observed. On July 31, 2008, NASA announced that Phoenix confirmed the presence of water ice at its landing site. During the initial heating cycle of a sample, the mass spectrometer detected water vapor when the sample temperature reached 0 °C. Liquid water cannot exist on the surface of Mars with its present low atmospheric pressure and temperature, except at the lowest elevations for short periods.

Perchlorate (ClO_4), a strong oxidizer, was confirmed to be in the soil. The chemical, when mixed with water, can lower the water freezing point in a manner similar to how salt is applied to roads to melt ice. It has been hypothesized that perchlorate may be allowing small amounts of liquid water to form on Mars today and may have formed visible gullies by eroding soil on steep slopes.

Additionally, during 2008 and early 2009, a debate emerged within NASA over the presence of "blobs" which appeared on photos of the vehicle's landing struts, which have been variously described as being either water droplets or "clumps of frost."

For about as far as the camera can see, the landing site is flat but shaped into polygons between 2 and 3 m in diameter and is bounded by troughs that are 20–50 cm deep. These shapes are due to ice in the soil expanding and contracting due to major temperature changes. The microscope showed that the soil on top of the polygons is composed of rounded particles and flat particles, probably a type of clay. Ice is present a few inches below the surface in the middle of the polygons, and along its edges, the ice is at least 3 cm deep. When the ice is exposed to the Martian atmosphere, it slowly sublimes.

Snow was observed to fall from cirrus clouds. The clouds formed at a level in the atmosphere that was around −65 °C, so the clouds would have to be composed of water ice, rather than carbon dioxide ice (CO_2 or dry ice) because the temperature for forming carbon dioxide ice is much lower than −120 °C. As a result of mission observations, it is now suspected that water ice (snow) would have accumulated later in the year at this location. The highest temperature measured during the mission, which took place during the Martian summer, was -−19.6 °C, while the coldest was −97.7 °C. So in this region, the temperature remained far below the freezing point (0 °C) of water.

8.3.9.9 Mars Exploration Rovers

The Mars Exploration rovers, Spirit and Opportunity, found a great deal of evidence for past water on Mars. The Spirit rover landed in what was thought to be a large lake bed. The lake bed had been covered over with lava flows, so evidence of past water was initially hard to detect. On March 5, 2004, NASA announced that Spirit had found hints of water history on Mars in a rock dubbed "Humphrey." Detailed inspection of the rock revealed a bright material filling internal cracks. Such material may have crystallized from water trickling through the volcanic rock. The amount of Mars once covered by ancient water remains unknown, as both rovers landed in regions thought likely to be once under water.

The Opportunity rover was directed to a site that had displayed large amounts of hematite from the orbit. Hematite often forms from water. The rover indeed found layered rocks and marble- or blueberry-like hematite concretions. Elsewhere on its traverse, Opportunity investigated aeolian dune stratigraphy in Burns Cliff in Endurance Crater. Its operators concluded that the preservation and cementation of these outcrops had been controlled by flow of shallow groundwater. In its years of continuous operation, Opportunity is still sending back evidence that this area on Mars was soaked in liquid water in the past (Figs. 8.13, 8.14, 8.15, and 8.16).

As Spirit traveled in reverse in December 2007, pulling a seized wheel behind, the wheel scraped off the upper layer of soil, uncovering a patch of white ground rich in silica. It must have been produced in one of two ways: one, hot spring deposits produced when water dissolved silica at one location and then carried it to another (i.e., a geyser), two, acidic steam rising through cracks in rocks stripped them of their mineral components, leaving silica behind. The Spirit rover also found evidence for water in the Columbia Hills of Gusev Crater. In the Clovis group of carbonate-rich rocks enriched in the elements phosphorus, sulfur, chlorine and bromine, the Mössbauer spectrometer (MB) detected goethite, which forms only in the presence of water. Goethite is a mineral with an iron oxyhydroxide with ferric iron (Fe^{3+}) α-FeO(OH). This means that various regions of the planet once harbored stable liquid water.

The MER rovers had been finding evidence for ancient wet environments that were very acidic. In fact, what Opportunity has mostly discovered, or found evidence for, was sulfuric acid, a harsh chemical for life. But in May 17, 2013, NASA announced that Opportunity found clay deposits that typically form in wet environments that are of near-neutral acidity. This finding provides additional evidence about a wet ancient environment possibly favorable for life.

Fig. 8.13 This image from the panoramic camera on NASA's Mars Exploration Rover Opportunity shows a rock called "Chocolate Hills," which the rover found and examined at the edge of a young crater called "Concepción." The rocks ejected outward from the impact that dug Concepción are chunks of the same type of bedrock Opportunity has seen at hundreds of locations since landing in January 2004: soft, sulfate-rich sandstone holding harder peppercorn-size dark spheres like berries in a muffin. The little spheres, rich in iron, gained the nickname "blueberries." Opportunity used tools on its robotic arm to examine the texture and composition of target areas on the rock with and without the unusual dark coating material. In some places, the layer of closely packed spheres lies between thinner, smoother layers. The rock is about 20 cm in size. Detailed inspection of the rock revealed a bright material filling internal cracks. Such material may have crystallized from water trickling through the volcanic rock in a region thought likely to once be under water. Long ago, water flowing through fractures could have dissolved the sandstone and liberated the blueberries that fell down into the fracture and packed together. This view is presented in "natural" color. Opportunity took the image during the 2,147th Martian day, or sol, of the rover's mission on Mars (http://www.nasa.gov/mission_pages/mer/images/pia12972.html – NASA/JPL-Caltech/Cornell University, February 6, 2010)

8.3.9.10 Mars Reconnaissance Orbiter

The Mars Reconnaissance Orbiter's High Resolution Imaging Science Experiment (HiRISE) instrument has taken many images that strongly suggest that Mars has had a rich history of water-related processes. A major discovery was finding evidence of ancient hot springs. If they have hosted microbial life, they may contain biosignatures. Research published in January 2010 described strong evidence for

Fig. 8.14 This sharp, closeup image taken by the microscopic imager on the Mars Exploration Rover Opportunity's instrument deployment device, or "arm," shows sedimentary layers and a hematite spherule within a rock target dubbed "Robert E," located on the rock outcrop at Meridiani Planum, Mars. Scientists are studying this area for clues about the rock outcrop's composition. This image measures 3 cm (1.2 in.) across – Opportunity rover (http://mars.jpl.nasa.gov/mer/gallery/press/opportunity/20040212a.html – NASA/JPL-Caltech/Cornell University/USGS/Texas A&M University, February 8, 2004)

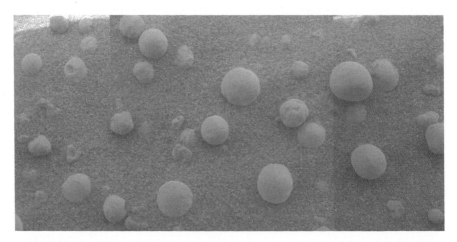

Fig. 8.15 This mosaic image shows an extreme close-up of round, blueberry-shaped hematite spherules in the Martian soil near a part of the rock outcrop at Meridiani Planum called Stone Mountain – Opportunity rover (http://mars.jpl.nasa.gov/mer/gallery/press/opportunity/20040212a.html – NASA/JPL-Caltech/Cornell University/USGS/Texas A&M University, February 10, 2004)

Fig. 8.16 Partly embedded spherules. This image is 1.3 cm (half an inch) across – Opportunity rover (http://photojournal.jpl.nasa.gov/catalog/pia05474 – NASA/JPL-Caltech/Cornell University/USGS/Texas A&M University, March 2, 2004)

sustained precipitation in the area around Valles Marineris. The types of minerals there are associated with water. Also, the high density of small branching channels indicates a great deal of precipitation.

Rocks on Mars have been found to frequently occur as layers, called strata, in many different places. Layers form by various ways, including volcanoes, wind, or water. Light-toned rocks on Mars have been associated with hydrated minerals like sulfates and clay. The orbiter helped scientists determine that much of the surface of Mars is covered by a thick smooth mantle that is thought to be a mixture of ice and dust.

The ice mantle under the shallow subsurface is thought to result from frequent, major climate changes. Changes in Mars' orbit and tilt cause significant changes in the distribution of water ice from polar regions down to latitudes equivalent to Texas. During certain climate periods, water vapor leaves polar ice and enters the atmosphere. The water returns to the ground at lower latitudes as deposits of frost or snow mixed generously with dust. The atmosphere of Mars contains a great deal of fine dust particles. Water vapor condenses on the particles, and then they fall down to the ground due to the additional weight of the water coating. When ice at the top of the mantling layer goes back into the atmosphere, it leaves behind dust, which insulates the remaining ice.

In 2008, research with the Shallow Radar on the Mars Reconnaissance Orbiter provided strong evidence that the lobate debris aprons (LDA) in Hellas Planitia and in mid- northern latitudes are glaciers that are covered with a thin layer of rocks. Its radar also detected a strong reflection from the top and base of LDAs, meaning that pure water ice made up the bulk of the formation. The discovery of water ice in LDAs demonstrates that water is found at even lower latitudes.

Research published in September 2009 demonstrated that some new craters on Mars show exposed, pure water ice. After a time, the ice disappears, evaporating into the atmosphere. The ice is only a few feet deep. The ice was confirmed with the compact imaging spectrometer (CRISM) on board the Mars Reconnaissance Orbiter.

8.3.9.11 Mars Science Laboratory and Curiosity Rover

The Mars Science Laboratory mission is a NASA project that launched, on November 26, 2011, the Curiosity rover, a nuclear-powered robotic vehicle, bearing instruments designed to assess past and present habitability conditions on Mars. The Curiosity rover landed on Mars on Aeolis Palus in Gale Crater, near Aeolis Mons (aka Mount Sharp), on August 6, 2012 (Various 2014) (Fig. 8.17).

Very early in its ongoing mission, NASA's Curiosity rover discovered unambiguous fluvial sediments on Mars. The properties of the pebbles in these outcrops suggested former vigorous flow on a streambed, with flow between ankle and waist

Fig. 8.17 View of Yellowknife Bay formation, with drilling sites. This mosaic of image from Curiosity's Mast Camera (Mastcam) shows geological members of the Yellowknife Bay formation and the sites where Curiosity drilled into the lowest-lying member, called Sheepbed, at targets "John Klein" and "Cumberland." The scene has the Sheepbed mudstone in the foreground and rises up through Gillespie Lake member to the Point Lake outcrop. These rocks record superimposed ancient lake and stream deposits that offered past environmental conditions favorable for microbial life. Rocks here were exposed about 70 Ma ago by removal of overlying layers due to erosion by the wind. The 50-cm scale bars at different locations in the image are about 20 in. long. The lower scale bar is about 8 m (26 ft) away from where Curiosity was positioned when the view was recorded. The upper scale bar is about 30 m (98 ft) from the rover's location. The scene is a portion of a 111-image mosaic acquired during the 137th Martian day, or sol, of Curiosity's work on Mars. The foothills of Mount Sharp are visible in the distance, upper left, southwest of camera position (http://www.nasa.gov/jpl/msl/mars-rover-curiosity-pia17595.html – NASA/JPL-Caltech/MSSS, December 24, 2012)

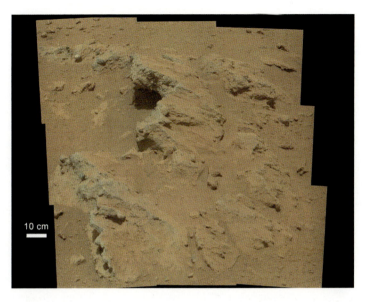

Fig. 8.18 "Hottah" rock outcrop – an ancient liquid water streambed discovered by the Curiosity rover team (3D version) (http://mars.jpl.nasa.gov/msl/multimedia/images/?ImageID=4722 – NASA/JPL-Caltech, September 14, 2012)

deep. These rocks were found at the foot of an alluvial fan system descending from the crater wall, which had previously been identified from the orbit (Various 2014) (Figs. 8.18, 8.19, and 8.20).

On October 2012, the first X-ray diffraction analysis of a Martian soil was performed by Curiosity. The results revealed the presence of several minerals, including feldspar, pyroxenes, and olivine, and suggested that the Martian soil in the sample was similar to the weathered basaltic soils of Hawaiian volcanoes. The sample used is composed of dust distributed from global dust storms and local fine sand. So far, the materials Curiosity has analyzed are consistent with the initial ideas of deposits in Gale Crater recording a transition through time from a wet to dry environment.

On December 2012, NASA reported that Curiosity performed its first extensive soil analysis, revealing the presence of water molecules, sulfur, and chlorine in the Martian soil. And on March 2013, NASA reported evidence of mineral hydration, likely hydrated calcium sulfate, in several rock samples including the broken fragments of "Tintina" rock and "Sutton Inlier" rock as well as in veins and nodules in other rocks like "Knorr" rock and "Wernicke" rock. Analysis using the rover's DAN instrument provided evidence of subsurface water, amounting to as much as 4 % water content, down to a depth of 60 cm (2.0 ft), in the rover's traverse from the Bradbury landing site to the Yellowknife Bay area in the Glenelg terrain (Various 2014) (Fig. 8.21).

Fig. 8.19 Rock outcrop on Mars, compared with a terrestrial (right) fluvial conglomerate, suggesting water "vigorously" flowing in a stream. The outcrop characteristics are consistent with a sedimentary conglomerate or a rock that was formed by the deposition of water and is composed of many smaller rounded rocks cemented together – Curiosity rover (http://mars.jpl.nasa.gov/msl/multimedia/images/?ImageID=4720 – NASA/JPL-Caltech/MSSS/PSI, September 2, 2012)

On September 26, 2013, NASA scientists reported that the Mars Curiosity rover detected abundant chemically bound water (1.5–3 wt%) in soil samples at the Rocknest region of Aeolis Palus in Gale Crater. In addition, NASA reported that the rover found two principal soil types: a fine-grained mafic type and a locally derived, coarse-grained felsic type (Meslin et al. 2013). The mafic type, similar to other Martian soils and Martian dust, was associated with hydration of the amorphous phases of the soil. Also, perchlorates, the presence of which may make detection of life-related organic molecules difficult, were found at the Curiosity rover landing site (and earlier at the more polar site of the Phoenix lander) suggesting a global distribution of these salts (Leshin and MSL Science Team 2013). NASA also reported that Jake M rock, a rock encountered by Curiosity on the way to Glenelg, was a mugearite and very similar to terrestrial mugearite rocks (Various 2014).

On December 9, 2013, NASA reported that the planet Mars had a large freshwater lake (which could have been a hospitable environment for microbial life) based on evidence from the Curiosity rover studying the plain Aeolis Palus near Mount Sharp in Gale Crater.

8 Mars Astrobiology: Recent Status and Progress

Fig. 8.20 The Mast Camera (Mastcam) on NASA's Mars rover Curiosity showed researchers interesting internal color in this rock called "Sutton Inlier," which was broken by the rover driving over it. The Mastcam took this image during the 174th Martian day, or sol, of the rover's work on Mars. The rock is about 12 cm (5 in.) wide at the end closest to the camera. This view is calibrated to estimate "natural" color or approximately what the colors would look like if we were to view the scene ourselves on Mars. The inside of the rock, which is in the "Yellowknife Bay" area of Gale Crater, is much less red than typical Martian dust and rock surfaces, with a color verging on grayish to bluish. As revealed by the Alpha Particle X-ray Spectrometer (APXS) on Curiosity's arm, the sedimentary rocks at Yellowknife Bay likely formed when original basaltic rocks were broken into fragments, transported, and redeposited as sedimentary particles and mineralogically altered by exposure to water. This rock appears to be a carbonatic rock (http://mars.jpl.nasa.gov/msl/multimedia/images/?ImageID=5152 – NASA/JPL-Caltech, January 31, 2013)

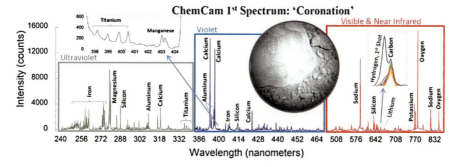

Fig. 8.21 First target, "Coronation" rock, on Mars of the ChemCam laser analyzer on the Curiosity rover (http://mars.jpl.nasa.gov/msl/multimedia/images/?ImageID=4541 – NASA/JPL-Caltech/LANL/CNES/IRAP, August 19, 2012)

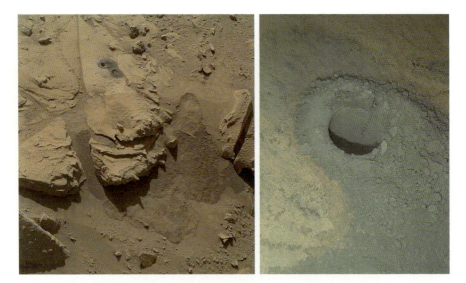

Fig. 8.22 (*Left*): This view from the Mars Hand Lens Imager (MAHLI) in NASA's Curiosity Mars Rover shows the sandstone rock target "Windjana" and its immediate surroundings. The open hole from sample collection is 1.6 cm (0.63 in.) in diameter and is about 2 cm (0.8 in.) deep. It was drilled on Sol 621 (May 5th, 2014) (http://mars.jpl.nasa.gov/msl/multimedia/images/?ImageID= 6237 – NASA/JPL-Caltech/MSSS, May 12th, 2014); and (*right*): NASA's Curiosity Mars rover used the Mars Hand Lens Imager (MAHLI) instrument on its robotic arm to illuminate and record this nighttime view of the sandstone rock target "Windjana". The rover had previously drilled a hole and collected sample powdered material from the interior of the rock, for future analysis by laboratory instruments inside the Curiosity rover. See the dark gray sandy material from the interior of the sedimentary rock, different from the reddish sand material which covers the Martian top surface. A vertical array of pits in the side of the hole resulted from using the laser-shooting Chemistry and Camera (ChemCam) instrument to assess composition at those points (http://mars.jpl.nasa.gov/msl/multimedia/images/?ImageID=6236 – NASA/JPL-CaltechSSS, May 13, 2014)

Curiosity's hammering drill collects powdered sample material from the interior of a rock, and then the rover prepares and delivers portions of the sample to laboratory instruments on board. The first two Martian rocks drilled and analyzed this way were mudstone slabs neighboring each other in Yellowknife Bay, about 4 km (2.5 miles) northeast of the rover's current location at a waypoint called "The Kimberley." Those two rocks yielded evidence last year of an ancient lake bed environment with key chemical elements and a chemical energy source that provided conditions billions of years ago favorable for microbial life (Various 2014) (Figs. 8.22 and 8.23).

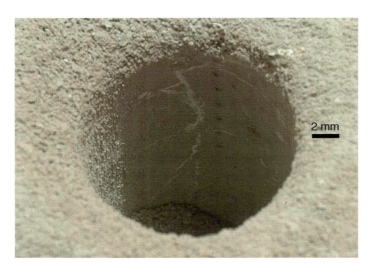

Fig. 8.23 The hole that NASA's Curiosity Mars rover drilled into target rock "John Klein" provided a view into the interior of the rock, as well as obtaining a sample of powdered material from the sedimentary rock. See the light gray sandy material in the interior of it, different from the reddish sand material which covers the Martian top surface. The rock is part of the Sheepbed mudstone deposit in the Yellowknife Bay area of Gale Crater. This image, taken by Curiosity's Mars Hand Lens Imager (MAHLI) camera, reveals gray colored cuttings, rock powder and interior wall. Notice the homogeneous, fine grain size of the mudstone, and the irregular network of sulfate-filled hairline fractures. A vertical array of pits in the side of the hole resulted from using the laser-shooting Chemistry and Camera (ChemCam) instrument to assess composition at those points. The MAHLI took this image during the 270th Martian day, or sol, of Curiosity's work on Mars (May 10th, 2013). The diameter of the hole is about 1.6 cm (0.6 in.) and is about 2 cm (0.8 in.) deep. The Sheepbed mudstone is interpreted to represent an ancient lake. It preserves evidence of an environment that would have been suited to support microbes that get their energy by eating chemicals in rocks. This wet environment was characterized by neutral pH, low salinity, and variable oxidation of iron- and sulfur-containing minerals. Carbon, hydrogen, oxygen, sulfur, nitrogen and phosphorus were measured directly as key elements for supporting possible life. These results highlight the biological viability of fluvial–lacustrine environments (streams and lakes) in the history of Mars after the earliest era of the Martian past, called the Noachian Era of planet Mars (4.1 Gyrs – 3.7 Gyrs ago) (http://mars.jpl.nasa.gov/msl/multimedia/images/? ImageID=5765 – NASA/JPL-Caltech/MSSS, May 10, 2013)

8.4 Timeline of Robotic Spacecraft to Study the "Red Planet"

Well, since October 10, 1960 – with the launch (failed) of the Soviet mission "Mars 1 M No. 1" – until nowadays, 2014, there have been 47 robotic missions to study Mars. Of them, 18 missions failed, 27 had success, and 2 are in route to Mars.

Below is a timeline of all missions to Mars with their present status. The colored logos represent the nations.

Mission (1960–1969)	Launch	Arrival at Mars	Termination	Elements	Outcome
Mars 1 M No. 1	10 October 1960		10 October 1960	Flyby	Launch failure
Mars 1 M No. 2	14 October 1960		14 October 1960	Flyby	Launch failure
Mars 2MV-4 No. 1	24 October 1962		24 October 1962	Flyby	Broke up shortly after launch
Mars 1	1 November 1962		21 March 1963	Flyby	Some data collected but lost contact before reaching Mars, flyby at approx. 193,000 km
Mars 2MV-3 No. 1	4 November 1962		19 January 1963	Lander	Failed to leave the Earth's orbit
Mariner 3	5 November 1964		5 November 1964	Flyby	Failure during launch ruined trajectory
Mariner 4	28 November 1964	14 July 1965	21 December 1967	Flyby	Success (21 images returned)
Zond 2	30 November 1964		May 1965	Flyby	Communication lost three months before reaching Mars
Mariner 6	25 February 1969	31 July 1969	August 1969	Flyby	Success
Mariner 7	27 March 1969	5 August 1969	August 1969	Flyby	Success
Mars 2 M No. 521	27 March 1969		27 March 1969	Orbiter	Launch failure
Mars 2 M No. 522	2 April 1969		2 April 1969	Orbiter	Launch failure

Mission (1970–1989)	Launch	Arrival at Mars	Termination	Elements	Outcome
Mariner 8	8 May 1971		8 May 1971	Orbiter	Launch failure
Kosmos 419	10 May 1971		12 May 1971	Orbiter	Launch failure
Mariner 9	30 May 1971	13 November 1971	27 October 1972	Orbiter	Success (first successful orbit)
Mars 2	19 May 1971	27 November 1971	22 August 1972	Orbiter	Success
			27 November 1971	Lander, rover	Crashed on the surface of Mars

(continued)

Mission (1970–1989)	Launch	Arrival at Mars	Termination	Elements	Outcome
Mars 3	28 May 1971	2 December 1971	22 August 1972	Orbiter	Success
			2 December 1971	Lander, rover	Partial success. First successful landing; landed softly but ceased transmission within 15 s
Mars 4	21 July 1973	10 February 1974	10 February 1974	Orbiter	Could not enter the orbit, made a close flyby
Mars 5	25 July 1973	2 February 1974	21 February 1974	Orbiter	Partial success. Entered the orbit and returned data but failed within 9 days
Mars 6	5 August 1973	12 March 1974	12 March 1974	Lander	Partial success. Data returned during descent but not after landing on Mars
Mars 7	9 August 1973	9 March 1974	9 March 1974	Lander	Landing probe separated prematurely; entered the heliocentric orbit
Viking 1	20 August 1975	20 July 1976	17 August 1980	Orbiter	Success
			13 November 1982	Lander	Success
Viking 2	9 September 1975	3 September 1976	25 July 1978	Orbiter	Success
			11 April 1980	Lander	Success
Phobos 1	7 July 1988		2 September 1988	Orbiter	Contact lost while en route to Mars
				Lander	Not deployed
Phobos 2	12 July 1988	29 January 1989	27 March 1989	Orbiter	Partial success: entered the orbit and returned some data. Contact lost just before deployment of landers
				Landers	Not deployed

(continued)

Mission (1990–1999)	Launch	Arrival at Mars	Termination	Elements	Outcome
Mars Observer	25 September 1992	24 August 1993	21 August 1993	Orbiter	Lost contact just before arrival
Mars Global Surveyor	7 November 1996	11 September 1997	5 November 2006	Orbiter	Success
Mars 96	16 November 1996		17 November 1996	Orbiter, lander, penetrator	Launch failure
Mars Pathfinder	4 December 1996	4 July 1997	27 September 1997	Lander, rover	Success
Nozomi (Planet-B)	3 July 1998		9 December 2003	Orbiter	Complications while en route; never entered the orbit
Mars Climate Orbiter	11 December 1998	23 September 1999	23 September 1999	Orbiter	Crashed on the surface due to metric-imperial mix-up
Mars Polar Lander	3 January 1999	3 December 1999	3 December 1999	Lander	Crash-landed on the surface due to improper hardware testing
Deep Space 2 (DS2)				Hard landers	

Mission (2000–2009)	Launch	Arrival at Mars	Termination	Elements	Outcome
2001 Mars Odyssey	7 April 2001	24 October 2001	Currently operational	Orbiter	Success
Mars Express	2 June 2003	25 December 2003	Currently operational	Orbiter	Success
Beagle 2			6 February 2004	Lander	Landing failure; fate unknown
MER-A Spirit	10 June 2003	4 January 2004	22 March 2011	Rover	Success
MER-B Opportunity	7 July 2003	25 January 2004	Currently operational	Rover	Success

(continued)

Mission (2000–2009)	Launch	Arrival at Mars	Termination	Elements	Outcome
Rosetta	2 March 2004	25 February 2007	Currently operational	Gravity assist en route to comet 67P/ Churyumov– Gerasimenko	Success
Mars Reconnais-sance Orbiter	12 August 2005	10 March 2006	Currently operational	Orbiter	Success
Phoenix	4 August 2007	25 May 2008	10 November 2008	Lander	Success
Dawn	27 September 2007	17 February 2009	Currently operational	Gravity assist to Vesta	Success

Mission (2010–2019)	Launch	Arrival at Mars	Termination	Elements	Outcome
Fobos-Grunt	8 November 2011		8 November 2011	Phobos lander, sample return	Failed to leave the Earth orbit. Fell back to Earth
Yinghuo-1			8 November 2011	Orbiter	
MSL Curiosity	26 November 2011	6 August 2012	Currently operational	Rover	Success
Mars Orbiter Mission	5 November 2013	24 September 2014	Currently operational	Orbiter	Launched successfully
MAVEN	18 November 2013	22 September 2014	Currently operational	Orbiter	Launched successfully

Since August 2012, there have been two scientific rovers on the surface of Mars beaming signals back to Earth (Opportunity of the Mars Exploration Rover mission and Curiosity of the Mars Science Laboratory mission), and three orbiters are currently surveying the planet: Mars Odyssey, Mars Express, and Mars Reconnaissance Orbiter. Two orbiters launched in November 2013, India's Mars Orbiter Mission and NASA's Mars Atmosphere and Volatile Evolution (MAVEN), are currently on their way to Mars. To date, no sample return missions have been attempted for Mars, and one attempted return mission for Mars' moon Phobos (Fobos-Grunt) has failed.

8.5 New Theoretical Models for Mars Astrobiology

8.5.1 Possible Biosignatures

8.5.1.1 Methane

Trace amounts of methane in the atmosphere of Mars were discovered in 2003 and verified in 2004. As methane is an unstable gas, its presence indicates that there must be an active source on the planet in order to keep such levels in the atmosphere. It is estimated that Mars must produce 270 t/year of methane, but asteroid impacts account for only 0.8 % of the total methane production. Although geologic sources of methane such as serpentinization are possible, the lack of current volcanism, hydrothermal activity, or hot spots is not favorable for geologic methane.

It has been suggested that the methane was produced by chemical reactions in meteorites, driven by the intense heat during entry through the atmosphere. Although research published in December 2009 ruled out this possibility, research published in 2012 suggests that a source may be organic compounds on meteorites that are converted to methane by ultraviolet radiation (Fig. 8.24).

The existence of life in the form of microorganisms such as methanogens is among possible, but as yet unproven sources. If microscopic Martian life is producing the methane, it probably resides far below the surface, where it is still warm enough for liquid water to exist.

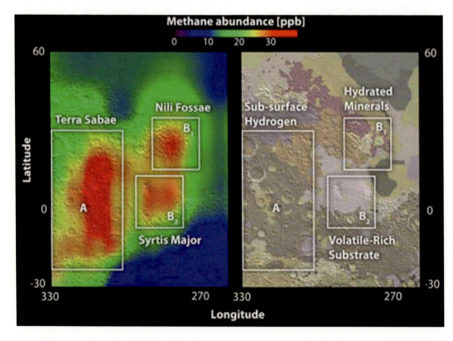

Fig. 8.24 Distribution of methane in the atmosphere of Mars in the northern hemisphere during summer (http://astrobiology.gsfc.nasa.gov/research1.html – NASA/JPL-Caltech/MSSS, 2009)

Since the 2003 discovery of methane in the atmosphere, some scientists have been designing models and in vitro experiments testing growth of methanogenic bacteria on simulated Martian soil, where all four methanogen strains tested produced substantial levels of methane, even in the presence of 1.0 wt% perchlorate salt. The results reported indicate that the perchlorates discovered by the Phoenix lander would not rule out the possible presence of methanogens on Mars.

It was suggested that both phenomena – methane production and degradation – could be accounted for by ecology of methane-producing and methane-consuming microorganisms.

In June 2012, scientists reported that measuring the ratio of hydrogen and methane levels on Mars may help determine the likelihood of life on Mars. According to the scientists, low H_2/CH_4 ratios (less than approximately 40) indicate that life is likely present and active (Oze et al. 2012). Other scientists have recently reported methods of detecting hydrogen and methane in extraterrestrial atmospheres.

In contrast to the findings described above, studies by Kevin Zahnle, a planetary scientist at NASA's Ames Research Center, and two colleagues conclude that there is as yet no compelling evidence for methane on Mars. They argue that the strongest reported observations of the gas to date have been taken at frequencies where interference from methane in Earth's atmosphere is particularly difficult to remove and are thus unreliable. Additionally, they claim that the published observations most favorable to interpretation as indicative of Martian methane are also consistent with no methane being present on Mars (Zahnle et al. 2011).

The Curiosity rover, which landed on Mars in August 2012, is able to make measurements that distinguish between different isotopologues of methane; but even if the mission is to determine that microscopic Martian life is the seasonal source of the methane, the life forms probably reside far below the surface, outside the rover's reach. The first measurements with the tunable laser spectrometer (TLS) in the Curiosity rover indicated that there is less than 5 ppb of methane at the landing site at the point of the measurement. On July 19, 2013, NASA scientists published the results of a new analysis of the atmosphere of Mars, reporting a lack of methane around the landing site of the Curiosity rover. On September 19, 2013, NASA again reported no detection of atmospheric methane with a measured value of 0.18 ± 0.67 ppbv corresponding to an upper limit of only 1.3 ppbv (95 % confidence limit) and, as a result, concluded that the probability of current methanogen microbial activity on Mars is reduced.

India's Mars Orbiter Mission, launched on November 5, 2013, will search for methane in the atmosphere of Mars using its Methane Sensor for Mars (MSM). The orbiter is scheduled to arrive at Mars on September 24, 2014. The Mars Trace Gas Mission orbiter planned to launch in 2016 would further study the methane, if present, as well as its decomposition products such as formaldehyde and methanol.

8.5.1.2 Formaldehyde

In February 2005, it was announced that the Planetary Fourier Spectrometer (PFS) on the European Space Agency's Mars Express Orbiter had detected traces of formaldehyde in the atmosphere of Mars. Vittorio Formisano, the director of the PFS, has speculated that the formaldehyde could be the by-product of the oxidation of methane and, according to him, would provide evidence that Mars is either extremely geologically active or harboring colonies of microbial life. NASA scientists consider the preliminary findings well worth a follow-up but have also rejected the claims of life.

8.5.1.3 Meteorites

NASA maintains a catalog of 34 Mars meteorites. These assets are highly valuable since they are the only physical samples available of Mars. Studies conducted by NASA's Johnson Space Center show that at least three of the meteorites contain potential evidence of past life on Mars, in the form of microscopic structures resembling fossilized bacteria (so-called biomorphs). Although the scientific evidence collected is reliable, its interpretation varies. To date, none of the original lines of scientific evidence for the hypothesis that the biomorphs are of exobiological origin (the so-called biogenic hypothesis) have been either discredited or positively ascribed to non-biological explanations.

In July 2014, it was published a paper with a new study co-funded by the NASA Astrobiology Institute, where it is shown that the ribosomal RNA core – which originated over 3 Gyrs ago before the Last Universal Common Ancestor (LUCA) of life – is essentially the same in all living systems, throughout the evolution of life on planet Earth (Petrov et al. 2014). So if meteorites containing encapsulate ribosomal RNA cores were ejected (by asteroidal/cometary impacts) from the Earth to Mars, or vice versa, there is a possibility that such molecular structures (ribosomal RNA, etc.) evolved into more complex biological molecules and, hypothetically, into initial living cells on Earth and on Mars, in a panspermia scenario (de Morais 2004).

Over the past few decades, seven criteria have been established for the recognition of past life within terrestrial geologic samples. Those criteria are:

1. Is the geologic context of the sample compatible with past life?
2. Is the age of the sample and its stratigraphic location compatible with possible life?
3. Does the sample contain evidence of cellular morphology and colonies?
4. Is there any evidence of biominerals showing chemical or mineral disequilibria?
5. Is there any evidence of stable isotope patterns unique to biology?
6. Are there any organic biomarkers present?
7. Are the features indigenous to the sample?

For general acceptance of past life in a geologic sample, essentially, most or all of these criteria must be met. All seven criteria have not yet been met for any of the Martian samples, but continued investigations are in progress.

As of 2014, reexaminations of the biomorphs found in the three Martian meteorites are underway with more advanced analytical instruments than previously available.

ALH84001

The ALH84001 meteorite was found in December 1984 in Antarctica, by members of the ANSMET project; the meteorite weighs 1.93 kg (4.3 lb). The sample was ejected from Mars about 17 Ma ago and spent 11,000 years in or on the Antarctic ice sheets. Composition analysis by NASA revealed a kind of magnetite that on Earth is only found in association with certain microorganisms (Fig. 8.25).

Then in August 2002, another NASA team led by Thomas-Keprta published a study indicating that 25 % of the magnetite in ALH84001 occurs as small, uniform-sized crystals that, on Earth, is associated only with biological activity and that the remainder of the material appears to be normal inorganic magnetite (Thomas-Keprta et al. 2009). The extraction technique did not permit the determination as to whether the possibly biological magnetite was organized into chains as would be expected. The meteorite displays indication of relatively low-temperature secondary mineralization by water and shows evidence of pre-terrestrial aqueous alteration. Evidence of polycyclic aromatic hydrocarbons (PAHs) has been identified with the levels increasing away from the surface.

Fig. 8.25 Mars meteorite ALH84001 (http://www.lpi.usra.edu/lpi/meteorites/The_Meteorite. shtml – NASA–JSC/Lunar and Planetary Institute, 1996)

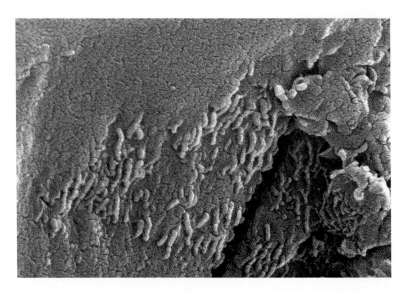

Fig. 8.26 Could the elongated forms in this electronic microscopic image of the interior of meteorite ALH84000 (which came from Mars' underground and felt on Earth's Antarctica), enlarged some 100,000 times, be microscopic fossils of past life on Mars? They bear a strong resemblance to simple organisms that were widespread on planet Earth nearly 3.5 Gyrs ago, though their terrestrial analogs were 100 times larger (http://www.lpi.usra.edu/lpi/meteorites/life.html – NASA–JSC/Lunar and Planetary Institute, 1996)

Some structures resembling the mineralized casts of terrestrial bacteria and their appendages (fibrils) or by-products (extracellular polymeric substances) occur in the rims of carbonate globules and pre-terrestrial aqueous alteration regions. The size and shape of the objects is consistent with Earthly fossilized nanobacteria, but the existence of nanobacteria itself is controversial (Figs. 8.26, 8.27, and 8.28).

In November 2009, NASA scientists reported after more detailed analyses that a biogenic explanation is a more viable hypothesis for the origin of the magnetite in the meteorite.

Nakhla Meteorite

The Nakhla meteorite fell on Earth on June 28, 1911, on the locality of Nakhla, Alexandria, Egypt. In 1998, a team from NASA's Johnson Space Center obtained a small sample for analysis. Researchers found pre-terrestrial aqueous alteration phases and objects of the size and shape consistent with Earthly fossilized nanobacteria, but the existence of nanobacteria itself is controversial. Analysis with gas chromatography and mass spectrometry (GC-MS) studied its high-molecular-weight polycyclic aromatic hydrocarbons in 2000, and NASA scientists concluded that as much as 75 % of the organic matter in Nakhla may not be recent terrestrial contamination.

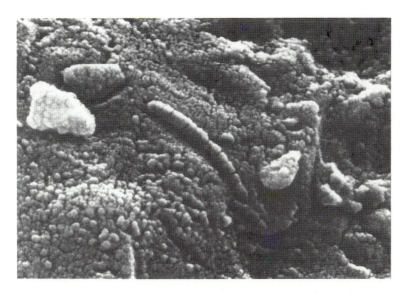

Fig. 8.27 An electron microscope reveals Archaean-like structures in meteorite fragment ALH84001 (http://www.lpi.usra.edu/lpi/meteorites/life.html – NASA–JSC/Lunar and Planetary Institute, 1996)

Fig. 8.28 Microscopic image of polycyclic aromatic hydrocarbon (PAH) molecules surrounded by magnetite crystals inside the ALH84001 Martian meteorite. PAHs and magnetite are strongly related to microbiological activity (http://www.nasa.gov/centers/johnson/pdf/403099main_GCA_2009_final_corrected.pdf – NASA–JSC, 1996)

This caused additional interest in this meteorite, so in 2006, NASA managed to obtain an additional and larger sample from the London Natural History Museum. On this second sample, a large dendritic carbon content was observed. When the results and evidence were published on 2006, some independent researchers claimed that the carbon deposits are of biological origin. However, it was remarked that since carbon is the fourth most abundant element in the universe, finding it in curious patterns is not indicative or suggestive of biological origin.

Shergotty

The Shergotty meteorite, a 4 kg Martian meteorite, fell on Earth on Shergotty, India, on August 25, 1865, and was retrieved by witnesses almost immediately. This meteorite is relatively young, calculated to have been formed on Mars only 165 Ma ago from volcanic origin. It is composed mostly of pyroxene and thought to have undergone pre-terrestrial aqueous alteration for several centuries. Certain features in its interior suggest remnants of a biofilm and its associated microbial communities. Work is in progress on searching for magnetite within alteration phases.

Yamato 000593

Yamato 000593 is the second largest meteorite from Mars found on Earth. Studies suggest the Martian meteorite was formed about 1.3 billion years ago from a lava flow on Mars. An impact occurred on Mars about 12 Ma ago and ejected the meteorite from the Martian surface into space. The meteorite landed on Earth in Antarctica about 50,000 years ago. The mass of the meteorite is 13.7 kg (30 lb) and has been found to contain evidence of past water movement. At a microscopic level, spheres are found in the meteorite that are rich in carbon compared to surrounding areas that lack such spheres. The carbon-rich spheres may have been formed by biotic activity according to NASA scientists.

8.5.1.4 Geysers on Mars

The seasonal frosting and defrosting of the southern ice cap results in the formation of spiderlike radial channels carved on 1-m-thick ice by sunlight. Then, sublimed CO_2 – and probably water – increases pressure in their interior producing geyser-like eruptions of cold fluids often mixed with dark basaltic sand or mud. This process is rapid, observed happening in the space of a few days, weeks, or months, a growth rate rather unusual in geology – especially for Mars.

A team of Hungarian scientists proposes that the geysers' most visible features, dark dune spots and spider channels, may be colonies of photosynthetic Martian microorganisms, which over winter are beneath the ice cap, and as the sunlight returns to the pole during early spring, light penetrates the ice; the microorganisms photosynthesize and heat their immediate surroundings. A pocket of liquid water, which would normally evaporate instantly in the thin Martian atmosphere, is trapped around them by the overlying ice. As this ice layer thins, the microorganisms show through gray.

When the layer has completely melted, the microorganisms rapidly desiccate and turn black, surrounded by a gray aureole. The Hungarian scientists believe that even a complex sublimation process is insufficient to explain the formation and evolution of the dark dune spots in space and time.

8 Mars Astrobiology: Recent Status and Progress

Fig. 8.29 Close-up of dark dune spots, probably created by cold geyser-like eruptions – Mars Global Surveyor (http://www.msss.com/moc_gallery/m07_m12/maps/M08/M0804688.gif – NASA/JPL-Caltech/MSSS, October 19, 1999)

A multinational European team suggests that if liquid water is present in the spiders' channels during their annual defrost cycle, they might provide a niche where certain microscopic life forms could have retreated and adapted while sheltered from solar radiation. A British team also considers the possibility that organic matter, microbes, or even simple plants might coexist with these inorganic formations, especially if the mechanism includes liquid water and a geothermal energy source. However, they also remark that the majority of geological structures may be accounted for without invoking any organic "life on Mars" hypothesis. It has been proposed to develop the Mars Geyser Hopper lander to study the geysers up close (Fig. 8.29).

8.5.2 Forward Contamination: Planetary Protection

Planetary protection of Mars aims to prevent biological contamination of the planet. A major goal is to preserve the planetary record of natural processes by preventing human-caused microbial introductions, also called forward contamination. There is abundant evidence as to what can happen when organisms from regions on Earth that have been isolated from one another for significant periods of time are introduced into each other's environment. Species that are constrained in one environment can thrive – often out of control – in another environment much to the detriment of the original species that were present. In some ways, this problem could be compounded if life forms from one planet were introduced into the totally alien ecology of another world.

The prime concern of hardware contaminating Mars derives from incomplete spacecraft sterilization of some hardy terrestrial bacteria (extremophiles) despite best efforts. Hardware includes landers, crashed probes, end of mission disposal of hardware, and hard landing of entry, descent, and landing systems. This has prompted research on radiation-resistant microorganisms including *Brevundimonas*, *Rhodococcus*, *Pseudomonas genera*, and *Deinococcus radiodurans* survival rates under simulated Martian conditions. Results from one of these experimental irradiation experiments, combined with previous radiation modeling, indicate that *Brevundimonas* sp. MV.7 emplaced only 30 cm deep in Martian dust could survive the cosmic radiation for up to 100,000 years before suffering 10^6 population reduction. Surprisingly, the diurnal Mars-like cycles in temperature and relative humidity affected the viability of *Deinococcus radiodurans* cells quite severely. In other simulations, *Deinococcus radiodurans* also failed to grow under low atmospheric pressure, under 0 °C, or in the absence of oxygen.

8.5.3 Reanalysis of the Viking Labeled Release Experiments

One of the designers of the Labeled Release experiment, Gilbert Levin, believes his results are a definitive diagnostic for life on Mars. Levin's interpretation is disputed by many scientists. A 2006 astrobiology textbook noted that with unsterilized terrestrial samples, though, the addition of more nutrients after the initial incubation would then produce still more radioactive gas as the dormant bacteria sprang into action to consume the new dose of food. This was not true of the Martian soil; on Mars, the second and third nutrient injections did not produce any further release of labeled gas. Other scientists argue that superoxides in the soil could have produced this effect without life being present. An almost general consensus discarded the Labeled Release data as evidence of life, because the gas chromatograph and mass spectrometer, designed to identify natural organic matter, did not detect organic molecules. The results of the Viking mission concerning life are considered by the general expert community, at best, as inconclusive.

In 2007, during a Seminar of the Geophysical Laboratory of the Carnegie Institution (Washington, D.C., USA), Gilbert Levin's investigation was assessed once more. Levin still maintains that his original data were correct, as the positive and negative control experiments were in order.

Moreover, Levin's team, on 12 April 2012, reported a statistical speculation, based on old data (mentioned above and below in the text) – reinterpreted mathematically through cluster analysis – of the Labeled Release experiments, that may suggest evidence of extant microbial life on Mars. Critics counter that the method has not yet been proven effective for differentiating between biological and non-biological processes on Earth so it is premature to draw any conclusions.

A research team from the National Autonomous University of Mexico headed by Rafael Navarro-González concluded that the GCMS equipment (TV–GC–MS) used by the Viking program to search for organic molecules may not be sensitive enough to detect low levels of organics. Klaus Biemann, the principal investigator of the GCMS experiment on Viking, wrote a rebuttal. Because of the simplicity of sample handling, TV–GC–MS is still considered the standard method for organic detection on future Mars missions, so Navarro-González suggests that the design of future organic instruments for Mars should include other methods of detection.

After the discovery of perchlorates on Mars by the Phoenix lander, practically the same team of Navarro-González published a paper arguing that the Viking GCMS results were compromised by the presence of perchlorates. A 2011 astrobiology textbook notes that while perchlorate is too poor as an oxidizer to reproduce the LR results (under the conditions of that experiment, perchlorate does not oxidize organics), it does oxidize, and thus destroy, organics at higher temperatures used in the Viking GCMS experiment (Plaxco and Gross 2011). Biemann has written a commentary critical of this Navarro-González paper as well, to which the latter have replied; the exchange was published in December 2011.

8.5.3.1 *Gillevinia straata*

The claim for life on Mars, in the form of *Gillevinia straata*, is based on old data reinterpreted as sufficient evidence of life, mainly by Gilbert Levin. The evidence supporting the existence of *Gillevinia straata* microorganisms relies on the data collected by the two Mars Viking landers that searched for biosignatures of life, but the analytical results were, officially, inconclusive.

In 2006, Mario Crocco, a neurobiologist at the Neuropsychiatric Hospital Borda in Buenos Aires, Argentina, proposed the creation of a new nomenclatural rank that classified the Viking landers' results as "metabolic" and therefore belonging to a form of life. Crocco proposed to create new biological ranking categories (taxa), in the new kingdom system of life, in order to be able to accommodate the genus of Martian microorganisms.

Crocco proposed the following taxonomical entry:

- Organic life system: Solaria
- Biosphere: Marciana
- Kingdom: Jakobia (named after neurobiologist Christfried Jakob)
- Genus and species: *Gillevinia straata*

As a result, the hypothetical *Gillevinia straata* would not be a bacterium (which rather is a terrestrial taxon) but a member of the kingdom "Jakobia" in the biosphere "Marciana" of the "Solaria" system.

The intended effect of the new nomenclature was to reverse the burden of proof concerning the life issue, but the taxonomy proposed by Crocco has not been accepted by the scientific community and is considered a single nomen nudum. Further, no Mars mission has found traces of biomolecules.

The only extraterrestrial life detection experiments ever conducted were the three which were components of the 1976 Viking Mission to Mars. Of these, only the Labeled Release experiment obtained a clearly positive response. In this experiment, ^{14}C-radiolabeled nutrient was added to the Mars soil samples. Active soils exhibited rapid, substantial gas release. The gas was probably CO_2 and, possibly, other radiocarbon-containing gases. We have applied complexity analysis to the Viking LR data. Measures of mathematical complexity permit deep analysis of data structure along continua including signal vs. noise, entropy vs. negentropy, periodicity vs. aperiodicity, order vs. disorder, etc. It employed seven complexity variables, all derived from LR data, to show that Viking LR active responses can be distinguished from controls via cluster analysis and other multivariate techniques.

Furthermore, Martian LR active response data cluster with known biological time series while the control data cluster with purely physical measures. It was concluded that the complexity pattern seen in active experiments strongly suggests biology while the different pattern in the control responses is more likely to be non-biological. Control responses that exhibit relatively low initial order rapidly devolve into near-random noise, while the active experiments exhibit higher initial order which decays only slowly. This suggests a robust biological response. These analyses support the interpretation that the Viking LR experiment did detect extant microbial life on Mars (Bianciardi et al. 2012).

Some scientists still believe the results were due to living reactions. No organic chemicals were found in the soil. However, dry areas of Antarctica do not have detectable organic compounds either, but they have organisms living in the rocks (Friedmann 1982).

8.5.4 New Forms of Hypothetical Life on Mars

The planet Mars has been studied for several years, and recently, with sophisticated spacecraft – orbiters and rovers – analyzing its atmosphere and surface, it has become clear that the "Red Planet" passed through great atmospheric and geological disturbances at some 2 Gyrs ago. Mars indeed had a significant quantity of liquid water flowing through its surface and subsurface at that time, and now it is an arid and cold planet with a very thin atmosphere. The lack of a moon with the size of Earth's Moon makes Mars to have its spin axis slowly oscillating, thus producing geological epochs with much differentiated climates.

The extensive volcanism at that time much possibly created subsurface cracks and caves (McKay et al. 2010) within different strata, and the liquid water could have been stored in these subterraneous places, forming large aquifers with deposits of saline liquid water, minerals, organic molecules, and geothermal heat – ingredients for life as we know it on Earth (de Morais 2004).

At the subsurface, iron-rich clays, such as montmorillonite and kaolinite, possibly catalyzed several organic chemical reactions. Clay minerals are excellent catalyzers and can catalyze peptide bond formation in fluctuating environments.

And it was also found that extensive mechanical distortion produced on freezing and drying of these clay minerals produces unusual luminescent phenomena, during wet/dry cycled reaction sequence. This dehydration-induced emission of bursts of UV and visible light, with peak emission $\lambda = 365$ nm (decaying monotonically in several hours and several days) from clays, is possibly related to its catalytic mechanism by means of electronic excitation, creating mobile or trapped holes and electrons in the lattice (Coyne et al. 1984, 1985). So I propose that with a possible biogeochemical evolution below Martian surface at around 2 Gyrs ago, using clays' catalytic properties and the light emitted from them, there is a possibility that life arose at Mars from biomolecules up to a very simple organism form – simpler than Earth's Archaea organisms – deep within subsurface aquifers.

Such kind of simple organisms, within extreme environments, might have evolved a two-way form to use energy – one at great depths, using geochemical energy by sulfur redox, and other near to the surface using the light emitted from clays at wet/dry cycles. At the surface, gravity would pull colonies of those microorganisms to underground, and when at depths, plumes of hot water/hydrated hot molten material would rise those microorganisms near to the surface again, in long-term periodic cycles. Here on Earth, there are studies of Archaean thermophile organisms, such as *Sulfolobus shibatae* virus-like particles (SSV1), which exhibit properties of double ways for the same function (as double proteinic coats, via DNA decoding) and like to grow at 89 °C with UV light (a strong stimulant for SSV1 production (Martin et al. 1984)) (Zillig et al. 1992, 1999), and other Archaean which use more than two sources for a common metabolic pathway as *Sulfolobus acidocaldarius* (growing best in the water of volcanic calderas at about 75 °C and at a pH range of 1–6), which oxidizes sulfur to sulfuric acid or can use Fe^{2+} or MnO_4^{2-} as electron acceptors while using glycolysis and the TCA cycle.

The idea of a "simple" microorganism using two different forms of energy sources is thus not so problematic. That hypothetical life could have lasted for some hundreds of millions years, 1 Gyr, and now be dormant or, much more possibly, fossilized inside sediment rocks at the subsurface of Mars.

Future robotic and manned missions to Mars can search for possible biogeochemical signatures of fossilized colonies of such hypothetical microorganisms, below the Martian surface, within locations with hydrothermal past (and possible present) volcanic activity, using sedimentary petrological microscopy and NMR – interesting to astrobiology.

8.6 Biogeochemical Parallels Between Earth and Mars

8.6.1 Geological Comparisons

Well, I make some parallels between what is known about carbonate diagenetic deposits and petrological studies of arenites, present at the Bauru Group, Brazil, South America, and what is known about intact rocks in contact with clays, rich in olivine, exposed at the surface of planet Mars containing carbonate mineral deposits.

The region where the Martian carbonate material was discovered is located within a system of small canyons known as Nili Fossae, at the northern hemisphere of Mars. The carbonates were discovered by NASA's Mars Reconnaissance Orbiter robotic spacecraft, in December 2008.

Plumes of CH_4 above that location was also discovered. The diagenetic alteration of the subsurface rocks was probably due to hydrothermal activity which might have existed in that region.

So I make specific comparisons among such diagenetic processes on Earth and on Mars to give a contribution – via the use of biased techniques and gained experience – for future exploration and understanding of that planet. By repeatedly studying geophysical-chemical properties of carbonate diagenetic deposits, it is possible to characterize their minerals each time better and, with that, to better determine the geological evolution of deposits of sediments, gradually altered by physicochemical-biological processes after their deposition (diagenesis), as, for example, the carbonates – the subject of this work (Leinz and Amaral 2003; Teixeira et al. 2008).

These practical carbonate sedimentology analysis and comparisons can be applied to more focused robotic and manned in situ search for biogeochemical signatures at past Earth and present Mars. The study of the arenites of the carbonate diagenetic deposits present at the Bauru Group shows evidence of sedimentary homogeneity at different depths of soundings and at different areas. The detritic properties of these arenites are constituted essentially by quartz and feldspars and, at smaller quantity, by lithic fragments and accessory minerals with great intergranular porosity. This is a good indicator of shallow eodiagenetic processes with low mechanical pressure, indicating past presence of subsurface waters in that location.

The principal diagenetic processes observed were dissolution of heavy minerals, lithic fragments, and aluminosilicates and cementing by microcrystalline calcite (Stradioto et al. 2008). I observed negative biaxial refringe and formation of clay minerals as montmorillonite and kaolinite.

Infrared images of Nili Fossae by CRISM–MRO show bedrock containing clays (principally smectites), exposed olivine, and carbonates at the same stratigraphic level. Much probably, the olivine was altered by hydrothermal action, possibly present at that location $\sim$3.6 Gyrs ago, transformed into the found carbonates. It is known that carbonates are formed in neutral or basic waters ($pH > 7$), which indicates that dissolution of the minerals was not only in acid environment, showing a great pH variation of the waters, interesting to the diagenetic evolution of hydrothermal, volcanic, and sedimentary regions on planet Mars, as this carbonatic one being studied here on planet Earth (Fig. 8.30).

Other minerals found by CRISM–MRO were aluminosmectite, iron/magnesium-smectite, hydrated silicates, minerals of the kaolinite group, and iron oxides. In 2004, the NASA's robotic rover Opportunity discovered the minerals hematite [iron (III) oxide (Fe_2O_3)] and jarosite [$KFe^{3+}_3(OH)_6(SO_4)_2$] in the Martian surface. Jarosite is formed by the oxidation of iron sulfides in the presence of liquid acid water, which shows (aside from several evidences, geomorphic and geologic ones by remote sensing) that Mars much probably had liquid water with relative abundance

8 Mars Astrobiology: Recent Status and Progress

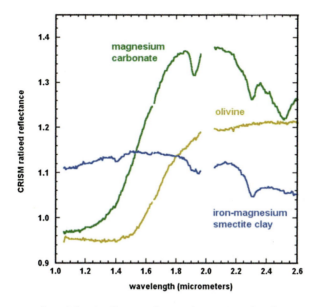

Fig. 8.30 Spectra collected by the Compact Reconnaissance Imaging Spectrometer for Mars (CRISM) indicate the presence of three distinct minerals. The graphed information comes from an observation of terrain in the Nili Fossae area of northern Mars. CRISM is one of six science instruments on NASA's Mars Reconnaissance Orbiter. Iron–magnesium smectite clay is formed through alteration of rocks by liquid water and is characterized by distinctive absorptions at 1.4, 1.9, and 2.3 μm due to water (H_2O) and OH in the atomic structure of the mineral. Olivine is an iron magnesium silicate and primary igneous mineral, and water is not in its structure. Its spectrum is characterized by a strong and broad absorption at 1.0 μm due to ferrous iron (Fe^{2+}). Carbonate is an alteration mineral identified by the distinctive paired absorptions at 2.3 and 2.5 μm. The precise band positions at 2.31 and 2.51 μm identify the carbonate at this location as magnesium carbonate. The broad 1.0-μm band indicates that some small amount of ferrous iron is also present and the feature at 1.9 μm indicates the presence of water. CRISM researchers believe the magnesium carbonate found in the Nili Fossae region was formed from alteration of olivine by water. The data come from a CRISM image catalogued as FRT00003E12. The spectra shown here are five-pixel-by-five-pixel averages of CRISM L – detector spectra taken from three different areas within the image that have then been ratioed to a five-pixel-by-five-pixel common denominator spectrum taken from a spectrally unremarkable area with no distinctive mineralogic signatures. This technique highlights the spectral contrasts between regions due to their unique mineralogy. The spectral wavelengths near 2.0 μm are affected by atmospheric absorptions and have been removed for clarity (http://www.nasa.gov/mission_pages/MRO/multimedia/carb-plot.html – NASA/JPL-Caltech/JHUAPL/Brown University, 2008)

in its surface and subsurface soils. This potentially demonstrates that, hypothetically, Mars might have had redox mechanisms associated with biogeochemical (bio meaning biomolecules) activity within subsurface hot springs, with stable liquid H_2O, at least by ∼3.6 Gyrs ago.

The study of carbonate diagenetic deposits uses a large spectrum of techniques, as optical/texture and microscopic/mineralogical (birefringence, pleochroism, MEV, etc.), to characterize a geological evolution of diagenesis of minerals found in sedimentary deposits, and this can be applied to understand such places – via geological comparisons – on planet Earth and on planet Mars. The more different complementary techniques are used in those analyses, the more accurate will be the research results outgiven.

So, I suggest more practically focused geological comparisons between selected locations on these planets. In the case of Mars, there are still missing many texture and microscopic analyses of minerals, to diminish the error bars related to the Martian hydrogeological diagenesis. Such studies will be future made by samples–return, in-situ and manned missions to that beautiful planet. The comparative evolution of carbonate sedimentary deposits on these two worlds, Earth and Mars, is very interesting to astrobiology, since we can better understand the origin and evolution of life within Earth, the possibility of that within Mars, and elsewhere in the cosmos. Nili Fossae has geochemical potential for which it could have had a beginning of biogeochemical processes there – interesting to the multidiscipline field of astrobiology.

8.6.2 Mathematical Turing Patterns in Estuarine Sediments by Microbiological Activity

I did field observations regarding ecological microbial mat distribution in marine estuarine locations, and by means of statistical mechanics, I discovered that we can mathematically show that distribution by using Turing mechanisms.

For better explaining about Turing patterns, I begin a naïve statistical description of a simple system by using an ensemble of small balls, but whose omission would perhaps make the paper a bit difficult to read. Imagine that all of them are equal except the colors, one group being red and the other white. If the balls do not have long-range interaction among everyone, only very short-range elastic shock interaction, and if all the balls have zero kinetic energy relative to each other, then after some time of continuous measurement of this system, it will remain in the same state.

But if, in this system, the balls have some kinetic energy, then after some time, it will be in a different state with red balls arranged in positions differently from the beginning of the measurement. Although clusters of red balls or white ones during the measurement can occur, it is not common. But if the balls have long-range interaction, then after some enough time of measurement, even if the balls begin with zero kinetic energy, they will form clusters of separated colors.

We can see now that the mechanical statistics of systems displaying long-range interactions leads to the formation of clusters within the system, setting apart from initial homogeneity boundary conditions. And such kind of non-Boltzmann–Gibbs statistics can be treated using the non-extensive statistical mechanics (Tsallis 1988; Tsallis et al. 1998; Tsallis and Gell-Mann 2004). Now we can use a different system, a group of just one kind of species of microorganisms, for instance, bacteria,

8 Mars Astrobiology: Recent Status and Progress

and their nutrients within a unit volume inside a homogeneously composed soil in the ground. The bacteria exhibit long-range interactions via signal molecules. We can think that after some enough time, the bacteria will be arranged in clusters. And this actually occurs. This can be mathematically well verified via the theoretical analysis of Turing structures (Murray 1989), which are stationary, spatially periodic morphological patterns (or clusters), with fractal boundaries behavior, resulting from the interplay between pure diffusion and nonlinear reaction kinetics mechanisms at nonequilibrium.

Interestingly, the first experimental observation of Turing patterns occurred almost 40 years after Turing's work (Turing 1952), in a chemical diffusion–reaction system apparatus (Castets et al. 1990; Ouyang and Swinney 1991).

There is a model which describes the dynamics of predator–prey systems, which is the Lotka–Volterra model (LV) (Murray 1989; Lotka 1920; Volterra 1926, 1936). And there is a description that it is a minimal complexity model, with mean field conservative dynamics which can be directly implemented on lattice and involves only two reactive species X_1 and X_2 (adsorbed on a lattice support) and the empty sites of the support S.

This description is named lattice Lotka–Volterra model (LLV) (Frachebourg et al. 1996; Provata et al. 1999). All reactive steps are bimolecular, and the reaction occurs via hard-core interactions. Schematically, the LLV model has the following form:

$$X_1 + X_2 \xrightarrow{k_s} 2\,X_2$$

$$X_2 + S \xrightarrow{k_1} 2S$$

$$S + X_1 \xrightarrow{k_2} 2X_1$$

In the LLV model, the kinetic rate equations are:

$$\frac{dx_1}{dt} = x_1\,(-k_s x_2 + k_2 s)$$

$$\frac{dx_2}{dt} = x_2\,(k_s x_1 - k_1 s)$$

$$\frac{ds}{dt} = s\,(-k_2 x_1 + k_1 x_2), \quad \text{with the entropies } k.$$

And the associated non-extensive q-entropy in the LLV model is:

$$S_q = 1 - \sum_{i=1}^{M} p_i^q(t)/q - 1, \quad S_1 = -\sum_{i=1}^{M} p_i(t)\ln\,p_i(t)$$

The LLV model was used to study such following problems: a biological system composed of three competing species, which gives rise to a stationary pattern of vortices and strings, clusters, in this bio-system (Tainaka 1989); the evolution of a system of N interacting biological species mimicking the dynamics of cyclic food (nutrients) chain, which gives rise to a spatial organization, clusters, in such system (Frachebourg et al. 1996; Saunders and Bazin 1974); and the dynamics of open reactive systems capable of giving rise to oscillatory behavior, which displays a result that in low dimensional supports, the system prefers specific oscillation frequencies, which gives rise to a spatial clustering mechanism (Provata et al. 1999).

Using mean field analysis and Monte Carlo simulations, it is found that spontaneous local clustering on lattice and homogeneous initial distribution turn into clustered structures and reactions among and between molecular and biological species occur within their interfaces adopting a fractal structure, i.e., besides clustering, there is fractality within the clusters' domains and borders (Tsekouras and Provata 2002); at the reaction limited regime, on square lattice, in the cyclic LLV model, there is the spontaneous development of dynamical patterns, in the form of consecutive stripes or rings, which give rise to clustering, and the borderlines between consecutive stripes are fractal (Provata and Tsekouras 2003); it is numerically shown that the LLV model, when realized on a square lattice support, gives rise to a finite production of the above mentioned non-extensive q-entropy S_q (Tsekouras and Provata 2004), and this evidence of non-extensivity is consistent with the spontaneous emergence of local domains of identical particles (which can be anything, from molecules to microorganisms, for instance), e.g., clustering, with fractal boundaries and competing interactions at long range (Tsallis and Gell-Mann 2004).

Schematically, the Turing mechanism has the following form:

$$\begin{cases} \dfrac{\partial u}{\partial t} = \nabla^2 u + uv - u - \alpha \\ \dfrac{\partial v}{\partial t} = d\nabla^2 v + \beta - uv \end{cases}$$

$$u_{ss} = \beta - \alpha; \ v_{ss} = \frac{\beta}{\beta - \alpha}$$

$$\alpha_c = \beta - \sqrt{\beta d} + \frac{d}{2}\left(\sqrt{1 + \frac{4\beta}{\sqrt{\beta d}}} - 1\right)$$

$$k^2 = \frac{(\alpha - \beta)^2 - \alpha d}{2d\,(\alpha - \beta)}$$

Based on the Jacobian matrix, there is also the activator–inhibitor and substrate–depletion Turing model. This model is called activator–inhibitor if the matrix is $\begin{pmatrix} + & - \\ + & - \end{pmatrix}$ and substrate–depletion if the matrix is $\begin{pmatrix} \pm & \pm \end{pmatrix}$.

8 Mars Astrobiology: Recent Status and Progress 211

By studying the following natural problems, it was shown that they exhibit the Turing nonequilibrium mechanisms of spatial patterns formations, with clustering and fractal boundaries behavior – the author has been studying since 2003 the subjects contained in items (3), (13), and (14):

1. Mutual interference between predators (Alonso et al. 2002)
2. The dynamics of nutrient cycling and food webs (DeAngelis 1992)
3. Kinetic modeling of microbial-driven redox chemistry of subsurface environments, with coupling transport, microbial metabolism, and geochemistry (Hunter et al. 1998)
4. Predator–prey dynamics in environments rich and poor in nutrients (McCauley and Murdoch 1990)
5. The occurrence and activity of sulfate-reducing bacteria (producing hydrogen sulfide gas, H_2S) in the bottom sediments of the Gulf of Gdansk (Mudryk et al. 2000)
6. The cycling of iron and manganese in surface sediments, with the coupled transport and reaction of carbon (C), oxygen (O_2), nitrogen (N_2), sulfur (S), iron (Fe), and manganese (Mn) (Van Capellen and Chang 1996)
7. Linear understanding of a huge aquatic ecosystem model using a group-collecting sensitive study (Köhler and Wirtz 2002)
8. Phenomenological pattern recognition in the dynamical structures of marine–estuarine tidal sediments from the German Wadden Sea (Kropp and Klenke 1997)
9. Global terrestrial distribution of nitrous oxide (N_2O) production and N inputs in freshwater (rivers, lakes, and ponds), estuarine, and coastal marine ecosystems (Seitzinger and Kroeze 1998)
10. Dynamic response of deep-sea sediments to seasonal variations (Soetaert et al. 1996)
11. A multicomponent reactive transport model of early diagenesis, with application to redox cycling in coastal marine sediments (Wang and Van Capellen 1996)
12. The control of biogeochemical cycling by mobility and metabolic strategies of microbes in the sediments (Wirtz 2003)
13. Instabilities and pattern formation in simple ecosystem models (Baurmann et al. 2003)
14. Stable squares and other oscillatory Turing patterns in a diffusion-reaction model (Yang et al. 2004)

We then can see clearly that the dynamical evolutions of the LLV models (which are based on nonlinear reactive and other processes between species on a lattice) lead toward structures with clustering and fractal boundaries behavior. And we know from above that the evolutions of the Turing structures (based on the formation of patterns in nonequilibrium mixing of the mechanisms of diffusion and nonlinear reactive kinetics, between natural objects) also lead toward structures with clustering and fractal boundaries behavior (Kropp et al. 1997). So, the LLV model is profoundly linked to the Turing patterns – which can exhibit spots, stripes, squares, hexagon, and other geometrical figures – mechanism, meaning that non-extensive statistical mechanics (with the use of the q-entropy S_q) can be applied to the study of the Turing patterns with fractal clustering behavior exhibited in nature, on Earth, and on Mars.

Thus, the use of LLV models and Turing mechanisms, also by means of the non-extensive statistical mechanics, can mathematically describe well the phenomena of clustering and their associated boundaries with fractal dimensionality, which occurs in various natural situations, among them, biogeochemical processes via microorganisms in estuarine and marine sediments on planet Earth.

An estuary is a semi-enclosed coastal body of water with one or more rivers or streams flowing into it and with a free connection to the open sea (Pritchard 1967). An estuary is typically the tidal mouth of a river, and estuaries are often characterized by sedimentation or silt carried in from terrestrial runoff and, frequently, from offshore. They are made up of brackish water. Estuaries are more likely to occur on submerged coasts, where the sea level has risen in relation to the land; this process floods valleys to form rias (a ria is a submerged marine coastal landform, often known as a drowned valley or drowned river valley; rias are almost always estuaries) and fjords (fjords are very long inlets from the sea with high steeply sloped walled sides; a fjord is a landform created during a period of glaciations). These can become estuaries if there is a stream or river flowing into them.

Estuaries and coastal marine waters are among the most biological productive ecosystems on Earth. And if there is enough liquid water on Mars in the past to produce such geological watery environments, then probably, biogeochemical evolution might have took place on such locations.

Sediments are characterized by heterogeneous distributions of nutrients and microorganisms which emerge as a result of the interaction between chemical and biological processes with physical transport. It was studied in a simplified model the dynamics of one population of microorganisms and its nutrients, taking into account that the considered bacteria possess an active as well as an inactive state, where activation is processed by signal molecules. Furthermore, the nutrients are transported actively by bio-irrigation.

It is shown that under certain conditions, Turing patterns can occur which yield heterogeneous spatial patterns of species. Furthermore, this model exhibits several stable coexisting spatial patterns. This phenomenon of multistability can still be observed when spatial patterns are externally imposed by considering a depth-dependent bio-irrigation. The influence of bio-irrigation on Turing patterns leads to the emergence of "hot spots," i.e., localized regions of enhanced bacterial activity. This above work did not take into account temporal behavior of the Turing patterns (the Turing mathematical mechanism also has a temporal degree of freedom within its space–time matrix metric dynamics).

I did an experimental analysis in fieldwork which took into account the spatial and temporal behavior of Turing patterns, in the form of microbial activity within estuarine subsurface sediments. The region of study was a flat region with a small tidal estuarine, a very shallow river (0.05 m $\leq$ depth ≤ 0.5 m) named Caraís

Lagoon, whose water is rich in iode (I), some meters ($\approx$10 m) distant from the sea water (which is very rich in phyto- and zooplankton), inside an environmental protected area, Setiba State Park, Brazil. I measured the temperatures of the subsurface sediments (mud) at different depths, ranging from 0 to -0.5 m, and took measurements at different locations in the lagoon area. I also measured those temperatures at different times during day 1, at 8 A.M., 12 A.M., 4 P.M., and 8 P.M., in February 2003 (summer season in Brazil).

I collected small quantities of the subsurface mud at those range in depth, analyzed the color and particulate consistency of it, and froze the samples at $-10\,°C$ (using a portable thermal insulator). The temperature of the subsurface sediments and their color, smell, and particulate consistency are simple and good indicators of possible microorganisms' catalytic activity. The vertical temperature profile shows that at 12 A.M., at the surface (0 m), $T = 20\ °C$, and at -0.05 m, $T = 25\ °C$. At -0.30 m, $T = 30\ °C$. Between and below these depths, there was a noticeable drop of temperature ($T = 18\ °C$) and the same occurring below 0.20 m.

The color of the sediments at -0.05 and -0.30 m are black–dark gray (mud), and the color of the sediments between and below those depths are gray. The odors of the collected material at -0.05 and -0.30 m are strongly of hydrogen sulfide (H_2S), and the odors of the material between and below those depths are lightly of H_2S.

The particulate consistency of the material collected at -0.05 and -0.30 m is of very wet clay (mud), and the material between and below those depths shows a silt consistency. All those above four biogeochemical characteristics united display a strong possible evidence of microbiological metabolic activity at -0.05 and -0.30 m. H_2S is known to be the by-product of sulfate-reducing bacteria, and the black color of the very wet clay also reinforces this behavior (Mudryk et al. 2000).

This vertical temperature profile is maintained at different horizontal places of the lagoon area. But the vertical T profile varies with time during the day (probably linked to the variation of the solar energy flux by insolation unit area during a day, since the Sun's infrared and ultraviolet energy are metabolized by microorganisms) in a cyclic pattern – T oscillates from a minimum recorded at the surface (0 m), $T = 18\ °C$ at 8 A.M., to a maximum recorded at 0 m, $T = 35\ °C$ at 4 P.M. And the depths of the subsurface material exhibiting the above characterized possible microbial metabolism (black, hot, clay mud with strong odor of H_2S) alternated by subsurface regions with these smaller physicochemical parameters also vary in a time-dependence cyclic behavior.

These sites with possible microbial mats are spatially (vertically and horizontally) arranged in the form of "hot spots," or "hot clusters," not scattered as small clusters but they are more concentrated in regions, and they exhibit temporal cyclic behavior. At the surface of the sediments (0 m), there were microscopic green algae, deposited there by the tidal sea water. Perhaps these photosynthetic microorganisms could be supplying O_2 and nutrients (S, P, others) to the possible subsurface bacteria communities.

And it was also noted that the boundaries between those sites exhibit scale-invariance of those above biogeochemical properties, i.e., there were ensembles of mud material near to ensembles of mud material near the boundaries at -0.05

and -0.3 m depths which had the same geochemical profile (T, color, odors, consistency). This is a fractal characteristic.

So, by this in-situ analysis, I found that possibly a fractal dimensionality could be used to better describe the spatial-temporal geochemical clustered patterns observed, with strong possible microbiological component. Thus, the in situ search for spatial-temporal geochemical clustered patterns and fractal dimensionality at the Martian surface and subsurface could facilitate the search of biosignatures on Mars (de Morais 2004).

Thus, concluding, we can possibly show that in such abovementioned estuarine location, one can find the characteristics of clustering and fractality which are present in the dynamical LLV model and Turing patterns mechanisms, and the non-extensive statistical mechanics could be used to find the q-entropy S_q and other nonequilibrium statistical parameters of the above estuarine (Caraís lagoon) subsurface biogeochemical system and the possibility of this on Mars (de Morais 2004).

Such kind of Turing patterns are geologically present at the Phoenix lander on the north polar region of Mars, in the form of those hexagon figures as mentioned earlier in this chapter. And such subsurface ecological systems, displaying fractality behavior, are also of interest to the field of astrobiology, mainly to the models for the origin of life by $\sim$4.2 Gy on the subsurface of planet Earth and for hypothetical biogeochemical evolution by $\sim$4.2 Gy at the subsurface of planet Mars (de Morais 2004) (Fig. 8.31).

8.7 New Experimental Equipment for Mars Astrobiology

8.7.1 Future Missions

- ExoMars is a European-led multi-spacecraft program currently under development by the European Space Agency (ESA) and the Russian Federal Space Agency scheduled for launch in 2016–2018. Its primary scientific mission will be to search for possible biosignatures on Mars, past or present. A rover with a 2-m (6.6 ft) core drill will be used to sample various depths beneath the surface where liquid water may be found and where microorganisms might survive cosmic radiation.
- Mars 2020 rover mission – The Mars 2020 rover mission is a Mars planetary rover mission concept under study by NASA with a possible launch in 2020. It is intended to investigate an astrobiologically relevant ancient environment on Mars and to investigate its surface geological processes and history, including the assessment of its past habitability and potential for preservation of biosignatures within accessible geological materials.
- Mars Sample Return Mission – The best life detection experiment proposed is the examination on Earth of a soil sample from Mars. However, the difficulty of providing and maintaining life support over the months of transit from Mars to Earth remains to be solved. Providing for still unknown environmental and

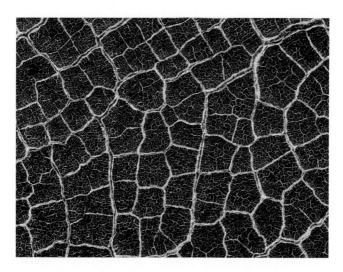

Fig. 8.31 Polygonal patterned ground. From a distance, the floor of this crater looks like a zoomed leaf or a giant honeycomb or a spider web. The intersecting shapes, or polygons, commonly occur in the northern lowlands (∼at 45° north) of Mars. The polygons in this "patterned ground" are easy to see because their edges are bound by troughs or ridges covered by bright frost relative to their darker, frost-free interiors. Patterned ground on Mars is thought to form as the result of cyclic thermal contraction cracking in the permanently frozen ground. Scientists study polygonal-patterned ground on Mars because the occurrence and physical characteristics of the polygons helps us to understand the recent and past distribution of ice (frozen water) in the shallow subsurface. These features also provide clues about climate conditions – Mars Reconnaissance Orbiter (http://hirise.lpl.arizona.edu/ESP_016641_2500 – NASA/JPL-Caltech/University of Arizona, February 13, 2010)

nutritional requirements is daunting. Should dead organisms be found in a sample, it would be difficult to conclude that those organisms were alive when obtained.

8.7.2 NMR Spectrometry

At Mars, some robotic missions have already analyzed its atmosphere and surface, using equipment with resolution down to millimetric scale. The NASA's Opportunity and Spirit rovers used microscope to study the subsurface of the "Red Planet" at millimetric depths in drilled holes on rocks. In 1996, as said earlier in this chapter, a NASA team announced the finding of organic molecules and morphological structures at nanometric scale, inside a meteorite which came from Mars. These possibly could be derived from an ancient Martian signature of biochemical activity, hypothetically, a fossilized "Archaean-type" microorganism.

In order to acquire better resolutions for the mineralogical study of samples of its surface, it is necessary to use nuclear magnetic resonance (NMR) spectrometers, with which one can obtain detailed astrobiological information below micrometer scale. NMR spectrometers are big equipment, but there are already miniature, lightweight, NMR spectrometers being developed at NASA–JPL which do not contain permanent magnets (they are designed to operate without applied magnetic fields); instead, they exploit the natural magnetic fields of the mineral phases (that contain iron) to be studied.

These fields give rise to nuclear magnetic resonance of the isotope ^{57}Fe at frequencies in the approximate range of 60–74 MHz in magnetite (Fe_3O_4). Such instrument has a mass of only 65 g – with a 5 V DC battery included – and consumes a power of only 0.5 W. It will be interesting the use of miniature NMR spectroscopy at Mars.

This instrument includes a marginal oscillator, the frequency of which is determined mainly by tuning capacitors, two varactors, and the NMR sample coil, in which a mineral specimen is placed. During operation, the frequency is swept slowly by use of one varactor and is modulated at a rate of 110 Hz by the use of the other varactor. The instrument also includes a digital-to-analog and an analog-to-digital converter and a microprocessor that communicates with an external laptop computer, generates frequency sweep and modulation signals, samples the output from the oscillator, and performs synchronous detection. Spectral data can be displayed on the computer screen. The software in the computer includes routines for processing data to estimate concentrations of mineral phases of interest. The performance of the instrument was demonstrated in NMR measurement experiments on two mineral specimens: one that contained magnetite in chlorite schist mixed with magnetically inert particles and one made of hematite ($a–Fe_2O_3$) (Kim et al. 2009).

Mössbauer (MB) spectrometers are very useful in planetary exploration, and such NMR spectrometer can be a useful tool for complementing data from MB ones. So, with the objective of the search for hypothetical extinct or extant life on the "Red Planet," I propose that in future robotic missions and a possible manned research at Mars, miniature NMR spectrometers be used.

Rovers can use their robotic arms to collect soil and rocky samples from below surfaces and put them inside the spacecraft where such those spectrometers can analyze the material. And astronauts can also use those miniature NMR spectrometers to do in situ very good quality research of surface and subsurface minerals, searching for biologically important minerals, organic molecules, and morphological structures relevant for astrobiology.

8.8 Hypothetical Future Life on the "Red Planet": Mars Colonization and Terraforming

Do little green men exist on Mars?

Well, no one knows the answer to this question. But, perhaps, this question could be reformulated as follows: will men exist on Mars? And the answer is

... much probably it will be we, humans. Various serious projects are being created, formulated, and analyzed, for the future real presence of persons on Mars. The manned colonization of planet Mars will require several intelligent and auto-sustainable steps. Among them, I mention here 2 great projects for the manned presence on the "Red Planet": the Caves of Mars Project and the Terraforming of Mars (de Morais 2004).

8.8.1 Caves of Mars Project

The Caves of Mars Project was a program funded through Phase II by the NASA Institute for Advanced Concepts to assess the best place to situate the research and habitation modules that a manned mission to Mars would require.

Caves and other underground structures, including lava tubes, canyon overhangs, and other Martian cavities, would be potentially useful for manned missions, for they would provide considerable shielding from both the elements and intense solar radiation that a Mars mission would expose astronauts to. They might also offer access to minerals, gases, ices, and any subterranean life that the crew of such a mission would probably be searching for.

The program also studied the designs for inflatable modules and other such structures that would aid the astronauts to build a livable environment for humans and earth creatures. This project can be reanalyzed again when more practical information is gathered on Mars' environments (de Morais 2004) (Figs. 8.32 and 8.33).

8.8.2 Terraforming of Mars

The terraforming of Mars is the hypothetical process by which Martian climate, surface, and known properties would be deliberately changed with the goal of making large areas of the environment more hospitable to human habitation, thus making human colonization much safer and more sustainable.

The concept relies on the assumption that the environment of a planet can be altered through artificial means. In addition, the feasibility of creating a planetary biosphere on Mars is undetermined. There are several proposed methods, some of which present prohibitive economic and natural resource costs and others may be currently technologically achievable.

8.8.2.1 Motivation and Ethics

Future population growth and demand for resources may necessitate human colonization of objects other than Earth, such as Mars, the Moon, and nearby planets.

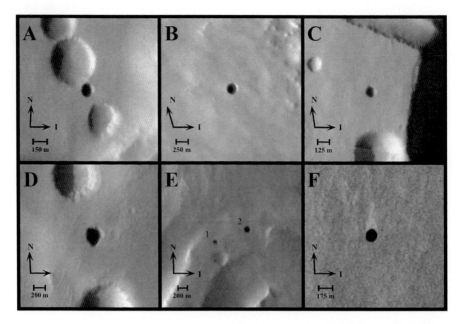

Fig. 8.32 THEMIS images of probable cave entrances on Arsia Mons. The pits have been informally named (A) Dena, (B) Chloe, (C) Wendy, (D) Annie, (E) Abby (left) and Nikki, and (F) Jeanne – Mars Odyssey (http://www.nasa.gov/mission_pages/odyssey/images/cave2.html – NASA/JPL-Caltech/Arizona State University/USGS, 2007)

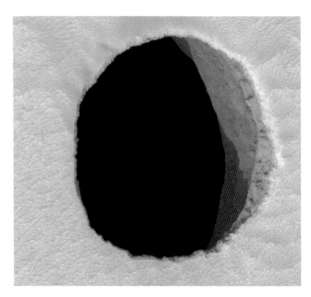

Fig. 8.33 HiRISE image of Mars hole "Jeanne," about 150 m (492 ft) across and at least 178 m (584 ft) deep – Mars Reconnaissance Orbiter (http://www.nasa.gov/mission_pages/MRO/multimedia/20070829-1.html – NASA/JPL-Caltech/University of Arizona, 2007)

Space colonization will facilitate harvesting the solar system's energy and material resources.

In many respects, Mars is the most Earth-like of all the other planets in the solar system. It is thought that Mars once did have a more Earth-like environment early in its history, with a thicker atmosphere and abundant water that was lost over the course of hundreds of millions of years. Given the foundations of similarity and proximity, Mars would make one of the most efficient and effective terraforming targets in the solar system.

Ethical considerations of terraforming include the potential displacement of indigenous life, even if microbial, if such life exists.

8.8.2.2 Challenges and Limitations

The Martian environment presents several terraforming challenges to overcome, and the extent of terraforming may be limited by certain key environmental factors.

Low Gravity

The surface gravity on Mars is 38 % of that on Earth. It is not known if this is enough to prevent the health problems associated with weightlessness. Additionally, the low gravity (and thus low escape velocity) of Mars may render it more difficult for it to retain an atmosphere when compared to the more massive Earth and Venus. Earth and Venus are both able to sustain thick atmospheres, even though they experience more of the solar wind that is believed to strip away planetary volatiles. Continuing sources of atmospheric gases on Mars might therefore be required to ensure that an atmosphere sufficiently dense for humans is sustained in the long term.

Countering the Effects of Space Weather (Cosmic Rays)

Mars lacks a magnetosphere, which poses challenges for mitigating solar radiation and retaining atmosphere. It is believed that fields detected on Mars are remnants of a magnetosphere that collapsed early in its history.

The lack of a magnetosphere is thought to be one reason for Mars' thin atmosphere. Solar-wind-induced ejection of Martian atmospheric atoms has been detected by Mars-orbiting probes. Venus, however, clearly demonstrates that the lack of a magnetosphere does not preclude a dense atmosphere.

Earth abounds with water because its ionosphere is permeated with a magnetosphere. The hydrogen ions present in its ionosphere move very fast due to their small mass, but they cannot escape to outer space because their trajectories are deflected by the magnetic field. Venus has a dense atmosphere, but only traces of water vapor (20 ppm) because it has no magnetic field. The Martian atmosphere also loses water to space. Earth's ozone layer provides additional protection. Ultraviolet light is blocked before it can dissociate water into hydrogen and oxygen. Because little water vapor rises above the troposphere and the ozone layer is in the upper stratosphere, little water is dissociated into hydrogen and oxygen.

The Earth's magnetic field is 31 μT. Mars would require a similar magnetic field intensity to similarly offset the effects of the solar wind at its distance further from the Sun. The technology for inducing a planetary-scale magnetic field does not currently exist.

The importance of magnetosphere has been brought into question. In the past, Earth has regularly had periods where the magnetosphere changed direction, yet life has continued to survive. A thick atmosphere similar to Earth's could also provide protection against solar radiation in the absence of a magnetosphere.

Advantages

According to modern theorists, Mars exists on the outer edge of the habitable zone, a region of the solar system where liquid water can exist, and so, where life (as we know it) can exist. Mars is on the border of a region known as the extended habitable zone where concentrated greenhouse gases could support the liquid water on the surface at sufficient atmospheric pressure. Therefore, Mars has the potential to support a hydrosphere and biosphere.

The lack of both a magnetic field and geologic activity on Mars may be a result of its relatively small size, which allowed the interior to cool more quickly than Earth's, though the details of such a process are still not well understood.

It has been suggested that Mars once had an environment relatively similar to that of Earth during its earlier stage of development. Although water appears to have once been present on the Martian surface, water appears to exist at the poles just below the planetary surface as permafrost. On September 26, 2013, NASA scientists reported that the Mars Curiosity rover detected abundant, easily accessible water (1.5–3 wt%) in soil samples at the Rocknest region of Aeolis Palus in Gale Crater.

The soil and atmosphere of Mars contain many of the main elements needed for life. Large amounts of water ice exist below the Martian surface, as well as on the surface at the poles, where it is mixed with dry ice and frozen CO_2.

Significant amounts of water are stored in the south pole of Mars, which, if melted, would correspond to a planet-wide ocean 11 m deep. Frozen carbon dioxide (CO_2) at the poles sublimates into the atmosphere during the Martian summers, and small amounts of water residue are left behind, which fast winds sweep off the poles at speeds approaching 400 km/h (250 mph). This seasonal occurrence transports large amounts of dust and water vapor into the atmosphere, giving potential for Earth-like cirrus clouds.

Most of the oxygen in the Martian atmosphere is present as carbon dioxide (CO_2), the main atmospheric component. Molecular oxygen (O_2) only exists in trace amounts. Large amounts of elemental oxygen can be also found in metal oxides on the Martian surface and in the soil, in the form of pernitrates. An analysis of soil samples taken by the Phoenix lander indicated the presence of perchlorate, which has been used to liberate oxygen in chemical oxygen generators. Electrolysis could be employed to separate water on Mars into oxygen and hydrogen if sufficient liquid water and electricity were available.

Proposed Methods and Strategies

Terraforming Mars would entail three major interlaced changes: building up the atmosphere, keeping it warm, and keeping the atmosphere from being lost to outer space. The atmosphere of Mars is relatively thin and has a very low surface pressure. Because its atmosphere consists mainly of CO_2, a known greenhouse gas, once Mars begins to heat, the CO_2 may help to keep thermal energy near the surface. Moreover, as it heats, more CO_2 should enter the atmosphere from the frozen reserves on the poles, enhancing the greenhouse effect. This means that the two processes of building the atmosphere and heating it would augment one another, favoring terraforming.

Comparison of dry atmosphere

	Mars	Earth
Pressure	0.6 kPa (0.087 psi)	101.3 kPa (14.69 psi)
Carbon dioxide (CO_2)	96.0 %	0.04 %
Argon (Ar)	2.1 %	0.93 %
Nitrogen (N_2)	1.9 %	78.08 %
Oxygen (O_2)	0.145 %	20.94 %

The tremendous air currents generated by the moving gases would create large, sustained dust storms, which would heat the atmosphere (by absorbing solar radiation).

Carbon Dioxide Sublimation

There is presently enough carbon dioxide (CO_2) as ice in the Martian south pole, absorbed by regolith (soil) on Mars that, if sublimated to gas by a climate warming of only a few degrees, would increase the atmospheric pressure to 30 kilopascals (0.30 atm), comparable to the altitude of the peak of Mount Everest, where the atmospheric pressure is 33.7 kilopascals (0.333 atm). Although this would not be breathable by humans, it is above the Armstrong limit and would eliminate the present need for pressure suits. Phytoplankton can also convert dissolved CO_2 into oxygen, which is important because Mars' low temperature will, by Henry's law, lead to a high ratio of dissolved CO_2 to atmospheric CO_2 in the flooded northern basin.

Importing Ammonia

Another more intricate method uses ammonia as a powerful greenhouse gas. It is possible that large amounts of it exist in frozen form on minor planets orbiting in the outer solar system. It may be possible to move these and send them into Mars' atmosphere. Because ammonia (NH_3) is mostly nitrogen by weight, it could also supply the buffer gas for the atmosphere. Sustained smaller impacts will also contribute to increases in the temperature and mass of the atmosphere.

The need for a buffer gas is a challenge that will face any potential atmosphere builders. On Earth, nitrogen is the primary atmospheric component, making up 78 % of the atmosphere. Mars would require a similar buffer gas component although not necessarily as much. Obtaining sufficient quantities of nitrogen, argon, or some other comparatively inert gas is difficult.

Importing Hydrocarbons

Another way to create a Martian atmosphere would be to import methane or other hydrocarbons, which are common in Titan's atmosphere (and on its surface). The methane could be vented into the atmosphere where it would act to compound the greenhouse effect.

Methane (or other hydrocarbons) could be helpful to increase atmospheric pressure. These gases also can be used to produce water and CO_2 for the Martian atmosphere:

$$CH_4 + 4Fe_2O_3 \rightarrow CO_2 + 2H_2O + 8FeO$$

This reaction could probably be initiated by heat or by Martian solar UV irradiation. Large amounts of the resulting products (CO_2 and water) are necessary for photosynthesis, which would be the next step in terraforming.

Importing Hydrogen

Hydrogen could be imported for atmosphere and hydrosphere engineering. For example, hydrogen could react with iron(III) oxide from the Martian soil, which would give water as a product:

$$H_2 + Fe_2O_3 \rightarrow H_2O + 2FeO$$

Depending on the level of carbon dioxide in the atmosphere, importation and reaction of hydrogen would produce heat, water, and graphite via the Bosch reaction. Alternatively, reacting hydrogen with the carbon dioxide atmosphere via the Sabatier reaction would yield methane and water.

Use of Fluorine Compounds

Because long-term climate stability would be required for sustaining a human population, the use of especially powerful fluorine-bearing greenhouse gases, possibly including sulfur hexafluoride or halocarbons such as chlorofluorocarbons (or CFCs) and perfluorocarbons (or PFCs), has been suggested. These gases are the most cited candidates for artificial insertion into the Martian atmosphere because they produce a strong effect as a greenhouse gas, thousands of times stronger than CO_2. This can conceivably be done relatively cheaply by sending rockets with payloads of compressed CFCs on collision courses with Mars. When the rockets crash onto the surface, they release their payloads into the atmosphere. A steady barrage of these "CFC rockets" would need to be sustained for a little over a decade while Mars changes chemically and becomes warmer.

In order to sublimate the south polar CO_2 glaciers, Mars would require the introduction of approximately 0.3 microbars of CFCs into Mars' atmosphere. This is equivalent to a mass of approximately 39 million metric tons. This is about three times the amount of CFC manufactured on Earth from 1972 to 1992 (when CFC production was banned by international treaty).

Mineralogical surveys of Mars estimate the elemental presence of fluorine in bulk composition of Mars at 32 ppm by mass vs. 19.4 ppm for the Earth. A proposal to mine fluorine-containing minerals as a source of CFCs and PFCs is supported by the belief that because these minerals are expected to be at least as common on Mars as on Earth, this process could sustain the production of sufficient quantities of optimal greenhouse compounds (CF_3SCF_3, $CF_3OCF_2OCF_3$, $CF_3SCF_2SCF_3$, $CF_3OCF_2NFCF_3$, $C_{12}F_{27}N$) to maintain Mars at "comfortable" temperatures, as a method of maintaining an Earth-like atmosphere produced previously by some other means.

Use of Orbital Mirrors

Mirrors made of thin aluminized PET film could be placed in orbit around Mars to increase the total insolation it receives. This would direct the sunlight onto the surface and could increase Mars' surface temperature directly. The mirror could be positioned as a statite, using its effectiveness as a solar sail to orbit in a stationary position relative to Mars, near the poles, to sublimate the CO_2 ice sheet and contribute to the warming greenhouse effect.

Albedo Reduction

Reducing the albedo of the Martian surface would also make more efficient use of incoming sunlight. This could be done by spreading dark dust from Mars' moons, Phobos and Deimos, which are among the blackest bodies in the solar system, or by introducing dark extremophile microbial life forms such as lichens, algae, and bacteria. The ground would then absorb more sunlight, warming the atmosphere.

If algae or other green life were established, it would also contribute a small amount of oxygen to the atmosphere, though not enough to allow humans to breathe. The conversion process to produce oxygen is highly reliant upon water. The CO_2 is mostly converted to carbohydrates. On 26 April 2012, scientists reported that lichen survived and showed remarkable results on the adaptation capacity of photosynthetic activity within the simulation time of 34 days under Martian conditions in the Mars Simulation Laboratory (MSL) maintained by the German Aerospace Center (DLR).

Asteroid Impact

Another way to increase the temperature would be to direct small asteroids onto the Martian surface. This could be achieved through the use of space-borne lasers to alter trajectories or other methods proposed for asteroid impact avoidance. The impact energy would be released as heat. This heat could sublimate CO_2 or, if there is liquid water present at this stage of the terraforming process, could vaporize it to steam, which is also a greenhouse gas. Asteroids could also be chosen for their composition, such as ammonia, which would then disperse into the atmosphere on impact, adding greenhouse gases to the atmosphere. Lightning may have built up nitrate beds in Mars' soil. Impacting asteroids on these nitrate beds would release additional nitrogen and oxygen into the atmosphere.

Thermodynamics of Terraforming

The overall energy required to sublimate the CO_2 from the south polar ice cap is modeled by raising the temperature of the poles by four Kelvin which would be necessary in order to trigger a runaway greenhouse effect. If using orbital mirrors, an estimated 120 MWe-years would be required in order to produce mirrors large enough to vaporize the ice caps. This is considered the most effective method, though the least practical.

If using powerful halocarbon greenhouse gases, an order of 1,000 MWe-years would be required to accomplish this heating. Although ineffectual in comparison, it is considered the most practical method. Impacting an asteroid, which is often considered a synergistic effect, would require approximately four 10-billion-ton ammonia-rich asteroids to trigger the runaway greenhouse effect, totaling an 8° increase in temperature.

8.8.2.3 New Proposals for Mars Terraforming

I make two proposals (suggestions) for Mars terraforming, and they are outlined in the form of research steps – firstly, aboard Earth's International Space Station

(ISS), and secondly, on the planet Mars' surface – which complement themselves. The human objective for Mars is its colonization for all nations' use, and it will only be satisfactorily accomplished through the union of the space agencies, research centers, universities, governments, and people who can contribute and efficiently work for this future, possible, and viable manned Martian exploration and, ultimately, colonization (de Morais 2004).

To achieve that, the future international manned exploration of planet Mars will require some independency of food and oxygen supplies to the crews on Mars. Vegetables and resistant microbes, growing inside secure vessels on the Martian ground, are the best candidates for supplying a quasi-continuous production of proteins, salt minerals, liquid water, and oxygen to the astronauts working in a semipermanent possible future living facility at Mars' surface (de Morais 2004).

To colonize Mars, it is necessary to change its atmosphere (mainly composed by carbon dioxide (CO_2) at very low pressure) into one more similar to Earth's (with oxygen at higher pressure), to be more habitable by humans – Mars terraforming. This can be done by introducing, within the Martian soil, very resistant microbes (and later, cacti) and heating the planet's atmosphere to liberate liquid water from subsurface permafrost (McKay et al. 1991a, b), for them to produce oxygen by photosynthesis (de Morais 2004).

Some microorganisms found on Earth can survive and grow in the present-day harsh conditions found on Mars – temperatures well below zero degrees centigrade; very dry soil; very thin CO_2 atmosphere, which makes ultraviolet (UV) radiation from the Sun strong; and global dust storms that take several Martian months (1 Martian month is equal to 0.9 terrestrial month, and there are 24 Martian months) blocking part of the Sun's infrared (IR) and visible light, which are four times less intense than on Earth's surface (de Morais 2004).

But before doing that, it is fundamentally necessary for a long-term (some years of) primary Earth study of the strong influences of such mentioned lower levels of IR and visible light and higher levels of ultraviolet, CO_2 atmosphere at low pressure, temperatures below zero degrees centigrade, very dry Mars-like soil and the lower gravity (as compared to Earth's one) field present on Mars, on microbes' and vegetables' growth to see if they can develop satisfactorily (de Morais 2004).

Ground experiments need to be done to simulate such influent characteristics found on Mars. The only one characteristic which cannot be simulated on the ground is the Martian 0.38 g gravity field. Studies of microgravity (near zero-g) environment aboard satellites in Earth's orbit, on microbes and plants, show strong effects on their growth and development (de Morais 2004).

Thus, there is a fundamental and strong necessity for a 0.38-g gravity simulation with vegetables and resistant microbes on Earth's orbit to get scientific information on which species will have optimal growth under the real Martian gravity environment, during the international manned missions on that planet (for two main objectives, the growth of plants for producing oxygen, food, and liquid water to the astronauts on Mars and the growth of microbes for producing oxygen by photosynthesis in a greenhouse for a controlled Martian terraforming) and to acquire technical and operational data by astronauts on how to accomplish these on Mars (de Morais 2004).

So, my proposals have two objectives – first, to suggest this gravity simulation in facilities on board the ISS, the only available place on Earth for a Mars gravity simulation, and second, to suggest the construction of a simple greenhouse facility on Mars for the growing of microbes and plants to produce oxygen and liquid water within it, for a possible future terraforming of that planet (de Morais 2004).

Regarding those facilities on board the ISS, there are already data and results available. These two suggestions are for the future, based on ground studies. Possibly, this initial international manned Martian exploration will give place to a future complex exploration and transformation of Martian characteristics for human use, a Mars terraforming (de Morais 2004).

As mentioned above, for Mars terraforming, there is the possibility of future introduction of some very resistant species of plants on Mars for the beginning of a future long (several decades), gradual change of the Martian environment (e.g., Martian soil and atmospheric temperature profiles close to the ground), increasing in some degrees the air and soil's temperatures for the transformation of sites with permafrost into liquid water and the release of oxygen molecules into the atmosphere, for future human use. To do this, we need first to understand how the production of oxygen is on Earth, since it can be somewhat analogous to a future terraforming of Mars (de Morais 2004).

Studies on the net production of oxygen by the phytoplankton on Earth's oceans show that the oceans contribute 2/3 of the oxygen content in Earth's atmosphere (Melillo et al. 1993), with a total oceanic primary production of 47.5 Pg C/year, with the global seasonal variation production data, March–May, 10.9 Pg C/year; June–August, 13.0 Pg C/year; September–November, 12.3 Pg C/year; and December–February: 11.3 Pg C/year (Behrenfeld and Falkowski 1997).

Then, with the knowledge of which photosynthetic microbes and plants for generating molecular free oxygen are resistant to present-day Mars, we will be able to begin studies on how to gradually change Mars to a developed state (liquid water and thicker oxygen atmosphere) capable of sustaining future human life. Good initial candidates for very resistant organisms can be cyanobacteria, microscopic algae, lichen, and cacti seeds (CAM-type plants, e.g., flowering plant species which utilize crassulacean acid metabolism for the fixation of CO_2) (de Morais 2004).

NASA–Ames Research Center has some studies on the two types of microscopic algae, which grow in sand in hard conditions, *Microcalens* sp. and *Escillatoria* sp., and their productivity are 100 mm CO_2/m^2/day. Equations (1) and (2) below (Benz et al. 1997) give the influence of radiation on photosynthesis – the influence of radiation on the photosynthetic partial rate assimilation process (r_Q) is described by a saturation function between the photocompensation point (Q_{MIN}), which depends on the ontogenetical (the genetic behavior of the organism during its life cycle) stage of the cultivation, and the saturation point (Q_{MAX}) of photosynthetic active radiation (Q_P); the influence of temperature on photosynthesis partial rate (r_T) depends on ontogenesis and equals to 1 if the value of phototemperature is at the optimum temperature (T_{OPT}) (de Morais 2004).

8 Mars Astrobiology: Recent Status and Progress

The minimum temperature (T_{MIN}), the maximum temperature (T_{MAX}), and the optimum temperature (T_{OPT}) depend on the ontogenetical stage of the cultivation, but the structure of the partial rate function is unchanged with the stage (de Morais 2004).

It is known by electromagnetism (Jackson 1999) that the quantity of received radiation follows an inverse square law of the distance from the source, in this case the Sun, to the object. So, by a simple calculation, it is shown that Mars' surface receives 0.4 Q_P of the one at Earth's surface. Such calculation is achieved by the following equations (de Morais 2004):

$$r_Q \begin{cases} \left[\dfrac{(Q_P(t)-Q_{MIN})^{x_1}}{(Q_P(t)-Q_{MIN})^{x_1}+Q_S} \right]^{x_2} & \text{if } Q_P(t) \leq Q_{MIN} \\ \dfrac{(Q_{MAX}(t)-Q_{MIN})^{x_1}}{(Q_{MAX}(t)-Q_{MIN})^{x_1}+Q_S} & \text{if } Q_{MIN} \leq Q_P(t) \leq Q_{MAX} \\ 1 & \text{if } Q_{MAX} \leq Q_P(t) \end{cases}$$

$$r_T = \begin{cases} 0 \\ \dfrac{R_{PH}(T_L, T_{OPT}) - R_{PH}(T_{MIN}, T_{OPT})}{1 - R_{PH}(T_{MIN}, T_{OPT})} \\ \dfrac{R_{PH}(T_L,T_{OPT})}{} \end{cases} \begin{cases} \text{if } T_L(t) \notin \{T_{MIN}, T_{MAX}\} \\ \text{if } T_L(t) - T_{OPT} < 0 \\ \text{if } T_L(t) - T_{OPT} \geq 0 \end{cases}$$

Nomenclature

Q_{MAX} = Saturation radiation for the influence of radiation on photosynthesis ($(kJ)/(m^2 d)$)

Q_{MIN} = Photocompensation point – minimal value for the influence of radiation on photosynthesis ($(kJ)/(m^2 d)$)

Q_P = Input of photosynthetic active radiation ($(kJ)/(m^2 d)$)

Q_S = Half photosaturation constant ($(kJ)/(m^2 d)$)

x_1 = Curving exponent to determine the influence of radiation on photosynthesis

x_2 = Combining exponent to determine the influence of radiation on photosynthesis

r_Q = Partial photosynthetic radiation rate

R_{ph} = Specific internal temperature function – describes the influence of temperature on ontogenesis

T_{COR} = Correction of optimum temperature ($(m^2 d)/(kJK)$)

T_L = Mean temperature during light period (°C)

T_{MIN} = Minimal temperature for cultivation development, depends on the ontogenetical stage (°C)

T_{MAX} = Maximal temperature for cultivation development, depends on the ontogenetical stage (°C)

T_{OPT} = Optimal temperature for cultivation development, depends on the ontogenetical stage (°C)

r_T = Partial photosynthetic rate by the influence of the phototemperature

Taking the influence of temperature (r_T) into the photosynthetic assimilation process, then the photosynthetic productivity on Mars will be below 0.4 r_Q, showing that although the organisms for cultivation will display a much lower metabolism, they will be able to grow and develop at that IR and visible light radiation levels found on Mars. But this is not the only factor to influence growth; gravity also strongly affects the photosynthetic organisms' growth and development; thus, it is fundamentally necessary to study gravity effects on them, firstly, simulating a Martian 0.38 g aboard the ISS for gaining data, results, and operational experience and, later, on real Mars environment (de Morais 2004).

There are no data on growth effects on microbes and plants within gravitational fields such as the 0.38 g of Mars. There is, however, a large body of research on the effects of 0 g on them from experiments flown aboard rockets, the Apollo program, satellites (such as the Cosmos series, the Long Duration Exposure Facility, the BIOPAN facility, etc.), the Space Shuttle, the MIR space station, and the International Space Station (ISS) (EXPOSE facility, etc.) (de Morais 2004).

Life Under Simulated Martian Conditions: Experiments

On 26 April 2012, scientists reported that an extremophile lichen survived and showed remarkable results on the adaptation capacity of photosynthetic activity within the simulation time of 34 days under Martian conditions in the Mars Simulation Laboratory (MSL) maintained by the German Aerospace Center (DLR). However, the ability to survive in an environment is not the same as the ability to thrive, reproduce, and evolve in that same environment, necessitating further study.

EXPOSE

EXPOSE was a multiuser facility mounted outside the International Space Station (ISS) dedicated to astrobiology. EXPOSE was developed by the European Space Agency (ESA) for long-term space flights and was designed to allow exposure of chemical and biological samples to outer space while recording data during exposure.

The results will contribute to our understanding of photobiological processes in simulated radiation climates of planets (e.g., early Earth, early and present Mars, and the role of the ozone layer in protecting the biosphere from harmful UVB radiation), as well as studies of the probabilities and limitations for life to be distributed beyond its planet of origin.

EXPOSE data support long-term in situ studies of microbes in artificial meteorites, as well as of microbial communities from special ecological niches. Some EXPOSE experiments investigated to what extent particular terrestrial organisms are able to cope with extraterrestrial environmental conditions. Others tested how organic molecules react when subjected for a prolonged period of time to unfiltered solar light.

Relevance

With the experiments on board of the EXPOSE facilities, various aspects of astrobiology were investigated that could not be sufficiently approached by the use of laboratory facilities on ground. The chemical set of experiments is designed to reach a better understanding of the role of interstellar, cometary, and planetary chemistry in the origin of life. Comets and meteorites are interpreted as exogenous sources of prebiotic molecules on the early Earth. All data achieved from the astrobiological experiments on both EXPOSE missions will add to the understanding of the origin and evolution of life on Earth and on the possibility of its distribution in space or origin elsewhere, as planet Mars.

The biology experiments used the full extraterrestrial spectrum of solar UV radiation and suitable cutoff filters to study both the role of the ozone layer in protecting our biosphere and the likelihood of resistant terrestrial microorganisms (extremophiles) to survive in outer space. The latter studies will provide experimental data to the lithopanspermia hypothesis, and they will provide basic data to planetary protection issues. To get better insight into the habitability of Mars, one set of samples was exposed to simulated Martian conditions (UV radiation climate, pressure, atmosphere), with and without a protective cover of simulated Martian soil. The biological test samples selected are hardy representatives of various branches of life.

General Mission Description

There were two EXPOSE facilities, EXPOSE-E and EXPOSE-R. The EXPOSE-E was launched on February 7, 2008, on board the Space Shuttle Atlantis and was mounted on the ISS European module Columbus to the European Technology Exposure Facility (EuTEF). EXPOSE-R was launched to the ISS on November 26, 2008, from Baikonur in Kazakhstan on board of a Progress capsule and was mounted on the ISS Russian module Zvezda. EXPOSE-E provided accommodation in three exposure trays for a variety of astrobiological test samples that were exposed to selected space conditions: either to space vacuum, to solar electromagnetic radiation at >110 nm and cosmic radiation (trays 1 and 3), or to simulated Martian surface conditions (tray 2) (Baqué et al. 2013). The different experiments consisted of exposing solid molecules, gas mixtures, or biological samples to solar ultraviolet (UV) radiation, cosmic rays, vacuum, and temperature fluctuations of the outer space as the ISS repeatedly passed between areas of direct sunlight and the cold darkness of Earth's shadow.

At the end of the exposition period, EXPOSE-E was brought back to the ground in September 2009 as part of the Space Shuttle Discovery mission STS-128. EXPOSE-R was brought back in 2011 by a Soyuz spacecraft. From the landing site in Kazakhstan, the trays were returned via Moscow and distributed to scientists for further analysis in their laboratory.

EXPOSE-R was equipped with three trays housing eight experiments and 3 radiation dosimeters. Each tray was loaded with a variety of biological organisms including plant seeds and spores of bacteria, fungi, and ferns that were exposed to the harsh space environment for about one and a half years. The ROSE (response of organisms to space environment) group of experiments is under the coordination of the German Aerospace Center (DLR) and has been composed of scientists from different European countries, from USA, and from Japan. In its 8 experiments of biological and chemical content, more than 1,200 individual samples were exposed to solar ultraviolet (UV) radiations, vacuum, cosmic rays, or extreme temperature variations. In their different experiments, the involved scientists are studying the question of life's origin on Earth, and the results of their experiments are contributing to different aspects of the evolution and distribution of life in the universe.

EXPOSE-E Results

PROCESS Experiment

The search for organic molecules at the surface of Mars is a top priority of Mars exploration space missions. Therefore, a key step in interpretation of future data collected by these missions is to understand the preservation of organic matter in the Martian environment. A 1.5-year exposure to Mars-like surface UV radiation conditions in space resulted in complete degradation of the organic compounds (glycine, serine, phthalic acid, phthalic acid in the presence of a mineral phase, and mellitic acid). Their half-lives were between 50 and 150 h for Martian surface conditions.

To understand the chemical behavior of organic molecules in the space environment, amino acids and a dipeptide in pure form and embedded in meteorite powder were exposed to space conditions for 18 months; the samples were returned to Earth and analyzed in the laboratory for reactions caused by solar UV and cosmic radiation. The results show that resistance to irradiation is a function of the chemical nature of the exposed molecules and the wavelengths of the UV light. The most altered compounds were the dipeptide, aspartic acid, and aminobutyric acid. The most resistant were alanine, valine, glycine, and aminoisobutyric acid. The results also demonstrate the protective effect of meteorite powder, which reemphasizes the importance of exogenic contribution to the inventory of prebiotic organics on early Earth.

ADAPT Experiment

Bacterial endospores of the highly UV-resistant *Bacillus subtilis* strain MW01 were exposed to low-Earth orbit and simulated Martian surface conditions for 559 days. It was clearly shown that solar extraterrestrial UV radiation ($\lambda \geq 110$ nm) as well as the Martian UV spectrum ($\lambda \geq 200$ nm) was the most deleterious factor applied; in some samples, only a few spore survivors were recovered from *Bacillus subtilis* MW01 spores exposed in monolayers. However, if shielded from solar irradiation, about 8 % of MW01 spores survived, and 100 % survived in simulated Martian conditions, compared to the laboratory controls.

PROTECT Experiment

Spore-forming bacteria are of particular concern in the context of planetary protection because their tough endospores may withstand certain sterilization procedures as well as the harsh environments of outer space or planetary surfaces. To test their hardiness on a hypothetical mission to Mars, spores of *Bacillus subtilis* 168 and *Bacillus pumilus* SAFR-032 were exposed for 1.5 years to selected parameters of space. It was clearly shown that solar extraterrestrial UV radiation ($\lambda \geq 110$ nm) as well as the Martian UV spectrum ($\lambda \geq 200$ nm) was the most deleterious factor applied; in some samples, only a few survivors were recovered from spores exposed in monolayers.

Spores in multilayers survived better by several orders of magnitude. All other environmental parameters encountered did little harm to the spores, which showed about 50 % survival or more. The data demonstrate the high chance of survival of spores on a Mars mission, if protected against solar irradiation. These results will have implications for planetary protection considerations.

The mutagenic efficiency of space was also studied in spores of *Bacillus subtilis* 168. The data show the unique mutagenic power of space and Martian surface conditions as a consequence of DNA injuries induced by solar UV radiation and space vacuum or the low pressure of Mars. Spores exposed to space demonstrated a much broader and more severe stress response than spores exposed to simulated Martian conditions.

A comparative protein analysis (proteomics) of *Bacillus pumilus* SAFR-032 spores indicated that proteins conferring resistant traits (superoxide dismutase) were present in higher concentration in space-exposed spores when compared to controls. Also, the first-generation cells and spores derived from space-exposed samples exhibited elevated ultraviolet C resistance when compared with their ground control counterparts. The data generated are important for calculating the probability and mechanisms of microbial survival in space conditions and assessing microbial contaminants as risks for forward contamination and in situ life detection.

LiFE Experiment

After 1.5 years in space, samples were retrieved, rehydrated, and spread on different culture media. The only two organisms able to grow were isolated from a sample exposed to simulated Mars conditions beneath a 0.1 % T Suprasil neutral density filter and from a sample exposed to space vacuum without solar radiation exposure, respectively. The two surviving organisms were identified as *Stichococcus* sp. (green algae) and *Acarospora glaucocarpa* sp. (lichened fungal genus). According to the researchers, the studies provide experimental information on the possibility of eukaryotic life transfer from one planet to another by means of rocks and of survival in Mars' environment.

Cryptoendolithic microbial communities and epilithic lichens have been considered as appropriate candidates for the scenario of lithopanspermia, which proposes a natural interplanetary exchange of organisms by means of rocks that have been impact ejected from their planet of origin. A 1.5-year exposure experiment in space was performed with a variety of rock-colonizing eukaryotic organisms.

Selected organisms are known to cope with the environmental extremes of their natural habitats. It was found that some – but not all – of those most robust microbial communities from extremely hostile regions on Earth are also partially resistant to the even more hostile environment of outer space, including high vacuum, temperature fluctuation, the full spectrum of extraterrestrial solar electromagnetic radiation, and cosmic ionizing radiation. Although the reported experimental period of 1.5 years in space is not comparable with the time span of thousands or millions of years believed to be required for lithopanspermia, the data provide the first evidence of the differential hardiness of cryptoendolithic communities in space.

SEEDS Experiment

The plausibility that life was imported to Earth from elsewhere was tested by subjecting plant seeds to 1.5 years of exposure to solar UV, solar and galactic cosmic radiation, temperature fluctuations, and space vacuum outside the International Space Station. Of the 2100 exposed wild-type *Arabidopsis thaliana* and *Nicotiana tabacum* (tobacco) seeds, 23 % produced viable plants after being returned to Earth. Germination was delayed in seeds shielded from solar light, yet full survival was attained, which indicates that longer space travel would be possible for seeds embedded in an opaque matrix. The team concluded that a naked, seed-like entity could have survived exposure to solar UV radiation during a hypothetical transfer from Mars to Earth, and even if seeds do not survive, components (e.g., their DNA) might survive transfer over cosmic distances.

The radiation dose during the mission was 1823.98 MJ m^{-2} for PAR, 269.03 MJ m^{-2} for UVA, 45.73 MJ m^{-2} for UVB, or 18.28 MJ m^{-2} for UVC. Registered sunshine duration during the mission was about 152 days (about 27 % of mission time). The surface of EXPOSE was most likely turned away from the Sun for considerably longer time. The highest daily averaged absorbed dose rate of 426 µGy per day came from the "South Atlantic Anomaly" (SAA) region of the inner radiation belt; galactic cosmic rays (GCR) delivered a daily absorbed dose rate of 91.1 µGy per day, and the outer radiation belt (ORB) source delivered 8.6 µGy per day.

EXPOSE-R Results

TROPI Experiment

TROPI, or "Analysis of a Novel Sensory Mechanism in Root Phototropism," is an experiment on the International Space Station (ISS) to investigate the growth and development of plant seedlings under various gravity and lighting combinations. It was launched on Space Shuttle Endeavour during the STS-130 mission and was performed on the ISS during Expedition 22. Frozen plant samples from the TROPI experiment were returned on the landing of the STS-131 mission on Space Shuttle Discovery.

Arabidopsis thaliana seeds (thale cress, the genome of which has been DNA sequenced as a reference organism for the study of plant biology and genetics) were germinated and grown under various lighting and gravity conditions, using centrifugal gravity simulation and LEDs of various wavelengths (colors) and intensities to model lighting conditions. The specific aim of this project was to investigate phototropism in plants grown in microgravity conditions without the complications of a 1-g environment. Experiments performed were used to explore the mechanisms of both blue-light- and red-light-induced phototropism in plants.

John Z. Kiss of Miami University (Oxford OH) is the principal investigator; Richard E. Edelmann of Miami University and Melanie J. Correll of the University of Florida are coinvestigators; Kenny Vassigh of NASA is the project manager, and Marianne Steele of NASA is the project scientist. The payload was developed by the NASA Ames Research Center, Moffett Field, California. The experiment was performed in the European Modular Cultivation System (EMCS) built by the European Space Agency (ESA). The Norwegian User Support Operation Centre (N-USOC), located in Trondheim, Norway, controlled the EMCS during the TROPI experiments on the ISS.

In the long term, the results from TROPI will help in the development of future space, Moon, and Mars life-support systems, in which plants are used to help remove carbon dioxide and generate oxygen via photosynthesis for maintenance of atmospheric and other conditions, reducing the need for very expensive resupply from Earth.

Zero-g strong effects became apparent when plants were germinated and grown in space: plants do grow under microgravity conditions but not normally and the problems become more severe the longer the duration. When plants are grown during space flight, roots are typically disoriented while shoots may orient to a light source. Root and shoot growth is often less in space, but the rate of development and aging may be faster. Deterioration and death of plants are common. Lignin and cellulose are reduced with thinner cell walls resulting. And there is altered flow of water and minerals around the roots and through the plant and out the leaves (Raven et al. 1986). But those studies are only through 0 g (de Morais 2004).

All influent characteristics of Mars, except its gravity, can be simulated on Earth's ground. There are no ways for doing these fundamentally necessary long-term studies with 0.38 g using the present-day space vehicles; the only capable one is the ISS, which accommodates two biological facilities built by ESA – the European Modular Cultivation System (EMCS) and the Biological Experiment Laboratory (BioLab) (de Morais 2004).

EMCS was launched to the ISS on the STS-121 mission in July 2006. BioLab was pre-installed inside the *Columbus* laboratory, and Space Shuttle *Atlantis* on ISS Assembly Flight 1E, mission STS-122, successfully delivered the *Columbus* module to the ISS on February 9, 2008.

One of the science objectives of the EMCS and the BioLab is the provision of a long-term study of growth and early development stability of biological samples. The EMCS and the BioLab are, respectively, automated and semi-automated hardware. Following it is a brief description of these two hardwares:

EMCS

The EMCS is basically composed by an incubator, an atmospheric control system of the incubator, and two centrifuges inside it both equipped with interfaces to four experiment containers each. The incubator has a thermal control system which provides selectable temperature range between +18 and +40 °C, which smaller value can be decreased to below 0 °C by a small modification of the electrical settings. The two centrifuges with diameter of 600 mm and variable gravity conditions ($0.00\times$ g up to $2\times$ g) are located inside the incubator. The following components can be accommodated in each centrifuge: four experiment containers, four life support system modules, two movable video cameras, and four movable mirrors for internal observations. The internal volume of EMCS available for experiment hardware and specimens is 0.58 l; the height is 160 mm. Liquids, nutrients, humidity, pressure, gas, and light systems are efficiently controlled. Electrical and data systems are simple and efficient.

BioLab

The BioLab is a single-rack multiuser science payload designed for use in the *Columbus* laboratory of the International Space Station (ISS). BioLab supports biological research on small plants, small invertebrates, microorganisms, animal cells, and tissue cultures. It includes an incubator equipped with centrifuges in which the preceding experimental subjects can be subjected to controlled levels of accelerations. These experiments help to identify the role that microgravity plays at all levels of an organism, from the effects on a single cell up to a complex organism including humans.

The BioLab is basically composed by two sections: automatic and manual. The automatic section allows performance of the experiments after manual loading by the crew. It is composed by an incubator, two centrifuges inside it, a robotic handling mechanism, and an automatic temperature-controlled stowage which can handle samples within temperatures ranging from −20 to +10 °C, an automatic ambient stowage, and analysis instruments.

The incubator has a thermal control system which provides selectable temperature range between +18 and +40 °C, in which smaller value can be decreased to below 0 °C by a small modification of the electrical settings. The two centrifuges with a diameter of 600 mm and variable gravity conditions ($0.001\times$ g up to $2\times$ g) are located inside the incubator. The manual section is mainly devoted to experiment control and the storage of biological samples such as for photosynthetic life growth.

It is equipped with the following units: temperature-controlled units, with temperatures ranging from −20 to +10 °C, a BioGloveBox, a crew commanding system – a laptop computer and video display for monitoring experiments, and the experiment containers, for development and study of biological samples as photosynthetic life and for the accommodation of experiment specific hardware. The internal volume of BioLab available for experiment hardware and specimens is 0.36 l, height is 60 mm. Liquids, nutrients, humidity, pressure, gas, and light systems are efficiently controlled. Electrical and data systems are also simple and efficient.

The engineering aspects for the development and operation of the (small) containers and related life support facilities can be made easily, using the experience gained with the operational work by astronauts during experiments already flown aboard space shuttles and nowadays in the ISS (de Morais 2004).

The first objective of the following proposal is that I suggest the study of effects of Mars' 0.38 g gravitational field on very resistant, primitive microbes – extremophiles (such as *Serratia liquefaciens* (Schuerger et al. 2013)), methanogens, cyanobacteria, the algae *Microcalens* sp., *Escillatoria* sp., and other such very resistant algae, lichens, and plant seeds (cacti, wheat, potatoes, and other vegetables), to determinate if they can develop normally at that field strength, by the use of variable centrifuges inside the EMCS hardware and inside the BioLab facility, or in any other similar hardware, on board the ISS Biological Research Facility. As stated above in the text, there is a strong necessity of that for the manned Mars exploration and its possible future colonization (de Morais 2004).

A Special Topic About Lichens

Lichens are great colonizers. Lichen is a composite organism in which a fungus is paired with a photosynthesizing partner (either green algae or cyanobacteria). The thallus of the lichen (which makes up the organism's body) is very different from either the fungal or algal components living on their own. The fungi surround and hold up the algae by sinking tendrils through the algal cell walls (in much the same manner parasitic fungi attack their hosts). By sharing the resources of the two different partners, the organism is capable of surviving extreme desiccation, and, when the lichen is again exposed to moisture, a flood of nutrients becomes available to both partners.

The partnership makes for an extraordinarily resilient organism which can be found everywhere on land from the rainforests to the deserts to the highest mountains to the harsh frozen rocks of Antarctica. The European Space Agency (ESA) explored the durability of lichen by putting living specimens in direct contact with outer space where the organisms were exposed to vacuum, wide fluctuations of temperature, and the complete spectrum of solar UV light and bombarded with cosmic radiation. During the Foton-M2 mission, which was launched into low-Earth orbit on May 31, 2005, the lichens *Rhizocarpon geographicum* and *Xanthoria elegans* were exposed for a total of 14.6 days before being returned to Earth. Analysis postflight showed a full rate of survival and an unchanged ability for photosynthesis.

Lichens' strange partnership also creates strange morphological forms. In many circumstances, these organisms resemble exotic corals, sponges, or plants. Additionally, many lichens are brightly colored. The result is often beautiful miniature landscapes.

Since it involves both algae and fungi, lichen reproduction can be complicated and takes many different forms depending on the species and the circumstance. Some lichens form *soredia*, small groups of algal cells surrounded by fungal filaments which are dispersed as a group by wind.

Others produce *isidia*, elongated outgrowths from the thallus which break away. During the dry season, certain lichens crumble into dusty flakes which are blown across the landscape. When the rains come, the flakes burst into full growths. In the most interesting and complicated pattern of reproduction, the fungal portion of the lichen produces spores (as a result of sexual exchange and meiosis); these spores are disseminated across the landscape, and then they must find compatible algae or cyanobacteria with which to partner.

Lichens are probably long lived, and what is certain is that they are one of life's most efficient colonizers: in areas such as the Atacama Desert and Antarctica, plants cannot grow unless lichen lived there previously.

On April 26, 2012, scientists reported that lichen survived and showed remarkable results on the adaptation capacity of photosynthetic activity within the simulation time of 34 days under Martian conditions in the Mars Simulation Laboratory (MSL) maintained by the German Aerospace Center (Deutsches Zentrum für Luft- und Raumfahrt – DLR) (Fig. 8.34) (Meeβen et al. 2013; de Vera et al. 2014).

And to better study the possible living conditions of future lichens on Mars, we need to – besides other organisms – perform simulations on the lichens' growth taxes under Martian gravity, 0.38 g, on board the International Space Station (ISS), as stated in this subchapter (de Morais 2004). These gravity simulators could be used for this objective during selected months-period shifts, to conduct the initial experiments using the NASA–Johnson Space Center's Mars Soil Analog (JSC-1), under 6×10^{-3} atm atmospheric pressure with pure CO_2, at 0 °C inside a secure chamber, during 1 year, or less, under about 800 μmol quanta$\cdot m^{-2}$ s^{-1} (1,370 Wm^2) of photosynthetic active radiation for the growing of the abovementioned bacteria, algae, lichens, and plant seeds, with ultraviolet radiation with wavelengths between 200 and 300 nm and with an energy of about 7×10^{-4} Jcm$^{-2} \cdot s^{-1}$, in a step-by-step approach, and simulating temperature and light variations of the Martian seasons (de Morais 2004).

These numbers (which can be altered for better simulation parameters) above are to simulate conditions found by robotic spacecraft on regions of Mars with good sites for the international manned bio-exploration and Martian terraforming: the northern hemisphere close to the equator, near to canyons, possible ancient liquid water quiet-flood valleys, depressions, caves, and volcanic systems are best places for astronauts to keep a semipermanent simple facility (with controlled closed growing of vegetables using Martian soil in a hydroponics system) to search for possible extinct (and hypothetical extant) microbial mats, in the form of Earth's

Fig. 8.34 The lichen *Pleopsidium chlorophanum*, which is an extremophile, collected from Antarctica were placed inside the German Aerospace Center (Deutsches Zentrum für Luft- und Raumfahrt – DLR)'s Mars Simulation Laboratory for 34 days. It was exposed to harsh Mars-like conditions in the laboratory and have been found to survive, preferring to cling to cracks in rocks and in gaps in the simulated Martian soil (http://www.dlr.de/dlr/en/desktopdefault.aspx/tabid-10081/151_read-3409/#/gallery/5681 – Image: DLR, CC-BY 3.0, 2012)

Archaean microfossil records (Schopf and Walter 1983; de Morais 2004), as the Earth's stromatolites analog (Walter 1983; Stoker 1996).

When Mars' environment was wetter than today, with possible liquid water flowing on its surface, or even small oceans (Head et al. 1999), perhaps it gave opportunity for microbial life to possibly have originated within Martian subsurface hydrothermal systems, as Earth's active Fe–S accumulations systems (Libes 1992), on distant-past Mars around 3–4 Gyrs, and became fossilized, or hypothetically is still dormant beneath possible ice-covered small lakes deep inside subsurface Martian ice (McKay 1993; Mitrofanov et al. 2003; de Morais 2004), as Mars dried up during its evolution (McKay and Stoker 1989; McKay et al. 1991a, b).

On Earth, there is a place which gets somewhat close, in terms of environmental conditions, to these possible Martian subsurface ice-covered small lakes. It is the McMurdo Dry Valley region of southern Victoria Land in Antarctica, the largest ice-free region of Antarctica and one of the coldest and driest deserts in the world. There are numerous meltwater lakes in this region thanks to two regional features, the Transantarctic Mountain Range that blocks the flow of ice from the polar plateau and eliminates precipitation and the very low mean air temperature (-15 to -30 °C) that provides perennial ice cover, 3–6 m thick, to the lakes.

"In November 2004, NIWA scientists Kay Vopel and Ian Hawes went under the ice to measure, for the first time, the rates of photosynthesis of microbial mats within one of the permanently ice-covered lakes of the McMurdo Dry Valleys.

Fig. 8.35 Underwater photograph of a microbial mat at 8 m depth in Lake Hoare, McMurdo Dry Valleys, Antarctica. It is not known why the microbes form this pinnacled surface, but it is speculated it could be an adaptation to low light and stagnant water (http://aut.researchgateway.ac.nz/bitstream/handle/10292/1449/1801.pdf?sequence=5 – NIWA/Image Copyright by: R. Ellwood, 2004)

Using underwater micro-sensors, they were able to prove that, during the Antarctic summer, mats in Lake Hoare are capable of using extremely low ambient light to fix carbon and to release oxygen into the already supersaturated lake water. With their measurements, those scientists have generated new insights into the physiology of photosynthesis under low-light and near-freezing conditions, and confirmed key aspects of photosynthesis previously based only on laboratory measurements. These findings greatly advance our understanding of the ecology of ice-covered polar lakes, and help illuminate the mats' secret geological record" (Vopel and Hawes 2006).

These findings are of great importance for the possibility of fossilized, dormant, or (hypothetically) living microbial mats deep underground below Mars' icy subsurface (Fig. 8.35).

The second objective regarding Mars terraforming in this chapter is the following proposal: I suggest it would be better to introduce greenhouse gases not directly into Mars' atmosphere but into an isolated Martian atmosphere by means of a robotic spacecraft putting those gases into a closed, simple 10-m-radius greenhouse with simple chambers (which can be easily fixed by astronauts), firmly attached to the ground. For an initial Mars terraforming, it is needed to liquefy water from the Martian subsurface ice (de Morais 2004). This heating of the atmosphere could be eased by the use of greenhouse gases – sulfur, fluor, and carbon molecules (Marinova et al. 2000). But since we know very few about present-day Mars, not talking about almost unknown past Mars, then, in terms of planetary protection (Rummel 2001), due to a hypothetical own Martian bio-protection for its preservation, this second proposal can be necessary for Mars exploration (de Morais 2004).

In order to protect places with possible biosignatures (microfossils and hypothetical extant microbes), the best place for putting this initial greenhouse for Martian terraforming must be far (several kilometers) from those places, at the equator – for the same best IR, luminosity, and UV from the Sun, on a flat soil with a surface permafrost for easier obtaining of liquid water for the culture of the bio-oxygen generators (cyanobacteria, the mentioned above algae, lichen, cacti) and of methanogens (de Morais 2004).

Two observations: (1) methanogens *Archaean* are not photosynthetic organisms; they derive their energy from CO_2, carbon monoxide, or hydrogen. Some hydrogen-consuming methanogens grow deep inside Earth's subsurface (Chapelle et al. 2002) and could survive the harsh conditions found on Mars, to liberate methane (a greenhouse gas) inside a secure and scientifically controlled greenhouse, warming it up, and interacting with the other cultivated organisms, for in situ studies of the evolution of a Martian terraforming; (2) cacti are necessary for increasing the auto-sustainability of such bio-system. Simple modified techniques and tools can be used by astronauts to perform the work (de Morais 2004).

The astronauts working on Mars will need to do three main things: to search for possible microfossils record (and hypothetical extant microbial life), to fix chambers for growing resistant vegetables and microbes for food and oxygen consumption, and (the second proposal in this chapter) to fix a simple 10-m-radius greenhouse facility with simple chambers to do a controlled study for Mars terraforming. To accomplish all these, it is strongly necessary (the first proposal in this chapter) for astronauts to perform controlled studies in simulated Martian 0.38 g gravity inside centrifuges built in ESA hardware, or in any other similar hardware, on board the ISS (de Morais 2004). These proposals, within an international cooperation, are possible and viable (de Morais 2004).

In conclusion, in my two proposals for the future Mars terraforming, we will need to:

- Obtain data and analyze results from the experiments I proposed in this chapter, to simulate Mars' 0.38 gravity, using the centrifuges of built biological facilities already on board the ISS (and future ones too), to obtain scientific information on which species of microbes and plants are best for growing on Mars and to get technical and operational data on how to future satisfactorily grow and develop them on its surface (de Morais 2004);
- Design, construct, and test on Earth a simple 10-m-radius greenhouse, to gather technical and operational data on how to fix it on the Martian ground, and to construct it on Mars' surface to begin a secure and scientifically controlled future Mars terraforming (de Morais 2004).

Mars is the known planet most similar to the Earth, it is not far, so an international manned trip to that planet is viable and necessary, and the ISS, with those future centrifuges aboard, can be of much valuable use for a solid planning for this trip and for a viable future international controlled Mars terraforming, for the firm and continuous future international manned exploration of that planet to, ultimately, colonize it for an equilibrated sustainable Human use (de Morais 2004).

8.9 Conclusions

We began this chapter about Mars astrobiology telling about the history of observations and studies of the "Red Planet"; analyzed several scientific and technological information regarding the mixture of physical, astronomical, mathematical, chemical, geological, biological, and engineering information on Mars; showed many informative images; described up-to-date data on the search for possible extinct or extant life at Mars, with the role of liquid water within that planet; and discussed the probable future manned presence on Mars (de Morais 2004).

Of all subjects in this quest for life on Mars. via several missions to it, I have three conclusions:

1. Missions' data and its analyses show there is no compelling strong evidence of biological activity, past or presently, on planet Mars. But it was confirmed that Mars had a large reservoir of liquid water in the past ($\sim$2 Gyrs ago) and it also confirmed the presence of a great quantity of water ice on its subsurface. It was also discovered biological-important hydrated minerals and carbonate salts present on Mars. Those associated with the planet's internal geochemical heat possibly contributed to the evolution from geochemical reactions into biogeochemical ones in subsurface permafrost and hot springs. It needed much more mathematical calculations and the greater reduction of statistical error bars in order to fit these above biogeochemical parameters (kinetics of reactions, entropy, thermodynamics of geochemical transformations, molecular quantum potentials, spin associations, biochemical functions, etc.) to the observed collected data (de Morais 2004);
2. Astrobiology is an interdisciplinary science. We are searching for biosignatures in this other planet surmising that the natural processes for the arisen of life, which took place here on planet Earth, are the same to the ones which hypothetically took place on Mars. This is an epistemological empirical thinking. But we know from the history of science that empirics is a mental tool which can lead to "better" or "not better" scientific conclusions about any phenomena. As an example, our empirical thinking on physics at the submolecular scale was proven to be completely wrong, and we had to formulate other set of "better" models – quantum physics. So, nowadays even if we are not finding any sign of biological-related organic chemistry on Mars, we should not discard the possibility that geochemical reactions within Mars might have followed other paths, e.g., via quantum physics (tunneling, etc.), evolving into biochemical reactions which we cannot detect with present focus and techniques. We could be open minded about the possibility that there could exist other kinds of metabolisms not involving water directly (de Morais 2004);

8 Mars Astrobiology: Recent Status and Progress

3. Mars is a beautiful planet, and it will indeed be very interesting for its further continuous exploration, with robotic spacecraft and manned missions, for the astrobiology of Mars (de Morais 2004).

Thus, our knowledge on the possibility of any life form on Mars is beginning to pass from a period of scientific blindness to a period of scientific hard work imagination brightness (de Morais 2004).

Acknowledgments I would like to acknowledge the most valuable comments, good advices for my international efforts, and sincere friendships by Dr. Adriana Ocampo, NASA Headquarters, and by Dr. Pascale Ehrenfreund (George Washington University), Dr. José Helayël-Neto (CBPF), Dr. Sebastião Dias (CBPF), Dr. Álvaro Nogueira (CBPF), Dr. João dos Anjos (ON), Dr. Joseph Boardman (Analytical Imaging and Geophysics, LLC.), Dr. Christopher McKay, Dr. Carol Stoker, Dr. Nathalie Cabrol, Dr. Heather Smith, Dr. Geoffrey Briggs (retired), and Dr. Scott Hubbard (Stanford University), NASA–Ames Research Center, since my studies and work at NASA–Ames in July 1999.

References

Alonso D, Bartumeus F, Catalan J (2002) Mutual interference between predators may give raise to Turing spatial patterns. Ecology 83(1):28–34

Baqué M, de Vera J-P, Rettberg P, Billi D (2013) The BOSS and BIOMEX space experiments on the EXPOSE-R2 mission: endurance of the desert *cyanobacterium Chroococcidiopsis* under simulated space vacuum, Martian atmosphere, UVC radiation and temperature extremes. Acta Astronaut 91:180–186

Baurmann M, Gross T, Feudel U (2003) Instabilities and pattern formation in simple ecosystem models. Berichte – Forschungszentrum Terramare 12:22–28

Behrenfeld MJ, Falkowski PG (1997) Photosynthetic rates derived from satellite-based chlorophyll concentration. Limnol Oceanogr 42(1):1–20

Benz J, Hoch R, Gabele T (1997) Documentation of mathematical models in ecology – an unpopular task ? Ecol Model 97:1–7

Bianciardi G, Miller JD, Straat PA, Levin GV (2012) Complexity analysis of the viking labeled release experiments. Int J Aeronaut Space Sci 13(1):14–26

Bibring J, Langevin Y, Poulet F, Gendrin A, Gondet B, Berthé M, Soufflot A, Drossart P, Combes M, Bellucci G, Moroz V, Mangold N, Schmitt B, OMEGA Team (2004) Perennial water ice identified in the South Polar Cap of Mars. Nature 428:627–630

Carr MH (1996) Water on Mars. Oxford University Press, New York, p 197

Castets V, Dulos E, Boissonade J, De Kepper P (1990) Experimental evidence of a sustained standing Turing-type non-equilibrium chemical pattern. Phys Rev Lett 64:2953

Chambers P (1999) Life on Mars: the complete story. Blandford Ed, London

Chapelle FH, O'Neill K, Bradley PM, Methé BA, Ciufo SA, Knobel LL, Lovley DR (2002) A hydrogen-based subsurface microbial community dominated by methanogens. Nature 415:312–315

Christensen PR (2006) Water at the poles and in permafrost regions of Mars. GeoSci World Elem 2(3):151–155

Coyne LM, Pollack G, Kloepping R (1984) Room-temperature luminescence from kaolin induced by organic molecules. Clays Clay Miner 32(1):58–66

Coyne LM, Lahav N, Lawless JG (1985) Characterization of dehydration–induced luminescence of kaolinite. Clays Clay Miner 33(3):207–213

de Morais A (2004) Use of gravity simulator in the international space station for Mars terraformation. In: El-Genk MS, Bragg MJ (eds) Space Technology and Applications International

Forum – STAIF 2004: conference on thermophysics in microgravity; commercial/civil next generation space transportation; 21st symposium on space nuclear power and propulsion; human space exploration; space colonization; new frontiers and future concepts, American Institute of Physics conference proceedings, Springer-Verlag, New York, vol 699, pp 961–966

de Vera J-P, Schulze-Makuch D, Khan A, Lorek A, Koncz A, Möhlmann D, Spohn T (2014) Adaptation of an Antarctic lichen to Martian niche conditions can occur within 34 days. Planet Space Sci 98:182–190

DeAngelis DL (1992) Dynamics of nutrient cycling and food webs. Chapman & Hall, New York

Dehant V, Lammer H, Kulikov YN, Grießmeier J-W, Breuer D, Verhoeven O, Karatekin O, Van Hoolst T, Korablev O, Lognonné P (2007) Planetary magnetic dynamo effect on atmospheric protection of early Earth and Mars. In: Geology and habitability of terrestrial planets, vol 24, Space sciences series of the International Space Science Institute (ISSI). International Space Science Institute, Bern, pp 279–300

Feldman WC, Prettyman TH, Maurice S, Plaut JJ, Bish DH, Vaniman DT, Mellon MT, Metzger AE, Squyres SW, Karunatillake S, Boynton WV, Elphic RC, Funsten HO, Lawrence DJ, Tokar RL (2004) Global distribution of near-surface hydrogen on Mars. J Geophys Res Planets 109(E9):253

Frachebourg L, Krapivsky PL, Ben-Naim E (1996) Spatial organization in cyclic Lotka-Volterra systems. Phys Rev E 54:6186

Friedmann EI (1982) Endolithic microorganisms in the Antarctic cold desert. Science 215:1045–1053

Gaidos E, Selsis F (2007) From protoplanets to protolife: the emergence and maintenance of life. In: Reipurth B, Jewitt D, Keil K (eds) Protostars and planets V. University of Arizona Press, Tucson, pp 929–944

Grotzinger JP (2014) Introduction to special issue – habitability, taphonomy, and the search for organic carbon on Mars. Science 343(6169):386–387

Grotzinger JP, MSL Science Team (2014) A Habitable Fluvio-Lacustrine environment at Yellowknife Bay, Gale Crater, Mars. Science 343:6169

Hassler DM, MSL Science Team (2014) Mars' surface radiation environment measured with the Mars science laboratory's curiosity rover. Science 343:6169

Head JW III, Hiesinger H, Ivanov MA, Kreslavsky M, Pratt S et al (1999) Possible ancient oceans on Mars: evidence from Mars orbiter laser altimeter data. Science 286:2134–2137

Hecht MH (2002) Metastability of liquid water on Mars. Icarus 156:373–386

Hunter KS, Wang Y, Van Capellen P (1998) Kinetic modelling of microbially–driven redox chemistry subsurface environments: coupling transport, microbial metabolism and geochemistry. J Hydrol 209:53–80

Jackson JD (1999) Classical electrodynamics. Wiley, New York, p 128

Kim SS, Mysoor N, Ulmer C (2009) Miniature NMR Spectrometers Without Magnets. NASA Tech Briefs

Köhler P, Wirtz KW (2002) Linear understanding of a huge aquatic ecosystem model using a group-collecting sensitivity analysis. Environ Model Software 17(7):613–625

Kropp J, Klenke T (1997) Phenomenological pattern recognition in the dynamical structures of tidal sediments from the German Wadden Sea. Ecol Model 103(2–3):151–170

Kropp J, Block A, von Bloh W, Klenke T, Schellnhuber HJ (1997) Multifractal characterization of microbially induced magnesian calcite formation in recent tidal flat sediments. Sediment Geol 109:37–51

Leinz V, Amaral SE (2003) Geologia Geral, 14th edn. Companhia Editora Nacional, São Paulo, pp 55–73

Leshin LA, Science Team MSL (2013) Volatile, isotope, and organic analysis of Martian fines with the Mars Curiosity Rover. Science 341:6153

Libes SM (1992) An introduction to marine biogeochemistry. Wiley, New York, pp 288–327

Lotka AJ (1920) Analytical note on certain rhythmic relations in organic systems. Proc Natl Acad Sci U S A 6(7):410–415

Marinova MM, McKay CP, Hashimoto H (2000) Warming Mars using artificial super-greenhouse gases. J Br Interplanet Soc 53:235–240

Martin A, Yeats S, Janekovic D, Reiter WD, Aicher W, Zillig W (1984) SAV 1, a temperate u.v.-inducible DNA virus-like particle from the archaebacterium *Sulfolobus acidocaldarius* isolate B12. EMBO J 3(9):2165–2168

McCauley E, Murdoch WW (1990) Predator–prey dynamics in environments rich and poor in nutrients. Nature 343:455–457

McKay CP (1993) Relevance of Antarctic microbial ecosystems to exobiology. In: Friedmann EI (ed) Antarctic microbiology. Wiley, New York, pp 593–601

McKay CP, Stoker CR (1989) The early environment and its evolution on Mars: implications for life. Rev Geophys 27(2):189–214

McKay CP, Friedmann EI, Wharton RA, Davis WL (1991a) History of water on Mars: a biological perspective. Adv Space Res 12(4):231–238

McKay CP, Toon OB, Kasting JF (1991b) Making Mars habitable. Nature 352:489–496

McKay CP, Williams KE, Toon OB, Head JW (2010) Do ice caves exist on Mars? Icarus 209(2):358–368

Meeßen J, Sánchez FJ, Brandt A, Balzer E-M, de la Torre R, Sancho LG, de Vera J-P, Ott S (2013) Extremotolerance and resistance of lichens: comparative studies on five species used in astrobiological research I. Morphological and anatomical characteristics. Origin Life Evol Biosph 43:283–303

Melillo JM, McGuire AD, Kicklighter DW, Moore B, Vorosmarty CJ, Schloss AL (1993) Global climate change and terrestrial net primary production. Nature 363:234–240

MEPAG Special Regions – Science Analysis Group, Beaty D, Buxbaum K, Meyer M, Barlow N, Boynton W, Clark B, Deming J et al (2006) Findings of the Mars special regions science analysis group. Astrobiology 6(5):677–732

Meslin PY, Gasnault O, Schröder S, Cousin A, Berger G, Clegg SM, Lause J, Maurice S, Sautter V, Madsen MB (2013) Soil diversity and hydration as observed by ChemCam at Gale Crater, Mars. Science 341:6153

Michalski JR, Cuadros J, Niles PB, Parnell J, Rogers AD, Wright SP (2013) Groundwater activity on Mars and implications for a deep biosphere. Nat Geosci 6:133–138

Mitrofanov IG, Zuber MT, Litvak ML, Boynton WV, Smith DE, Drake D, Hamara D, Kozyrev AS, Sanin AB, Shinohara C, Saunders RS (2003) CO_2 snow depth and subsurface water-ice abundance in the Northern Hemisphere of Mars. Science 300:2081–2084

Mudryk ZJ, Podgórska B, Ameryk A, Bolałek J (2000) The occurrence and activity of sulphate–reducing bacteria in the bottom sediments of the Gulf of Gdańsk. Oceanologia 42(1):105–117

Murray JD (1989) Mathematical biology. Springer, Heidelberg

Ouyang Q, Swinney HL (1991) Transition from a uniform state to hexagonal and striped Turing patterns. Nature 352:610–612

Oze C, Jones LC, Goldsmith JI, Rosenbauerd RJ (2012) Differentiating biotic from abiotic methane genesis in hydrothermally active planetary surfaces. Proc Natl Acad Sci U S A 109(25):9750–9754

Petrov AS, Bernier CR, Hsiao C, Norris AM, Kovacs NA, Waterbury CC, Stepanov VG, Harvey SC, Fox GE, Wartell RM, Hud NV, Williams LD (2014) Evolution of the ribosome at atomic resolution. Proc Natl Acad Sci U S A 111(28):10251–10256

Plaxco KW, Gross M (2011) Astrobiology: a brief introduction, 2nd edn. Johns Hopkins University Press, Baltimore, pp 285–286

Pollack JB (1987) The case for a wet, warm climate on early Mars. Icarus 71:203–224

Pritchard DW (1967) What is an Estuary: Physical Viewpoint. In: Lauf GH (ed) Estuaries, vol 83. American Association for the Advancement of Science, Washington, DC, pp 3–5

Provata A, Tsekouras GA (2003) Spontaneous formation of dynamical patterns with fractal fronts in the cyclic lattice Lotka–Volterra model. Phys Rev E 67:056602

Provata A, Nicilis G, Baras F (1999) Oscillatory dynamics in low-dimensional supports: a lattice Lotka–Volterra model. J Chem Phys 110:8361–8368

Raven PH, Evert RF, Eichhorn SE (1986) Biology of plants, 4th edn. Worth Publishers, New York

Rummel JD (2001) Planetary exploration in the time of astrobiology: protecting against biological contamination. Proc Natl Acad Sci U S A 98(5):2128–2131

Sagan CE (1980) Cosmos, 1st edn. Francisco Alves Editora, Rio de Janeiro, pp 195–216

Saunders PT, Bazin MJ (1974) On the stability of foodchains. J Theor Biol 52:121–142

Schopf WJ, Walter MR (1983) Archean microfossils: New evidence of ancient microbes. In: Schopf JW (ed) Earth's earliest biosphere – it's origin and evolution. Princeton University Press, Princeton, pp 214–239

Schuerger AC, Golden DC, Ming DW (2012) Biotoxicity of Mars soils: dry deposition of analog soils on microbial colonies and survival under Martian conditions. Planet Space Sci 72(1):91–101

Schuerger AC, Ulrich R, Berry BJ, Nicholson WL (2013) Growth of *Serratia liquefaciens* under 7 mbar, 0 °C, and CO_2–enriched anoxic atmospheres. Astrobiology 13(2):115–131

Seitzinger SP, Kroeze C (1998) Global distribution of nitrous oxide production and N inputs in freshwater and coastal marine ecosystems. Global Biogeochem Cycles 12:93–113

Soetaert K, Herman PMJ, Middelburg JJ (1996) Dynamic response of deep-sea sediments to seasonal variations: a model. Limnol Oceanogr 41(8):1651–1668

Stoker CR (1996) Science strategy for human exploration of Mars. In: Stoker CR, Emmart C (eds) Strategies for mars: a guide to human exploration, vol 86. American Astronautical Society, Univelt, Inc., San Diego, pp 536–560

Stradioto MR, Kiang CH, Caetano-Chang MR (2008) Caracterização petrográfica easpectos diagenéticos dos arenitos do Grupo Bauru na região sudoeste do Estado de São Paulo. Geociências, Revista Escola de Minas, Ouro Preto 61(4):433–441

Summons RE, Amend JP, Bish D, Buick R, Cody GD, Des Marais DJ, Dromart G, Eigenbrode JL, Knoll AH, Sumner DY (2011) Preservation of Martian organic and environmental records: final report of the Mars Biosignature Working Group. Astrobiology 11(2):157–181

Tainaka KI (1989) Stationary pattern of vortices or strings in biological systems: lattice version of the Lotka–Volterra model. Phys Rev Lett 63:2688–2691

Teixeira W, Toledo MCM, Taioli F, Fairchild TR (2008) Decifrando a Terra, 3rd edn. Companhia Editora Nacional, São Paulo, pp 285–304

Thomas-Keprta KL, Clemett SJ, McKay DS, Gibson EK, Wentworth SJ (2009) Origins of magnetite nanocrystals in Martian meteorite ALH84001. Geochim Cosmochim Acta 73(21):6631–6677

Tsallis C (1988) Possible generalization of Boltzmann-Gibbs entropy. J Stat Phys 52:479

Tsallis C, Gell-Mann M (eds) (2004) Nonextensive entropy-interdisciplinary applications. Oxford University Press, New York

Tsallis C, Mendes RS, Plastino AR (1998) The role of constraints within generalized nonextensive statistics. Phys A 261(3):534–554

Tsekouras GA, Provata A (2002) Fractal properties of the lattice Lotka-Volterra model. Phys Rev E 65(016204):016201–016208

Tsekouras GA, Provata A, Tsallis C (2004) Nonextensivity of the cyclic lattice Lotka-Volterra model. Phys Rev E 69(016120):161201–161207

Turing AM (1952) The chemical basis of morphogenesis. Philos Trans R Soc Lond B Biol Sci 237(641):37–72

Van Capellen P, Chang Y (1996) Cycling of iron and manganese in surface sediments; a general theory for the coupled transport and reaction of carbon, oxygen, nitrogen, sulfur, iron, and manganese. Am J Sci 296(3):197–243

(Various) (2014) Special issue – curiosity rover on Mars – exploring Martian habitability. Science 343:6169

Volterra V (1926) Fluctuations in the abundance of a species considered mathematically. Nature 118:558–560

Volterra V (1936) Lecons sur la Theorie Mathematique de la Lutte pour la Vie. Gauthier-Villars, Paris

Vopel KC, Hawes I (2006) Photosynthetic performance of benthic microbial mats in Lake Hoare, Antarctica. Limnol Oceanogr 51(4):1801–1812

Walter MR (1983) Archaean stromatolites: evidence of the Earth's earliest benthos. In: Schopf JW (ed) Earth's earliest biosphere – it's origin and evolution. Princeton University Press, Princeton, pp 187–213

Wang Y, Van Capellen P (1996) A multicomponent reactive transport model for early diagenesis: application to redox cycling in coastal marine sediments. Geochim Cosmochim Acta 60:2993–3014

Westall F, Loizeau D, Foucher F, Bost N, Betrand M, Vago J, Kminek G (2013) Habitability on Mars from a microbial point of view. Astrobiology 13(9):887–897

Wirtz KW (2003) Control of biogeochemical cycling by mobility and metabolic strategies of microbes in the sediments: an integrated model study. Fed Eur Microbiol Soc (FEMS) Microbiol Ecol 46:295–306

Yang L, Zhabotinsky AM, Epstein IR (2004) Stable squares and other oscillatory Turing patterns in a reaction-diffusion model. Phys Rev Lett 92:198303

Zahnle K, Freedman RS, Catling DC (2011) Is there methane on Mars? Icarus 212(2):493–503

Zillig W, Schleper C, Kubo K (1992) The particle SSV1 from the extremely thermophilic archaeon Sulfolobus is a virus: demonstration of infectivity and of transfection with viral DNA. Proc Natl Acad Sci U S A Microbiol 84(16):7645–7649

Zillig W, Stedman KM, Schleper C, Rumpf E (1999) Genetic requirements for the function of the archaeal virus SSV1 in *Sulfolobus solfataricus*: construction and testing of viral shuttle vectors. Genet Soc Am Genet 152:1397–1405

Chapter 9
Classical Physics to Calculate Rotation Periods of Planets and the Sun

Sahnggi Park

Abstract The rotation period of the Earth was calculated from the fundamental quantities, mass, distance, and radius, of the Earth-Moon system by almost an exact number $24^h3^m5^s$. The rotation periods of Mars, Jupiter, Saturn, Uranus, Neptune, and the Sun were also calculated by the same equation. The Earth spin axis which is inclined by 23.45° with respect to the Earth orbit was derived from the gravitation of the Sun acting on the Earth and calculated by almost an exact number, 23.487°. An optical experiment to measure the reaction torque on the Earth acted by the Moon is proposed and discussed.

Keywords Planetary rotation • Spin • Earth's rotation • Reaction torque

9.1 Introduction

Cosmogonists believe that planet's rotation is determined by a combination of processes that occur during and after its accretion from a protoplanetary disk. A considerable progress toward understanding the origin of terrestrial planet rotations has been made over the last two decades (Lissauer et al. 2000; Hughes 2003). Most conventional models of planet formation begin with a protoplanetary disk gas and dust orbiting a central protostar and proceed to collisional coagulation of dust particles into 10^{12}–10^{18} g (kilometer-sized) planetesimals. A swarm of 10^{10}–10^{16} planetesimals accumulate gravitationally into 10^{26}–10^{27} g (Mars-sized) planetary embryos. Near 1 AU, in a narrow accumulation zone of semimajor axis width $\Delta a \approx 0.05$ AU, formation of a planetary embryo is characterized by runaway growth of the largest body in the zone on a timescale of the order 10^5 years. Several independent research groups have corroborated this finding by using distinct numerical methods: gas dynamic statistical simulations or direct N-body orbital integrations. The final stage is described by giant impacts between embryos that

---

S. Park (✉)
Electronics and Telecommunications Research Institute (ETRI), 218 Gajong-ro, Yusong-Gu, Daejon 305-700, South Korea
e-mail: sahnggi@etri.re.kr

© Springer-Verlag Berlin Heidelberg 2015
S. Jin et al. (eds.), *Planetary Exploration and Science: Recent Results and Advances*,
Springer Geophysics, DOI 10.1007/978-3-662-45052-9_9

result in full-sized 10^{27}–10^{28} g terrestrial planets (Kortenkamp et al. 2000; Dones and Tremaine 1993; Wetherill 1990; Lissauer 1993; Lissauer and Stewart 1993).

Until almost 1900, cosmogonist believed that planets accumulated from the material in Keplerian orbits would have retrograde rotation, because the material closer to the Sun moves faster in Keplerian orbits. As the rotation directions of most planets were known to be prograde, various "solutions" to this dilemma were proposed. Chamberlin (1897) was apparently the first to realize that the retrograde rotation depended on the assumption of circular orbits and that accumulation of material on eccentric orbits could produce prograde rotation (Lissauer et al. 2000; Chamberlin 1897).

If the planet and planetesimals are all on circular orbits prior to approaching one another, the planet accumulates enough spin angular momentum to rotate at roughly the rates of Earth and Mars, but in the retrograde direction (Giuli 1968). There is a small range of eccentricities that produce prograde rotation at a comparable rate, but for a realistic distribution of planetesimal eccentricities, a planet accreting from either a uniform or nonuniform surface density disk of planetesimals rotates much slower than does Earth and Mars, mainly because a high degree of cancellation occurs between prograde and retrograde accumulations (Ida and Nakazawa 1990; Greenberg et al. 1997; Ohtsuki and Ida 1998; Lissauer and Kary 1991). It suggests that the random component of planetary spin imparted by large impactors must provide most of the spin angular momentum of the terrestrial planets. Cosmogonists generally accept that the final stage of terrestrial planet formation is the giant impact stage, where terrestrial planets obtain spin angular momentum from the relative motions of colliding protoplanets (Lissauer et al. 2000; Dones and Tremaine 1993; Kokubo and Ida 2007).

Six of the nine planets in our solar system have an obliquity (the angle between the rotational angular momentum vector and orbital angular momentum vector) of less than $30°$, a distribution that would have only a 10^{-5} probability of occurring randomly (Lissauer and Kary 1991). The spin angular momentum imparted by giant impacts should produce an isotropic distribution of obliquities with both prograde and retrograde rotations, which has been used to argue against giant impacts as the source of the terrestrial planet's spin (Lissauer et al. 2000; Kokubo and Ida 2007).

Cosmogonists believe that primordial rotation periods of planets have been changed significantly by tidal frictions between planet and satellites and Sun. The Earth spun faster in the past, and the Moon was closer to the Earth. Obliquities of planets have also changed by spin-orbit coupling to make the planet's spin axis precess like the Earth's precession. For the Earth, a torque to precess it is exerted not only by the Sun but also by the Moon. Orbits of the planets also precess as a result of their mutual gravitational interactions (Lissauer et al. 2000).

Four gas giant planets rotate faster than terrestrial planets, the fastest being Jupiter with a spin period of only 0.387 days. The origin of gas giant rotation has been less studied and far from general acceptance, more interested in the formation scenarios of the gas giants. One scenario, the core accretion hypothesis (Podolak 2007; Pollack et al. 1996), argues that the planetesimals in the outer solar system

9 Classical Physics to Calculate Rotation Periods of Planets and the Sun

were gradually accreted into a planetary core on the order of 15 Earth masses. Such a massive core could then attract a large amount of hydrogen and helium from the surrounding disk to form a gas giant planet, marked by runaway gas accretion, starting when the solid and gas masses are about equal. The other scenario, disk instability hypothesis, argues that the gas disk itself was unstable and that density fluctuations became large enough that some portion of the disk collapsed under its own gravity. This collapsing clump would eventually evolve into a gas giant planet. Some of the solids that were in the gas would eventually settle into a core, but that core is expected to be small, on the order of a few Earth masses (Boss 2000; Stevenson 2006; Giampieri et al. 2006).

Here it is shown that planetary spins including the spin of the Sun can be calculated accurately from an extension of classical Newtonian physics into the planetary rotations especially between a planet and its satellites (Park 2008, 2013). A planet acts as a torque on its satellites to rotate them synchronously. There should be a reaction torque on the planet, of which magnitude and direction with respect to the center of moment of inertia (CMI) are the same as those of the torque, which is analogous to Newton's third law in the linear motion. The spins of the Earth and the Moon are driven by the torque and the reaction torque in such a way that the rotation period of the Earth can be calculated by almost exact accuracy from the fundamental quantities of the Earth and Moon system. It is also shown that the Earth spin axis which is tilted by 23.45° with respect to the Earth orbit can be derived from the gravitation of the Sun acting on the Earth in a similar way that the gravitation of the Earth acts on the Moon to rotate it synchronously. The rotation period of the Sun is also calculated by the same way as the Earth by reducing the many-body system into a two-body system. The theory and equations can also be applied to the planets, Mars, Jupiter, Saturn, Uranus, and Neptune, which have satellites and rotate fast, to calculate the periods of their spins in a very consistent way. Planets including the Sun take two different kinds of rotations. One is the near synchronous rotations by the influence of the gravitation of the Sun. The other is the fast rotations which are driven by the planet and its satellites. All spins in the latter can be calculated accurately and consistently by the theory and equations in this article. Finally, a possible experiment to measure the reaction torque on the Earth is discussed.

9.2 The Rotation Period of the Earth

It is well known in mechanical physics that there is a good correspondence in quantities and equations between linear motions and angular motions, mass and moment of inertia, velocity and angular velocity, acceleration and angular acceleration, force and torque, etc. The correspondence works for the first and second Newton's laws; a body rotating with a certain velocity tends to keep rotating with the same velocity if there is no torque acted on the body, and a torque acted

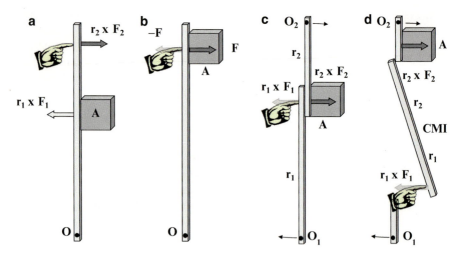

Fig. 9.1 Torque and reaction torque. (**a**) Reaction torque in opposite direction to the torque, (**b**) a specific case, $\mathbf{r}_1 = \mathbf{r}_2$, (**c**) and (**d**) reaction torque in the same direction with the torque

on a body should be mathematically equal to the moment of inertia times angular acceleration of the body. For Newton's third law, however, a reaction torque is not defined in physics to lead to a complete correspondence.

Considering Fig. 9.1, there should be a reaction torque by exactly the same reason as the reaction force in Newton's third law which is simply a special case that the radius vectors of torque and reaction torque are equal in magnitude and direction. Let us suppose that the rod is rigid and its mass is negligible. As a person in Fig. 9.1a acts as a torque, $\mathbf{r}_2 \times \mathbf{F}_2$, on an object via the rod to angularly accelerate the object with respect to the pivot O, the person will feel a reaction torque, $\mathbf{r}_1 \times \mathbf{F}_1$, acted by the object, where the torque and reaction torque have the same magnitudes and opposite directions.

Newton's third law is a specific case of $\mathbf{r}_1 = \mathbf{r}_2$, as shown in Fig. 9.1b. Since a torque is a product of two vectors, as shown in Fig. 9.1c, d, torque and reaction torque may have the same directions depending on the positions of pivots. If all the bodies are in an isolated system and in a stationary state initially, the person in Fig. 9.1a rotates in opposite direction to the object to conserve the total angular momentum before and after the person acts the torque, $\mathbf{J} = \mathbf{J}_1 + \mathbf{J}_2, \mathbf{J} = 0, \mathbf{J}_1 = -\mathbf{J}_2$. On the other hand, since the torque and reaction torque are in the same directions in Fig. 9.1c, d, there should be two different kinds of rotations, spin and orbital angular momenta, to conserve the total angular momentum. The person rotates with a spin angular momentum $\mathbf{S}_1$ with respect to the pivot O_1, and the object rotates with $\mathbf{S}_2$ with respect to the pivot O_2. To conserve the total angular momentum, $\mathbf{J} = 0$, the whole system rotates with orbital angular momentum $\mathbf{L} = -(\mathbf{S}_1 + \mathbf{S}_2)$, $\mathbf{J} = \mathbf{L} + \mathbf{S}_1 + \mathbf{S}_2$, in the direction indicated by arrows in O_1 and O_2. The pivot of the rotation will be a point, center of moment of inertia (CMI), about which moment of

Fig. 9.2 A rotational system of the Earth and the Moon bound by the gravitational force

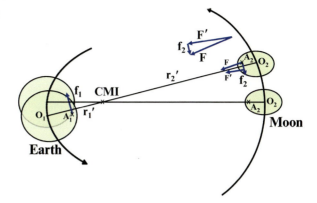

inertia of the system is symmetric, similar to the center of mass in linear motion. It can be represented mathematically as, for a system consisting of *n* bodies,

$$\sum_{i=1}^{n} m_i r_i^2 \mathbf{i} = 0, \qquad (9.1)$$

where **i** is a unit vector directing a body from CMI, r_i is a radial distance from CMI to the body, and m_i is a mass of the body. The rotations of the Earth and the Moon can be derived from the same fundamental physics, torque and reaction torque, as in Fig. 9.1d.

Figure 9.2 shows a rotating system of the Earth and the Moon. The gravitational force of the Earth acting on the Moon is highest at the point A_2 which is the nearest point of the Moon to the Earth. The radial distance from A_2 to O_2 is largest due to the different gravitational force. If the Moon does not spin, as the Moon moves along its orbit, the point A_2 moves away and the new point comes in to the nearest point. The highest point is not exactly on the line connecting two centers of the Earth and the Moon, but it is shifted as shown in Fig. 9.2. Here, it is assumed that Moon's radial response to the tidal force is retarded by a certain time. The gravitational force on the shifted highest point induces a force $\mathbf{f}_2$, which tends to make the spin of the Moon synchronous to the period of orbital motion.

Since the force $\mathbf{f}_2$ makes a torque with respect to CMI, there should be a reaction torque on the Earth, $\mathbf{r}_1' \times \mathbf{f}_1 = \mathbf{r}_2' \times \mathbf{f}_2$, as shown in Fig. 9.2, where $\mathbf{r}_1'$ and $\mathbf{r}_2'$ are radial vectors from CMI to the points A_1 and A_2. The force $\mathbf{f}_1$ on the Earth is magnified by the ratio of distances $\mathbf{r}_2'/\mathbf{r}_1'$. As the force $\mathbf{f}_2$ makes the spin of the Moon synchronous to the period of orbital motion, the force $\mathbf{f}_1$ will make the Earth rotate by about ten times stronger force than the force $\mathbf{f}_2$. As soon as the Moon rotates synchronously to the period of orbital motion, the forces $\mathbf{f}_1$ and $\mathbf{f}_2$ decrease to zero, and the rotation of the Earth will reach to the speed accelerated by the torque and the reaction torque. To conserve total angular momentum before and after torque and reaction torque were acted, orbital angular momentum must decrease by increase of total spin angular momentum, $\Delta \mathbf{L} = -(\mathbf{S}_1 + \mathbf{S}_2)$, $\Delta \mathbf{J} = \Delta \mathbf{L} + \mathbf{S}_1 + \mathbf{S}_2 = 0$.

The Moon will make the same kind of torque on the Earth and reaction torque on the Moon. Since the tidal force of the Moon is much smaller, roughly, by the mass ratio of the Earth and Moon, than the tidal force of the Earth, the torque and reaction torque acted by the Moon are much smaller than those acted by the Earth in Fig. 9.2, and these torques will be neglected here. If the Earth rotates faster than the period of orbital motion, the torque and the reaction torque acted by the Moon are opposite in directions to the torques in Fig. 9.2 and may retard the speed of the Earth's rotation slowly.

Torques and reaction torques acting on the points A_1 and A_2 with respect to the CMI are

$$\mathbf{r}'_1 \times \mathbf{f}_1 = \mathbf{r}'_2 \times \mathbf{f}_2 \tag{9.2}$$

$$\mathbf{r}'_1 \times \Delta m_1 \mathbf{a}_1 = \mathbf{r}'_2 \times \Delta m_2 \mathbf{a}_2, \tag{9.3}$$

where the masses, Δm_1 and Δm_2, are the mass increments in the points, A_1 and A_2, which have a relation $\Delta m_1/\Delta m_2 = m_1/m_2$, where m_1 and m_2 are masses of the Earth and the Moon, respectively. The forces, $\mathbf{f}_1$ and $\mathbf{f}_2$, induce tangential velocities $\mathbf{v}_1$ and $\mathbf{v}_2$ in the points, as shown in Fig. 9.3, which are transformed into spin velocities of the Earth and the Moon, where the tangential velocities $\mathbf{v}_1$ and $\mathbf{v}_2$ are the relative velocities of points A_1 and A_2 with respect to O_1 and O_2, respectively. By integrating Eq. 9.3 from initial velocities $\mathbf{v}_{10}$, $\mathbf{v}_{20}$ to final velocities $\mathbf{v}_1$, $\mathbf{v}_2$, the velocities have a relation:

$$\mathbf{r}'_1 \times \Delta m_1 (\mathbf{v}_1 - \mathbf{v}_{10}) = \mathbf{r}'_2 \times \Delta m_2 (\mathbf{v}_2 - \mathbf{v}_{20}) \tag{9.4}$$

It is assumed that initial velocities $\mathbf{v}_{10}$, $\mathbf{v}_{20}$ which had not been driven by the forces $\mathbf{f}_1$ and $\mathbf{f}_2$ will decrease to zero by tidal friction or are negligibly small because of a

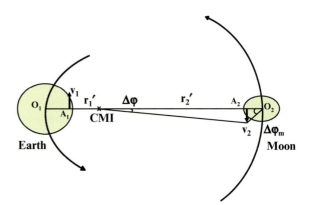

Fig. 9.3 Angles $\Delta \varphi$ and $\Delta \varphi_m$ defined by the relative velocity $\mathbf{v}_2$

9 Classical Physics to Calculate Rotation Periods of Planets and the Sun 253

high degree of cancellation (Lissauer et al. 2000; Ida and Nakazawa 1990), and only the driven components of spins remain to current speeds.

$$\mathbf{r}_1' \times \Delta m_1 \mathbf{v}_1 = \mathbf{r}_2' \times \Delta m_2 \mathbf{v}_2. \tag{9.5}$$

Since the Earth and the Moon consist of rigid materials, the tangential relative velocities of the mass increments in the points, A_1 and A_2, can determine the spin velocities of the whole bodies. A planet consisting of gas materials, however, may have different spin velocities along the radial distance or along the latitude. Since radial vectors are perpendicular to the velocities $\mathbf{v}_1$ and $\mathbf{v}_2$, Eq. 9.5 can be written as, with a relation of $\Delta m_1 / \Delta m_2 = m_1 / m_2$,

$$m_1 \mathbf{r}_1' \mathbf{v}_1 = m_2 \mathbf{r}_2' \mathbf{v}_2 \tag{9.6}$$

From Fig. 9.3, the velocity $\mathbf{v}_2$ is

$$\mathbf{v}_2 = \frac{\mathbf{r}_2' \Delta \phi}{\Delta T}, \tag{9.7}$$

where T is a period of point A_2 if the point A_2 rotates around CMI with the tangential velocity of $\mathbf{v}_2$. To calculate the spin period of the Moon, the velocity $\mathbf{v}_2$ must be expressed by the variables, R_m, $\Delta\varphi_m$, ΔT_m, which are not independent of $\mathbf{r}_2'$, $\Delta\varphi$, ΔT, where T_m is the spin period of the Moon, R_m is the equatorial radius of the Moon, and $\Delta\varphi_m$ is the angle shown in Fig. 9.3. The relation, $\mathbf{r}_2' \Delta\varphi = R_m \Delta\varphi_m$, is obvious from Fig. 9.3. The period, ΔT_m, must be inversely proportional to $\Delta\varphi_m$ and ΔT because the spin period is small as $\Delta\varphi_m$ is large. As the point A_2 moves from CMI to O_2, the period T_m decreases, while T increases with $\mathbf{v}_2$ fixed. It leads to a relation:

$$\mathbf{v}_2 = \frac{\mathbf{r}_2' \Delta\varphi}{\Delta T} \propto R_m \Delta\varphi_m \Delta T_m, \tag{9.8}$$

$$\mathbf{v}_2 = C R_m \Delta\varphi_m \Delta T_m, \tag{9.9}$$

where C is a proportional constant.

For an average velocity during one period,

$$\bar{v}_2 = C 2\pi R_m T_m. \tag{9.10}$$

By the same way,

$$\bar{v}_1 = C 2\pi R_e T_e, \tag{9.11}$$

where T_e is the spin period of the Earth and R_e is the equatorial radius of the Earth.

For the average velocities during one period, Eq. 9.6 becomes

$$m_1 (\mathbf{r}_1 - R_e) R_e T_e = m_2 (\mathbf{r}_2 - R_m) R_m T_m, \qquad (9.12)$$

where $\mathbf{r}_1$ and $\mathbf{r}_2$ are radial distances from CMI to the centers of the Earth and the Moon. Since T_m is the same as the period of orbital motion, the spin period of the Earth can be calculated by Eq. 9.12. From Eq. 9.1, the position of CMI of two bodies is easily found by two equations:

$$m_1 \mathbf{r}_1^2 = m_1 \mathbf{r}_2^2, \qquad (9.13)$$

$$d = \mathbf{r}_1 + \mathbf{r}_2 \qquad (9.14)$$

Although the Earth spin rate decreases due to the tidal force of the Moon, which makes the orbital period and the distance between the Earth and the Moon increase to conserve the total angular momentum, the ratio of spin periods given by Eq. 9.12 must always be true, because the tidal force of the Moon leads to the same equation, Eq. 9.12, by the torque and the reaction torque acted by the Moon.

In Figs. 9.2 and 9.3, it is assumed that the planes of equators of the Earth and the Moon are in the same plane as their orbits, which leads to a subsequent assumption that the velocities $\mathbf{v}_1$ and $\mathbf{v}_2$ of points A_1 and A_2 are in the same directions as the forces $\mathbf{f}_1$ and $\mathbf{f}_2$. If the planes of equators of the Earth and the Moon are inclined by θ_1 and θ_2 from their orbital planes as shown in Fig. 9.4, then the points A_1 and A_2 are accelerated toward the directions different, by the angles, θ_1 and θ_2, from the forces that are always in the plane of orbit. Since the directions of accelerations are different from the forces, the inertial masses, Δm_1 and Δm_2, at the points A_1 and A_2 should be taken larger by cosine factors, $\Delta m_1 \rightarrow \Delta m_1/\cos\theta_1$, $\Delta m_2 \rightarrow \Delta m_2/\cos\theta_2$. Since the velocity shown in Fig. 9.4 is $\mathbf{v}_1 \rightarrow \mathbf{v}_1\cos\theta_1$, the factor in Eq. 9.8 becomes $R_e \Delta\varphi_e \rightarrow R_e \Delta\varphi_e \cos\theta_1$, $R_m \Delta\varphi_m \rightarrow R_m \Delta\varphi_m \cos\theta_2$. It is easy to prove that Eq. 9.12 is unchanged. Although the planes of equators of the Earth and the Moon are inclined by θ_1 and θ_2 from their orbital planes, these angles do not affect the periods of their spins. If θ_1 and θ_2 are large to approach to 90°, however, then they might affect the calculation of periods.

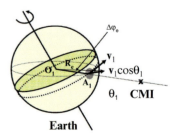

Fig. 9.4 The rotation axis of the Earth inclined by θ

9 Classical Physics to Calculate Rotation Periods of Planets and the Sun 255

Using Eqs. 9.13 and 9.14, and the observed values, $d = 3.844 \times 10^8$ m, $m_1 = 5.9736 \times 10^{24}$ kg, $m_2 = 7.349 \times 10^{22}$ kg, $R_e = 6.3781 \times 10^6$ m, $R_m = 1.7381 \times 10^6$ m (Allen 1973; Williams 2010), r_1 and r_2 are calculated, $r_1 = 3.8379 \times 10^7$ m, $r_2 = 3.4602 \times 10^8$ m. Using the spin period of the Moon $T_m = 27.3217$ day (Williams 2010), the spin period of the Earth is calculated:

$$T_e = 23^h 39^m 2^s$$

Compare with the observation $T_e = 23^h 56^m 4^s$ (Williams 2010).

To consider the different densities of the Earth along the radial distance, if a root-mean-square radius of the Earth defined as $\overline{R}_e = \left\{ \left(\int R^2 \cdot dm \right) / m \right\}^{1/2}$ is used, using the density distributions, 17 g/cm^3 for 1,300 km thick inner core, 10 g/cm^3 for 2,200 km thick outer core, 4.5585 g/cm^3 for 2,843.1 km thick mantle, and 3.3 g/cm^3 for 35 km thick shell (Smith and Jacobs 1973), where the density and the thickness for mantle were adjusted to match with the mass and the radius of the Earth, the root-mean-square radius of the Earth is calculated, $\overline{R}_e = 4.6157 \times 10^6$ m. Using the fact that the density of the Moon is uniform, approximately 3.35 g/cm^3 (Allen 1973), the root-mean-square radius of the Moon is calculated, $\overline{R}_m = 1.3481 \times 10^6$ m. Refer to Park (2013) for Mathcad calculations in detail. Using these radii, T_e is calculated, $T_e = 24^h 3^m 5^s$, which is closer to the observation. If the density distribution of the Moon is known accurately, the calculation will yield a closer value to the observation. The root-mean-square radius is a radius of a spherical shell of which mass is uniformly distributed on the shell and the moment of inertia to rotate the shell is the same as that of the Earth.

9.3 Calculation of the Earth Spin Axis

The Earth spin axis which is inclined by 23.45° with respect to the Earth orbit can be derived from the gravitation of the Sun acting on the Earth. As shown in Fig. 9.5, the Sun acts the strongest gravitational force on the point A of the Earth and induces the force $\mathbf{F}_s$, which is exactly in the reverse direction of tangential spin velocity $\mathbf{v}_e$.

The horizontal component $\mathbf{F}_{sx}$ of the force makes slower the horizontal component $\mathbf{v}_{ex}$ of the spin velocity because the force $\mathbf{F}_{sx}$ is in the opposite direction to the velocity $\mathbf{v}_{ex}$. The vertical component $\mathbf{F}_{sy}$, however, may increase the vertical component $\mathbf{v}_{ey}$ of the spin velocity by tilting the Earth spin axis into larger angles. If the vertical component $\mathbf{v}_{ey}$ becomes larger than the horizontal component $\mathbf{v}_{ex}$, then the same process will occur with the two components switched with each other. Eventually, both components converge to the same magnitude. It may also be explained by statistical equipartition of velocity components.

$$\frac{\mathbf{v}_{ex}}{\mathbf{v}_{ey}} = C, \ C \to 1 \tag{9.15}$$

Fig. 9.5 The Sun acts a different gravitational force on the point A of the Earth

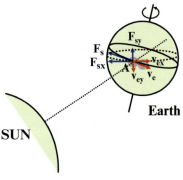

Fig. 9.6 A spherical coordinate on the Earth

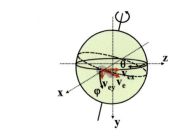

With reference to the spherical coordinate shown in Fig. 9.6, θ measures the angle from positive z axis to a concerned point and φ measures the angle from positive x axis to a concerned point. The point on the equator circle crossed by x axis is $(\varphi, \theta) = (0°, 90°)$. The horizontal component of the spin velocity is $v_{ex} = -R_e \dot{\theta}$ and the vertical component of the spin velocity is $v_{ey} = R_e \dot{\varphi} \sin\theta$.

From Eq. 9.15, the ratio is

$$\frac{-\dot{\theta}}{\dot{\varphi} \sin\theta} = C \tag{9.16}$$

The differential equation can be solved by integration:

$$\int \frac{d\theta}{\sin\theta} = -C \int d\varphi \tag{9.17}$$

$$\theta = 2\tan^{-1}\left(e^{-C\varphi + \text{const}}\right), \tag{9.18}$$

where const denotes an integration constant which is determined to zero as the equator point of the Earth passes through the point $(\varphi, \theta) = (0°, 90°)$. If $C = 1$, then Eq. 9.18 gives a final form as

$$\theta = 2\tan^{-1}\left(e^{-\varphi}\right) \tag{9.19}$$

For the angle $\varphi = 90°$ or $-90°$, the tilted angle of Earth axis is obtained by almost exact number, 23.487°.

9 Classical Physics to Calculate Rotation Periods of Planets and the Sun

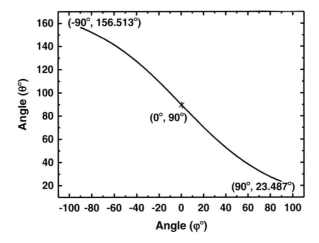

Fig. 9.7 A curve of Earth equator obtained theoretically

The graph in Fig. 9.7 shows a curve obtained from Eq. 9.19, which represents a trace of Earth equator inclined by 23.487° from the orbital plane of the Earth. Observed angles of planets inclined from their orbits are listed in Table 9.1 (Williams 2010). The angles are dominantly dispersed near 23.5° and 0° which are gravitationally stable or equilibrium angles as expected from Fig. 9.5. The vertical component of F_s vanishes at 0° so that the gravitation of the Sun to tilt the spin axis of the planet might be negligible. As shown in Table 9.1, the Earth is very close to the case of $C \approx 1$. Uranus and Pluto might be regarded as exceptional cases.

9.4 The Rotation Period of the Sun

The Earth and the Moon are an ideal two-body system where the spin period and the spin axis of the Earth can be calculated by an accuracy of almost exact numbers. A system involving many bodies such as the solar system or Jupiter and its satellites, however, implies very complicacy. The spin of the Sun must be affected by various motions of all the planets and satellites in very complicate ways. The calculation of the spin may be approximated by reducing the many-body system into a two-body system on the basis that the weighted collective contributions from planets determine the spin of the Sun. It is assumed that the period of spin of the Sun can be calculated by an imaginary planet which rotates synchronously as the case of the Earth.

Figure 9.8 shows an imaginary planet which has the same mass and the same moment of inertia as those of total sums of all planets. The Sun is denoted by m_s, the imaginary planet by m_o, and planets by m_i. CMI means the center of moment of inertia of the solar system. The distance from CMI to the Sun or planets is denoted by r_s or r_i.

Table 9.1 Planetary data

	Mercury	Venus	Earth	Mars	Jupiter	Saturn	Uranus	Neptune	Pluto
Spin axis	0.01°	2.64° (177.36°)	23.45°	25.19°	3.13°	26.73°	97.77°	28.32°	122.53°
Mass ($\times 10^{24}$ kg)	0.3302	4.8685	5.9736	0.64185	1898.6	586.46	86.832	102.43	0.0125
Density (g/cm)	5.427	5.243	5.515	3.933	1.326	0.687	1.27	1.638	1.75
Semimajor axis ($\times 10^6$ m)	57.91	108.21	149.60	227.92	778.57	1,433.53	2,872.42	4,495.06	5,906.38

9 Classical Physics to Calculate Rotation Periods of Planets and the Sun

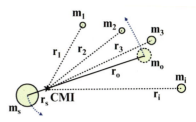

Fig. 9.8 An imaginary planet and the solar system

The equations to find relevant parameters are given as

$$m_s r_s^2 = m_1 r_1^2 + m_2 r_2^2 + \cdots + m_i r_i^2 \tag{9.20}$$

$$m_o r_o^2 = m_1 r_1^2 + m_2 r_2^2 + \cdots + m_i r_i^2 \tag{9.21}$$

$$m_o = m_1 + m_2 + \cdots + m_i. \tag{9.22}$$

$$d_i = r_i + r_s \tag{9.23}$$

Some manipulations of the above equations provide the distance r_s:

$$(m_s - m_o) r_s^2 + 2C_1 r_s - C_2 = 0, \tag{9.24}$$

where C_1, C_2 are constants consisting of observed quantities:

$$C_1 = m_1 d_1 + m_2 d_2 + \cdots + m_i d_i \tag{9.25}$$

$$C_2 = m_1 d_1^2 + m_2 d_2^2 + \cdots + m_i d_i^2 \tag{9.26}$$

Using the numbers in Table 9.1, the distance r_s is calculated, $r_s = 4.921 \times 10^{10}$ m, and using Eqs. 9.21, 9.22, and 9.23, the distance r_o is calculated, $r_o = 1.344 \times 10^{12}$ m. The imaginary planet has a mass $m_o = 2.668 \times 10^{27}$ kg from Eq. 9.22. To find the radius of the imaginary planet, weighted mass densities are considered as

$$\rho_o = \rho_1 \frac{m_1}{m_o} + \rho_2 \frac{m_2}{m_o} + \cdots + \rho_i \frac{m_i}{m_o}. \tag{9.27}$$

From the mass and the density given by Eqs. 9.22 and 9.27, the radius of the imaginary planet is calculated, $R_o = 8.057 \times 10^7$ m with $\rho_o = 1.218$ g/cm³, where the imaginary planet is assumed to be a sphere having an average mass density given by Eq. 9.27.

Now calculate the spin period of the Sun using Eq. 9.12. As the numbers are put into the appropriate variables in Eq. 9.12, $T_s = T_o/232.528$. To find the spin period of imaginary planet which is assumed to be the same as the period of orbital motion, the classical equation is used:

$$T_o = \frac{2\pi}{\sqrt{G\,(m_s + m_0)}} r_{cm}^{3/2},$$ (9.28)

where G is the gravitational constant and r_{cm} is a distance between the imaginary planet and the center of mass of the Sun and the imaginary planet, which is calculated, $r_{cm} = 1.391 \times 10^{12}$ m. Using $T_o = 28.356$ year from Eq. 9.28, the calculation results in $T_s = 44.5$ day. Refer to Park (2013) for Mathcad calculations in detail. Considering 10^{30} order of astronomical numbers involved in the equations, the calculated number is considerably close to the observation ~ 25 days. Two main approximations are involved in the calculation. First, the parameters of imaginary planet are dominated by Jupiter and Saturn that consist of gas. As shown in the calculation of the Earth spin, the rigid part having an average density of 5.515 g/cm^3 for the Earth and 3.341 g/cm^3 for the Moon is a major part to decide the spin rate accurately. The radial point of the imaginary planet to give 25 days of spin on the equator surface of the Sun is 4.53×10^7 m or $0.56 \times R_0$. Second, the imaginary planet is assumed to rotate synchronously with its orbital motion as the case of the Moon. Considering the spin of Mercury and Venus, this assumption may result in difference between calculation and observation. Nonvanished initial spin of the Sun which was not driven by the reaction torque may also contribute to the current spin of the Sun.

9.5 Spins of Mars, Jupiter, Saturn, Uranus, and Neptune

The spin rates of planets Mars, Jupiter, Saturn, Uranus, and Neptune are also calculated by the same theory and equations as those of the Earth and the Sun with a slight modification due to the fact that the CMIs locate inside the planets. Although the spin rates of the planets look irregularly distributed, it is possible to derive them in a consistent way with considerable accuracies. Refer to Park (2013) for Mathcad calculations in detail.

Figure 9.9 takes Jupiter, as an example, to describe a planet system of which CMI locates inside the planet. The point CMI about which moment of inertia of the system is symmetric is calculated, approximately, by taking the same volume B as volume A and assuming that the whole mass of volume C locates in the center O. The volume A is given by equation

$$V = \frac{2}{3}\pi R_J^3 - \left(\pi R_J^2 r_J - \frac{1}{3}\pi r_J^3\right),$$ (9.29)

where r_J is a distance between the center O and CMI. The moment of inertia of volume C is equal with that of the satellites as

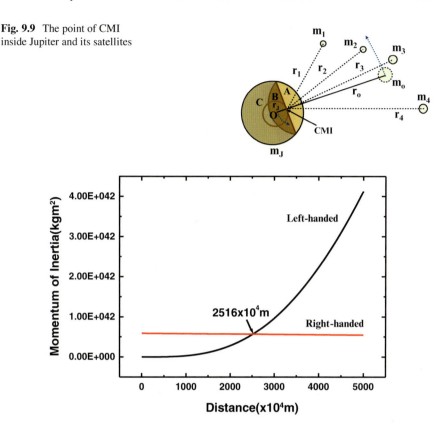

Fig. 9.9 The point of CMI inside Jupiter and its satellites

Fig. 9.10 Graphical solution to find the point of CMI inside Jupiter

$$(m_J - 2V\rho_J) r_J^2 = m_1 r_1^2 + m_2 r_2^2 + \cdots + m_i r_i^2, \quad (9.30)$$

$$d_i = r_i + r_J, \quad (9.31)$$

where ρ_J is the mean density of Jupiter and d_i is a distance between Jupiter and the ith satellite.

Now the distance r_J can be decided by Eqs. 9.30 and 9.31, graphically. Figure 9.10 shows the graphs of left-handed and right-handed sides of Eq. 9.30, where four main satellites of Jupiter, Io, Europa, Ganymede, Callisto, were taken into account. The graphs provide $r_J = 2.516 \times 10^7$ m. The equatorial radius of Jupiter is 7.1492×10^7 m. As soon as the distance r_J is decided, variables of imaginary planet can be decided by Eqs. 9.21, 9.22, 9.23, and 9.27, $r_0 = 1.201 \times 10^9$ m, $m_0 = 3.931 \times 10^{23}$ kg, $R_0 = 3.393 \times 10^6$ m. As that of the Sun, Fig. 9.11 shows a two-body system consisting of Jupiter and an imaginary satellite which has the same mass and the same momentum of inertia as those of total sums of the four satellites. Comparing Fig. 9.11 with Fig. 9.2, the force $\mathbf{f}_1$ in the reaction torque acts on the inertial mass Δm_J which is just inside CMI.

Fig. 9.11 Forces acting on Jupiter and the imaginary satellite

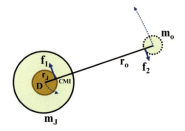

To proceed as in Eq. 9.5, the inertial mass Δm_0 is replaced by m_0. The inertial mass Δm_J has to be replaced not by m_J but by the mass inside the radius of r_J, because the inertial mass is bound to Jupiter by the gravitational force of the mass inside the radius of r_J. It is easy to prove that the gravitational force of the shell measured at the point CMI vanishes, where the inner and outer radii of the shell are r_J and R_J, respectively. Now it is ready to use Eq. 9.12. Since the sphere D has the radius of $R_D = r_J$, Eq. 9.12 does not work. Instead of R_D, root-mean-square radius $\overline{R}_D = \left\{ \left(\int R^2 \cdot dm \right) / m \right\}^{1/2}$ giving an accurate value of the Earth spin is used. For a uniform density, the root-mean-square radius is related by $\overline{R}_D = (3/5)^{1/2} r_J$ and $\overline{R}_0 = (3/5)^{1/2} R_0$.

The sphere D is a core volume of Jupiter and its density is much higher than the mean density of Jupiter. Since the core density of Jupiter is not known, it is more meaningful to obtain the core density that provides the exact value of the Jupiter spin using Eq. 9.12 rather than guessing a core density. The obtained core density should not only be within a reasonable boundary, but core densities of all five planets should have a consistency within their boundaries.

To find the rotational period of the imaginary satellite of Jupiter, Eq. 9.28 is used to obtain $T_0 = 8.773$ day with the fact that the four satellites of Jupiter are rotating synchronously. Putting $r_J = 2.516 \times 10^7$ m, $r_0 = 1.201 \times 10^9$ m, $\overline{R}_D = 1.949 \times 10^7$ m, $\overline{R}_0 = 2.628 \times 10^6$ m, and $m_0 = 3.931 \times 10^{23}$ kg into Eq. 9.12,

$$T_J = \frac{m_0 (r_0 - \overline{R}_0) \overline{R}_0}{m_D (r_J - \overline{R}_D) \overline{R}_D} T_0,$$

the exact number of the Jupiter spin, $T_J = 9.925$ h, is obtained from the core density of the sphere, $\rho_D = 3.563$ g/cm^3 and $m_D = 4\pi R_D^3 \rho_D/3 = 2.377 \times 10^{26}$ kg. Considering the densities of the Earth, 17 g/cm^3 for core and 5.515 g/cm^3 for mean, a reasonable boundary of the core density would be around 3 times the mean density of a planet. The number, 3.563 g/cm^3 as a core density of Jupiter, is definitely within a reasonable boundary, considering the mean density of Jupiter, $\rho_J = 1.326$ g/cm^3.

The spin periods of Uranus and Neptune can also be calculated by the same way as that of Jupiter. The spin period of Uranus, $T_U = 17.24$ h, is obtained by Eq. 9.12, and the densities of Uranus are 3.1 g/cm^3 for core and 1.27 g/cm^3 from the

9 Classical Physics to Calculate Rotation Periods of Planets and the Sun

Table 9.2 Densities to give the exact periods of spin

	Periods of spin (hours)	Density of core sphere (g/cm^3)	Density of volume A + B (density of planets) (g/cm^3)	Mass of volume A + B (mass of planet) $(\times 10^3)$	Radius of core sphere (radius of planet) $(\times 10^4 \text{ m})$
Mars	24.623	15.289	3.968 (3.933)	6.402 (6.418)	3.732 (339.6)
Jupiter	9.925	3.563	1.326 (1.326)	10,020 (18,990)	2,516 (7,149.2)
Saturn	10.657	1.827	0.6 (0.687)	2,263 (5,685)	2,513 (6,028.6)
Uranus	17.245	3.1	1.27 (1.27)	524 (868.3)	717.5 (2,555.9)
Neptune	16.11	3.224	1.638 (1.638)	531.9 (1,024)	840.4 (2476.4)

observed mean density. Neptune is orbiting with Triton as if it is a two-body system, because other satellites of Neptune are so small to be ignored. The spin period of Neptune, $T_N = 16.11$ h, is obtained by Eq. 9.12, and the densities of Neptune are 3.677 g/cm^3 for core and 1.638 g/cm^3 from the observed mean density. The numbers are summarized in Table 9.2. Core densities of the three planets, Jupiter, Uranus, and Neptune, are closely distributed around 3.5 g/cm^3, though the spin periods are largely distributed.

The spin periods of Saturn and Mars are also obtained by the same way as above with small changes of their mean densities. For Saturn, the mass, 2×10^{26} kg, of volumes A and B in Fig. 9.9 is smaller than the half of the Saturn mass, 5.685×10^{26} kg, which means that the density of volumes A and B can be smaller than the mean density, 0.687 g/cm^3, of Saturn. A combination to give the spin period, 10.656 h, of Saturn can be chosen 0.6 g/cm^3 for volumes A and B and 1.827 g/cm^3 for core. If the mean density 0.687 g/cm^3 for volumes A and B is taken into account, then the density for core to give the spin period is 1.55 g/cm^3. The procedure to calculate the spin of Mars is in the opposite way to Saturn. The mass of the volumes A and B, 6.402×10^{23}, is close to the mass, 6.418×10^{23}, of Mars, which means that the density of volumes A and B can be a little larger than the mean density of Mars, because the core is included in the volumes A and B. The best combination to give the rotation period, 24.623 h, of Mars may be chosen 3.968 g/cm^3 for volumes A and B and 15.289 g/cm^3 for core. If the mean density, 3.933 g/cm^3, of Mars is used in the density of volumes A and B, then the density for core to give the spin period is unreasonably large, 119.5 g/cm^3.

The masses of the volumes A and B for Jupiter, Uranus, and Neptune are a little larger than half of their masses, for Jupiter 1.002×10^{28} kg/1.899×10^{28} kg, for Uranus 5.24×10^{25} kg/8.683×10^{25} kg, and for Neptune 5.319×10^{25} kg/10.24×10^{25} kg, so that the mean densities of the planets can be used for volumes A and B. As summarized in Table 9.2, rotation periods of planets are calculated accurately in a very consistent way using the densities for volumes A and B and densities for cores, which proves that the theory and equations to obtain spins of the planets including the Sun are valid and universal.

Table 9.3 All numbers of satellite data used in the article

		Mass ($\times 10^{21}$ kg)	Distance ($\times 10^7$)	Mean density (g/cm^3)	Orbital period (days)	
Mars	Phobos	10.6×10^{-6}	0.99378	1.9	0.31891	S
	Deimos	2.4×10^{-6}	2.3459	1.75	1.26244	S
Jupiter	Io	89.32	42.16	3.53	1.769	S
	Europa	48.0	67.06	3.01	3.551	S
	Ganymede	148.19	107.04	1.94	7.154	S
	Callisto	107.59	188.27	1.83	16.689	S
Saturn	Mimas	0.0379	18.552	1.15	0.942	S
	Enceladus	0.108	23.802	1.61	1.370	S
	Tethys	0.618	29.466	0.96	1.888	S
	Dione	1.1	37.740	1.47	2.737	S
	Rhea	2.31	52.704	1.23	4.518	S
	Titan	134.55	122.183	1.88	15.945	S
	Hyperion	0.0055	148.11	0.57	21.277	C
	Iapetus	1.81	356.13	1.09	79.330	S
Uranus	Miranda	0.066	12.939	1.2	1.413	S
	Ariel	1.35	19.102	1.67	2.520	S
	Umbriel	1.17	26.630	1.4	4.144	S
	Titania	3.52	43.591	1.71	8.706	S
	Oberon	3.01	58.352	1.63	13.463	S
Neptune	Proteus	0.05	11.7647	–	1.122	
	Triton	21.4	35.476	2.05	5.877	S
	Nereid	0.03	551.34	–	360.136	

Since Triton is a dominant satellite in Neptune, the spin of Neptune might be calculated by the satellite alone in a good approximation. If the mass and the moment of inertia of Triton are subtracted by those of two other biggest satellites shown in Table 9.3 to consider the effect of small satellites, the core density of Neptune to produce the exact spin period increases to 4.418 from 3.677 g/cm^3. To maintain a stable orbit between a planet and a satellite, torque and reaction torque with respect to CMI should be in the same direction as in Fig. 9.2. If torque and reaction torque with respect to CMI is in the opposite direction as in Neptune, the orbit might be unstable until the torque decreases to zero because orbital angular accelerations and angular velocities of the planet and its satellite induced by the torque and the reaction torque are in the opposite directions.

Pluto has a dominant satellite, Charon, to be a binary system. It is easy to find that Eq. 9.12 predicts the same periods of rotations if masses and densities of two bodies are equal. Tidal forces of Pluto and Charon might lock their spins synchronous to their orbital rotation, though the mass of Pluto is about 8 times larger than Charon.

9.6 Experiment to Measure the Reaction Torque

As stated in Fig. 9.2, the force $\mathbf{f}_1$ in the reaction torque is a key to the Earth spin, which is about ten times larger than the force $\mathbf{f}_2$ on the Moon. Since the magnitude and the direction of the force in the reaction torque are dependent on the position along the radius and the latitude of the Earth, it is expected that the force will be an important factor for a lot of Earth activities, flow of seawater, tides, air flows, volcanoes, earthquakes, etc. The possibility to measure the force is discussed in Fig. 9.12. Since the force is largest at the point A_1, a pendulum in a chamber as shown in the magnification in Fig. 9.12 is located at the point A_1, where the chamber is just for blocking air flow to prevent any shift of pendulum by air.

If the Earth surface at the chamber is not accelerated by the force $\mathbf{f}_1$, then the pendulum will move by a certain degree from the perpendicular direction to balance with the sine component of gravitation. An accurate measurement of the pendulum displacement induced by the libration of the Moon will provide information on the reaction torque. Magnitude and direction of the force in the reaction torque might be decided by measuring the displacement of pendulum, the position of the Moon relative to the Earth, and eccentricity of the Moon.

For the libration of the Moon facing 58 % of its surface to the Earth, the largest possible shift of pendulum at the equator is calculated to be 0.3043 nm with 1 m pendulum which can be magnified to 152.2 µm by a 5 km laser beam as a 5 m pendulum and a 5 cm mirror are used as shown in Fig. 9.13. From Eq. 9.3 and Fig. 9.13a,

$$r'_1 \times \Delta m_1 a_1 = r'_2 \times \Delta m_2 i_2 \left(\frac{2Gm_1 R_m}{d^3} \right) \sin \delta°, \qquad (9.32)$$

where i_2 is a unit vector directing a_2 and the quantity in the parentheses is a magnitude of a_2 by the Earth's tidal force on the surface of the Moon multiplied by $\sin \Delta°$ from 4 % (7°) shift of the highest point with assumption that the highest point moves by the same angle of 7° to calculate the largest possible shift of pendulum.

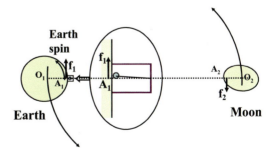

Fig. 9.12 A possible experiment to measure the reaction torque

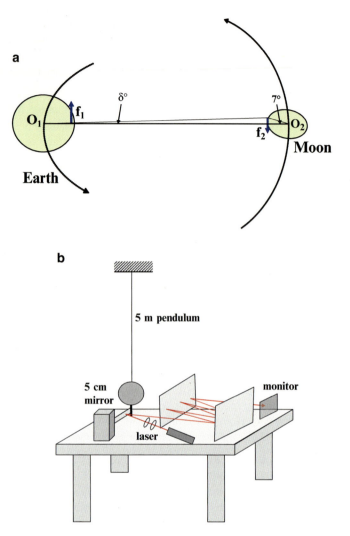

Fig. 9.13 The largest possible shift of pendulum at the equator is magnified to 152.2 μm

The force in the reaction torque has a ratio, a_1/g, with the gravitational force at the Earth surface:

$$\frac{a_1}{g} = \frac{[r'_2 \Delta m_2 \left(2Gm_1 R_m/d^3\right) \sin \delta°]}{[r'_1 \Delta m_1 \left(Gm_1/R_e^2\right)]} \quad (9.33)$$

$$d \sin \delta° = R_m \sin 7° \quad (9.34)$$

The ratio a_1/g is calculated as 1.522×10^{-10} to give a 0.3043 nm shift with 1 m pendulum for 58 % of Moon face. A relation $\Delta m_1/\Delta m_2 = m_1/m_2$ is used.

9 Classical Physics to Calculate Rotation Periods of Planets and the Sun

Figure 9.13b shows a possible experiment to measure a 0.3043 nm shift. A pendulum of 5 m length and a laser beam with multiple reflection mirrors are set on an optical table. The total path length of laser beam is 5 km. A monitor displays a shift of laser beam magnified to 152.2 μm. The whole system is enclosed by a chamber to block air flow.

9.7 Conclusion

As the Earth acts a torque to rotate the Moon synchronously with the orbital motion, there should be a reaction torque on the Earth, and the torque and the reaction torque should have the same magnitude and direction with respect to CMI. The rate of the Earth spin can be calculated by almost an exact number using the fundamental quantities of the Earth-Moon system. The calculated numbers of the Earth spin are $23^h39^m2^s$ without taking into account the density distribution and $24^h3^m5^s$ with taking into account the density distribution.

It is also shown that the Earth spin axis which is inclined by 23.45° with respect to the Earth orbit can be derived from the gravitation of the Sun acting on the Earth. The ratio of the horizontal component to the vertical component of the spin velocity approaches to unity. The calculated number, 23.487°, is close to the observation, 23.45°. The obliquities of planets to their orbits are dominantly dispersed near 23.45° and 0° which are stable or equilibrium angles.

The spin of the Sun was also calculated by the same way as the Earth by reducing the many-body system into a two-body system. It is based on the fact that the weighted collective contributions determine a single value of spin in the equator area of the Sun, which is assumed to be affected by a single imaginary planet. The rotation rates of planets Mars, Jupiter, Saturn, Uranus, and Neptune were also calculated by the same theory and equations as those of the Earth and the Sun with a slight modification due to the fact that the CMIs locate inside the planets. All the periods of the spins of the planets were calculated in a very consistent way using the densities for cores and volumes A and B. Considering $10^{24\sim30}$ order of astronomical numbers involved in the equations, the calculated values are considerably close to the observations to validate calculations. Finally, a possible experiment to measure the reaction torque in the Earth is discussed.

Appendix

Validity of the Initial Velocities Being Zero

The assumption in Eq. 9.4 that initial velocities $\mathbf{v}_{10}$, $\mathbf{v}_{20}$ which had not been driven by the forces $\mathbf{f}_1$ and $\mathbf{f}_2$ will be damped out may be validated by estimating the time to take for an initial velocity to decrease to a hundredth of its initial value. The increased radius, ΔR_e, of the Earth by the tidal force of the Moon can be inferred by an equation:

$$-\frac{Gm_1}{(R_e + \Delta R_e)} \approx -\frac{Gm_1}{R_e} + \left(\frac{Gm_2}{d - R_e} - \frac{Gm_2}{d}\right), \tag{9.35}$$

where the gravitational potential at the increased radius is equal to the gravitational potential increased by the tidal force of the Moon. The increased radius is calculated $\Delta R_e = 21.6$ m. The volume of the increased radius is given by $\Delta V \approx (4\pi R_e^2 \Delta R_e)$ (1/3) (1/2), where the factor 1/3 comes from an increase in one direction among three perpendicular directions and the factor 1/2 from a rough average of the increased radius. The mass is given by $\Delta m \approx \Delta V \rho_e$ with the average density ρ_e of the Earth. For more accuracy, the radius of the Earth can be replaced by a mass-averaged radius $R_e' = [\int r \rho_e (4\pi r^2 dr)]/m \approx (3/4)R_e$ to use the average density ρ_e of the Earth instead of the surface density of the Earth. As the Earth rotates, gravitational potential of the mass Δm increases and decreases twice a rotation by height $\Delta R_e' \approx \zeta(3/4)\Delta R_e = \zeta \times 16.2$ m, where ζ is a constant representing the difference between the real increased radius and the calculated one here. The whole volume of increased radius per rotation may be approximated by $2 \times \Delta V$, because there are two perpendicular directions on the equator plane. The potential energy loss per rotation is, approximately, $\Delta W \approx \eta 2 \cdot 2 \cdot \Delta m \cdot g \cdot \Delta R_e'$, which must be a loss of the kinetic energy of the Earth per rotation. The constant η is a nonconservative potential energy loss rate as the mass Δm moves by the increased radius. The number of rotations to lose all the kinetic energy of spin is $E_k/\Delta W \approx (1/2)I\omega^2/\Delta W$. If the initial velocity v_{10} makes 100 rotations per year, then the number is 5.86×10^6 for $\zeta = 1$ and $\eta = 1$. It takes 2.7×10^5 years, many orders less than the Earth age, for the initial spin to be damped to one rotation per year, where $R = R_0\exp(-Ct)$, $N_0 - 100 = N_0\exp(-Ct_0)$, $R_0 = 100$ rotations per year, and $R = 1$ rotation per year to calculate the time, $N_0 = 5.86 \times 10^6$ and $t_0 = 1$ year to calculate the constant C, are used. Even for $\zeta = 0.1$ and $\eta = 0.1$, it takes 2.7×10^8 years, still an order less than the Earth age.

As the same calculations are made with the reduced two-body systems of planets, for $\zeta = 1$ and $\eta = 1$, it takes 1.77×10^7 years for Jupiter, 4.55×10^7 years for Saturn, 6.98×10^7 years for Uranus, and 1.93×10^7 years for Neptune, for 100 rotation per year of a initial spin to be damped to one rotation per year. For the Sun, it takes 1.93×10^{13} years, but if the imaginary planet were in a distance of Mercury in early stage of solar system, it would take 5.63×10^7 years. For Mars, it takes 1.46×10^{12} years calculated from its satellites. If the tidal force of the Sun was taken into account in much closer distance at the early stage, it may take much shorter time by many orders, which may apply to other planets. Although it is a very rough calculation, the assumption that if they had not been driven by torques, initial velocities were damped out to negligibly small values may be validated.

One further validity may be inferred from the spins of satellites in Table 9.3 where all major satellites but Hyperion are rotating synchronously to their orbital periods. It proves that if their initial spins were much faster than the synchronous spins, kinetic energy losses of satellites due to gravitations of mother planets were large enough for the spins of the satellites to decrease to the synchronous spins. If it is the same for the planets, the remaining initial spins of the planets are, at most, smaller than the synchronous spins.

Influence of the Fast Rotating Planets on the Spin of the Sun

If the Sun acts torques on the fast rotating planets, Earth, Mars, Jupiter, Saturn, Uranus, Neptune, in the direction to decrease the speeds of spins, the reaction torques would make the Sun rotate faster than the planets in the opposite direction to the current spin of the Sun. A possible reason why it does not happen may be explained by a small or negligible magnitude of the torques being acted by the Sun. With reference to Fig. 9.2, for the gravitation of the Sun to make a torque on the Earth, the volume of increased radius must be in a certain angle between $0°$ and $90°$ or $-90°$ with respect to the line connecting two centers of the Sun and the Earth. As the Earth rotates so fast, the increased radius may appear in an angle of ineffective direction or the volume of increased radius may spread over all $360°$, due to the retarded response of the Earth materials. In this case, the Sun would not make a realistic torque on the Earth, which would be the same with the other fast rotating planets.

To explain the reaction torque on the Sun to hold the current spin of the Sun, the fast rotating planets would have some elliptic shape of orbital trajectories with their own satellites due to the tidal forces of the Sun. As the planets rotate around the Sun, the long axis of the elliptic orbit would always direct toward the Sun just as the synchronous rotation of the Moon. If the long axis of elliptic orbit deflects from the line connecting two centers of the Sun and the planet, then the gravitational force of the Sun would act a torque to make the elliptic orbit rotate synchronously with the orbital motion of the planet around the Sun. From this torque, there should be a reaction torque on the Sun to hold the current spin.

References

Allen CW (1973) Astrophysical quantities, 3rd edn. Athlone Press, London
Boss AP (2000) Possible rapid gas giant planet formation in the solar nebular and other protoplanetary disks. Astrophys J 536:L101–L104
Chamberlin TC (1897) A group of hypotheses bearing on climate change. J Geol 5:653–683
Dones L, Tremaine S (1993) Why does the Earth spin forward? Science 259:350–354
Giampieri G, Dougherty MK, Smith EJ, Russell CT (2006) A regular period for Saturn's magnetic field that may track its internal rotation. Nature 441:62–64
Giuli RT (1968) On the rotation of the Earth produced by gravitational accretion of particles. Icarus 8:301–323

Greenberg R, Fischer M, Valsecchi GB, Carusi A (1997) Sources of planetary rotation: mapping planetesimals' contributions to angular momentum. Icarus 129:384–400

Hughes DW (2003) Planetary spin. Planet Space Sci 51:517–523

Ida S, Nakazawa K (1990) Did rotation of the protoplanets originate from the successive collisions of planetesimals? Icarus 86:561–573

Kokubo E, Ida S (2007) Formation of terrestrial planets from protoplanets. II. Statistics of planetary spin. Astrophys J 671:2082–2090

Kortenkamp SJ, Kokubo E, Weidenschilling SJ (2000) Formation of planetary embryos. In: Canup RM, Righter K (eds) Origin of the earth and moon. The University of Arizona Press, Tucson, Arizona, US, pp 85–100

Lissauer JJ (1993) Planet formation. Annu Rev Astron Astrophys 31:129–174

Lissauer JJ, Kary DM (1991) The origin of the systematic component of planetary rotation I. Planet on a circular orbit. Icarus 94:126–159

Lissauer JJ, Stewart GR (1993) Growth of planets from planetesimals. In: Levy EH, Lunine JI (eds) Protostars and planets III. The University of Arizona Press, Tucson, Arizona, US, pp 1061–1088

Lissauer JJ, Dones L, Ohtsuki K (2000) Origin and evolution of terrestrial planet rotation. In: Canup RM, Righter K (eds) Origin of the earth and moon. The University of Arizona Press, Tucson, Arizona, US, pp 101–112

Ohtsuki K, Ida S (1998) Planetary rotation by accretion of planetesimals with nonuniform spatial distribution formed by the planet's gravitational perturbation. Icarus 131:393–420

Park S (2008) Why are the Earth spin and its axis 24 hours, tilted by 23.5 degree?. APS April Meeting and HEDP/HEDLA Meeting 53(5)

Park S (2013) MathcadText.pdf. http://sites.google.com/site/sahnggi

Podolak M (2007) The case of Saturn's spin. Science 37:1330–1331

Pollack JB, Hubickyj O, Bodenheimer P, Lissauer J, Podolak M, Greenzweig Y (1996) Formation of the giant planets by concurrent accretion of solids and gas. Icarus 124:62–85

Smith EP, Jacobs KC (1973) Introductory astronomy and astrophysics. W. B. Saunders, Philadelphia

Stevenson DJ (2006) A new spin on Saturn. Nature 441:34–35

Wetherill GW (1990) Formation of the Earth. Annu Rev Earth Planet Sci 18:205–256

Williams DR (2010) Planetary fact sheet – metric. http://nssdc.gsfc.nasa.gov/planetary/factsheet/index.html. Accessed 17 Nov 2010

Chapter 10
Estimates of the Size of the Ionosphere of Comet 67P/Churyumov–Gerasimenko During Its Perihelion Passage in 2014/2015

Wing-Huen Ip

Abstract The Rosetta mission will begin its comet rendezvous and lander mission in mid 2014. The plasma instruments onboard Rosetta will provide detailed measurements of the plasma environment of comet 67P/Churyumov–Gerasimenko and its responses to solar wind interaction. In this chapter the basic scale lengths connected to the dimension of the radial-expanding magnetic field-free ionosphere and to the boundary separating photoelectrons cooled by collisional interaction with water molecules, respectively, are explained. Their variations at different heliocentric distances along the orbit of comet 67P are described. It is found that the radii of the diamagnetic ionospheric cavity ($\sim$35 km at perihelion) and that of the cold photoelectron zone ($\sim$350 km at perihelion) are all much smaller than those of comet Halley. These theoretical estimates will be tested by the upcoming in situ plasma measurements and remote-sensing observations onboard the Rosetta spacecraft.

Keywords Rosetta mission • Comet 67P • Solar wind interaction

10.1 Introduction

The Rosetta mission for comet rendezvous and nucleus landing measurements of comet 67P/Churyumov–Gerasimenko provides a unique opportunity to study the structure, composition, and dynamics of the neutral gas, dust, and plasma emitted from the nucleus surface. The Rosetta spacecraft will start to approach the target comet after being successfully reactivated from deep-space hibernation on January 20, 2014. It will arrive at the flight-formation orbit around comet 67P in May in preparation of the Philae lander deployment in November. After accomplishment of

---

W.-H. Ip (✉)
Institute of Astronomy, National Central University, Taoyuan, Taiwan

Space Science Institute, Macau University of Science and Technology, Macau, China
e-mail: wingip@astro.ncu.edu.tw

© Springer-Verlag Berlin Heidelberg 2015   271
S. Jin et al. (eds.), *Planetary Exploration and Science: Recent Results and Advances*,
Springer Geophysics, DOI 10.1007/978-3-662-45052-9_10

Fig. 10.1 The particle-and-field experiments included in the Rosetta Plasma Consortium (RPC). ICA is Ion Composition Analyser, IES is Ion and Electron Sensor, LAP is Langmuir Probe, MAG is Fluxgate Magnetometer, MIP is Mutual Impedance Probe, and PIU is Plasma Interface Unit (Courtesy of European Space Agency)

this exciting phase of the mission operation, the spacecraft will continue to move around the comet to perform remote-sensing observations and in situ measurements of the gas and dust coma.

The Rosetta Plasma Consortium (RPC) package with a full complement of particle-and-field experiments (see Fig. 10.1) will carry out unprecedented study of the process of comet–solar wind interaction. It is unprecedented because never before did we have the opportunity to monitor the cometary plasma environment as a comet moves from large heliocentric distance ($r > 3$ AU) to $r \sim 1.5$ AU at perihelion, and beyond, at its vicinity.

To prepare for the plasma measurements, several types of theoretical models of high level of sophistication have been developed over the year. These include MHD computations (Benna and Mahaffy 2006; Rubin et al. 2012) and hybrid kinetic simulations (Motschmann and Kuehrt 2006; Koenders et al. 2013). The MHD models can provide accurate descriptions of the global behavior of cometary plasma dynamics and physical parameters (ion density, flow velocity, and temperature) as long as the finite gyroradius effect of the cometary ions can be ignored. If the ion gyroradii are comparable to the length scale of the solar wind interaction region, for example, when comet 67P is at $r > 2$ AU, hybrid kinetic simulations with the ions

10 Estimates of the Size of the Ionosphere of Comet 67P/Churyumov–... 273

taken as individual test particles and electrons as fluid will provide a more realistic picture (Motschmann and Kuehrt 2006). As the cometary outgassing rate increases when r decreases below 2 AU, both models would give similar results according to the recent works by M. Rubin and C. Koenders (private communication, 2014).

10.2 Diamagnetic Ionospheric Cavity

All cometary ions have their source from the neutral gas coma via photoionization, solar wind charge exchange, and electron impact ionization. In the inner region of the cometary ionosphere, the plasma dynamics is controlled by collisional effect with the neutral molecules in radial expansion from the central nucleus. It is so much so that the ions have the same radial velocity as the neutral gas up to a boundary where they will be decoupled from the gas stream because of the Lorentz force. What is the size of the ionospheric cavity containing the unmagnetized plasma? That was the question asked prior to the Giotto mission to comet Halley (Ip and Axford 1982).

A simple approximation can be obtained by considering that the collisional drag force on the cometary ions from the neutral gas is counteracted by the curvature force of the bent magnetic field lines, namely, $j \times B/c = k \, m \, n_i \, n_g \, v$, where k ($\sim 2 \times 10^{-9}$ cm^3 s^{-1}) is the rate coefficient of ion–molecule reaction (e.g., $H_2O^+ + H_2O \rightarrow H_3O^+ + OH$), m is the mass of the water molecule, n_i is the ion number density, n_g is the neutral gas number density, and v (~ 1 km s^{-1}) is the expansion speed of the neutral gas. For steady-state and spherically symmetric condition, n_g at cometocentric distance R can be expressed in terms of the gas production rate (Q) as $n_g = Q/4\pi v R^2$. In turn, the ion number density will be determined by equating the ion production rate via photoionization to the ion loss rate via electron dissociative recombination. As a result, $n_i = \sqrt{(n_g/\alpha \tau_i)}$, where α ($=10^{-7}$ cm^3 s^{-1}) is the electron dissociative recombination rate and τ_i is the photoionization rate taken to be $\tau_i = 10^6 \, r^2$ s with the heliocentric distance in AU to account for the radial variation of the ionizing photon flux.

At the radial position ($R = R_{max}$) where B reaches the maximum value (B_{max}), the force balance equation can be written as

$$\frac{B_{max}^2}{4\pi R_{max}} = Km \, n_i n_g v, \tag{10.1}$$

from which we obtain

$$R_{max} \sim 1.34 \times 10^{-17} \left(Q^{3/4}/B_{max}/\sqrt{r} \right) \text{ cm,} \tag{10.2}$$

with B_{max} in nT and r in AU. For comet Halley near the Giotto encounter, $Q \sim 8 \times 10^{29}$ H_2O s^{-1} and $B_{max} \sim 90$ nT, we can find $R_{max} \sim 4,000$ km.

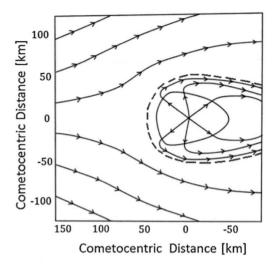

Fig. 10.2 A schematic view of the Rubin et al. (2012) model of the structure of the cometary ionosphere of comet 67P near perihelion

The magnetometer measurements on Giotto (Neubauer et al. 1986) showed that the theoretical estimate was nearly correct. Some subsequent first-order one-dimensional model calculations (Cravens 1986; Ip and Axford 1987) demonstrated that the basic structure of the diamagnetic ionospheric boundary such as the sharp drop-off of the magnetic field to near zero can be well described by the following equation (Fig. 10.2) (Ip and Axford 1987):

$$B\left(\frac{R}{R_{max}}\right) = B_{max} \frac{[1 + 2\ \ln(R/R_{max})]^{1/2}}{[R/R_{max}]} \quad (10.3)$$

If we use Table 6 of comet 67P's gas production rates of H_2O and CO_2 at different values of r given in Snodgrass et al. (2013) and the pressure balance condition, $B_{max}^2/8\pi \sim mn_{sw}v_{sw}^2$, where m is the proton mass and n_{sw} and v_{sw} are, respectively, the solar wind density and velocity ($\sim$400 km s^{-1}), at r with $n_{sw} = 5/r^2$ (protons cm^{-3}), we can find the size of the diamagnetic ionospheric cavity as a function of r from the application of Eq. 10.2.

Figure 10.3 shows that at $r \sim 3$ AU where the Rosetta spacecraft is to deliver the lander, $R_{max} \sim 10$ km. This means that the solar wind should penetrate all the way to close distances to the comet nucleus according to this simplified treatment. At perihelion, R_{max} will increase to 35 km. This result is in agreement with MHD model calculations (Rubin et al. 2012). Thus, if the Rosetta spacecraft can move to a cometocentric distance of about 30–50 km, the RPC instrument package will have the possibility to transverse the critical region where the ionospheric plasma is decoupled from the neutral gas outflow and be stagnated.

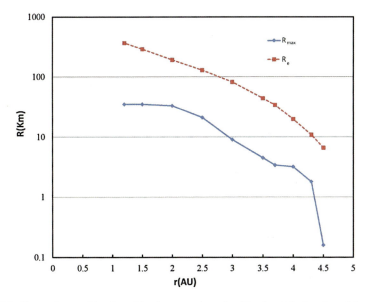

Fig. 10.3 Comparisons of the size of the diamagnetic cavity (R_{max}) and the location of the electron temperature discontinuity (R_e), respectively, at different heliocentric distances of comet 67P

10.3 Ion Pile-Up Region

The exact profile of the magnetic field ahead of the diamagnetic cavity is determined by a number of factors. The spatial and temporal variations of the draped interplanetary magnetic field (IMF) could lead to the formation of thin current sheets and probably flux ropes also. A case in point is the so-called ion tail disconnection events. It was suggested that encounter of a cometary ionosphere with the sector boundary of the IMF where the magnetic polarity reverses by 180° could lead to magnetic reconnection and hence the detachment of the "old" ion tail from the cometary coma (Niedner and Brandt 1978; Yi et al. 1994).

Other mechanisms such as plasma instabilities generated by interaction with solar wind events (i.e., CMEs) have also been investigated (Ip and Mendis 1978; Wang 1991; Wegmann 2000; Voelzke 2005). Small-scale structures like the formation of ion rays could also be related to solar wind disturbances (Bonev and Jockers 1994). Based on the detection of an ion pile-up region at comet Halley by the plasma instruments of the Giotto spacecraft (Balsiger et al. 1986) and its subsequent interpretation as the signature of a sharp electron temperature discontinuity (from $T_e \sim 100$ K to about 10^4 K) in the cometary ionosphere (Ip et al. 1987; Gan and Cravens 1990), it has been proposed (Ip 1994) that expansion or contraction of the boundary of the electron temperature discontinuity could also lead to the ion source strength in cometary ion tail. This is because the molecular ion loss rates of H_2O^+, CO^+, and CO_2^+ are determined by the electron dissociative recombination rate which is proportional to $T_e^{-\beta}$ with $\beta \sim 0.5$.

During steady-state condition, the cometocentric distance of the electron temperature discontinuity (R_{tec}) could be scaled to a critical neutral gas density of $n \sim 10^6$ cm^{-3} according to Giotto observation at comet Halley and ground-based observations (Ip et al. 1988; Bouchez et al. 1999; Ip 2004). If the expansion speed can be fixed to be 1 km s^{-1}, we have $R_e \sim 9.1 \times 10^{-7}$ Q(H$_2$O)$^{1/2}$ cm. The expected R_e values of comet 67P at different heliocentric distances are shown in Fig. 10.3. It can be seen that $R_e \sim 370$ km at perihelion.

10.4 Discussion

Using physical arguments, we show that some important length scales of comet 67P (and any comets for this matter) can be derived analytically. It is found that the diamagnetic cavity of this weakly outgassing comet with peak outgassing rate given to be about 2×10^{27} H$_2$O s^{-1} at the perihelion distance of 1.25 AU (Snodgrass et al. 2013) should be small with $R_{max} < 35$ km most of the time. The ionospheric region with cold electrons because of the strong cooling effect of the water molecules is also small in size ($R_{tec} \sim 370$ km at perihelion). These estimates are derived by assuming steady state and spherical symmetry in the neutral coma structure. However, we expect large temporal variations of the cometary plasma environment under the influence of the cometary gas outflow (Rubin et al. 2012) and the ever-changing interplanetary magnetic field and solar wind disturbances. Should magnetic reconnection occur either at the front side (Niedner and Brandt 1978) or in the ion tail (Ip 2004; Russell et al. 1986), enhanced ionization effect might result (Ip 1979). Comprehensive measurements by the RPC plasma instruments would allow us to investigate these many intriguing processes. Furthermore, insights will be provided by intercomparison of the in situ particle-and-field observations with the remote-sensing imaging experiments such as the OSIRIS camera system and the MIRO microwave spectrometer. The MIRO observations will give key information on the gas production rates, while the OSIRIS observations could be used to monitor the ion emissions also.

10.5 Summary

The Rosetta mission to comet 67P/Churyumov–Gerasimenko is expected to bring many breakthroughs to our understanding of the origin and structures of comets. It will also give us an excellent opportunity to test our theoretical models of solar wind–comet interaction developed since the first space mission to comet Halley in 1986. Because of the large difference in the outgassing rates of comet Halley and comet 67P, the plasma instruments onboard Rosetta will explore a complete new regime of cometary plasma physics of which finite gyroradius effects will probably dominate. From simple analytic approximations, we find that the radius of the diamagnetic ionospheric cavity of comet 67P could be as small as 30–40 km at perihelion. The region of cold photoelectrons, or the boundary of the ion pile-up region discovered at comet Halley, could be no more than a few hundred km. This means that the plasma environment of comet 67P could be very turbulent reflecting the time variability of the cometary outgassing rate and the interplanetary condition.

Acknowledgment I would like to thank Prof. Shuanggen Jin for giving the opportunity of producing this paper. The useful discussions and comments from Prof. Monio Kartalev, Prof. Vladimir Baranov, Prof. Susan McKenna-Lawlor, Dr. Martin Rubin, and Dr. Christoph Koenders, at the ISSI team meetings on Modeling cometary environments in the context of the heritage of the Giotto mission to comet Halley and of forthcoming new observations at Comet 67P/Churyumov–Gerasimenko, are gratefully acknowledged. This work was partially supported by NSC 102-211-M-008-014 and Project 019/2010/A2 of the Macau Science and Technology Development Fund: MSAR No. 0166.

References

Balsiger H et al (1986) Ion composition and dynamics at Comet Halley. Nature 321:330–334
Benna M, Mahaffy PR (2006) New multi-fluid MHD model of comet 26P/Grigg-Skjellerup: extrapolation to comet 67P/Churyumov-Gerasimenko. Geophys Res Lett 33, L10103. doi:10.1029/2006GL026197
Bonev T, Jockers K (1994) H2O+ ions in the inner plasma tail of comet Austin 1990 V. Icarus 107:335–357
Bouchez AH, Brown ME, Spinrad H, Misch A (1999) Observations of the ion pile-up in comets Hale-Bopp and Hyakutake. Icarus 137:62–68
Cravens TE(1986) The physics of the cometary contact surface, in 20th ESLAB symposium on the exploration of Halley's Comet. In: Batrick B et al (eds) ESA SP-250, Noordwijk, pp 241–246
Gan L, Cravens TE (1990) Electron energetics in the inner coma of Comet Halley. J Geophys Res 95:6285–6303
Ip W-H (1979) Currents in the cometary atmosphere. Planet Space Sci 27:121–125
Ip W-H (1994) On a thermodynamic origin of the cometary ion rays. Astrophys J Lett 432:L143–L145
Ip W-H (2004) Global solar wind interaction and ionospheric dynamics. In: Festou MC, Keller HU, Weaver HA (eds) Comets II. LPI, Arizona, pp 605–629
Ip W-H, Axford WI (1982) Theories of physical processes in the cometary comae and ion tails. In: Wilkening LL (ed) Comets. Arizona University Press, Tucson, pp 588–636
Ip W-H, Axford WI (1987) The formation of a magnetic field free cavity at Comet Halley. Nature 325:418–419
Ip W-H, Mendis DA (1978) The flute instability as the trigger mechanism for disruption of cometary plasma tails. Astrophys J 223:671–675
Ip W-H, Schwenn R, Rosenbauer H et al (1987) An interpretation of the ion pile-up region outside the ionospheric contact surface. Astron Astrophys 187:132–136
Ip W-H, Spinrad H, McCarthy P (1988) A CCD observation of the water ion distribution in the coma of Comet P/Halley near the Giotto encounter. Astron Astrophys 206:129–132

Koenders C, Glassmeier K-H, Richter I et al (2013) Revisiting cometary bow shock positions. Planet Space Sci 87:85–95

Motschmann U, Kuehrt E (2006) Interaction of the solar wind with weak obstacles: hybrid simulations for weakly active comets and for Mars. Space Sci Rev 122:197–208

Neubauer FM et al (1986) Giotto magnetic field results on the boundaries of the pile-up region and the magnetic cavity. Astron Astrophys 187:73–79

Niedner MB Jr, Brandt JC (1978) Interplanetary gas. XXIII. Plasma tail disconnection events in comets: evidence for magnetic field line reconnection at interplanetary sector boundaries. Astrophys J 223:655–670

Rubin M, Hansen KC, Combi MR et al (2012) Kelvin-Helmholtz instabilities at the magnetic cavity boundary of comet 67P/Churyumov-Gerasimenko. J Geophys Res 117, A06227. doi:10.1029/2011JA017300

Russell CT, Saunders MA, Phillips JL, Feder JA (1986) Near-tail reconnection as the cause of cometary tail disconnections. J Geophys Res 91:1417–1423

Snodgrass C, Tubiana C, Bramich DM et al (2013) Beginning of activity in 67P/Churyumov-Gerasimenko and predictions for 2014–2015. Astron Astrophys 557:A33. doi:10.1051/0004-6361/201322020

Voelzke MR (2005) Disconnection events processes in cometary tails. Earth Moon Planets 97:399–409

Wang S (1991) A mechanism for the formation of knots, kinks and disconnection events in the plasma tail of comets. Astron Astrophys 243:521–530

Wegmann R (2000) The effect of some solar wind disturbances on the plasma tail of a comet: models and observations. Astron Astrophys 358:759–775

Yi Y, Caputo FM, Brandt JC (1994) Disconnection events (Des) and sector boundaries: the evidence from Comet Halley 1985-1986. Planet Space Sci 42:705–720

Chapter 11
Photometric and Spectroscopic Observations of Exoplanet Transit Events

Liyun Zhang and Qingfeng Pi

Abstract Firstly, we review simply the exoplanet transit events and introduce our ongoing optical wavelength project. Then, we present new photometric observations of the transiting planets (HAT-P-10b/WASP-11b, HAT-P-19b, WASP-43b, ...) using an 85-cm telescope of Xinglong Station of the National Astronomical Observatories of China in 2012. New transit curves are modeled using the JKTEBOP code and adopting the quadratic limb-darkening law. The preliminary parameters (the orbital inclination, relative radius ......) of these star-planet systems are re-obtained. We also revise the ephemeris with a transit epoch and discuss a period change by the observed minimum minus calculated transit times. Finally, our future plans in China are given and discussed.

Keywords Extrasolar planet • Photometry • Spectroscopy • Transit timing variation

11.1 Introduction

Astronomers have discovered 973 extrasolar planets up to 10 November 2013 by ground and space survey (catalogue from http://www.openexoplanetcatalogue. com). More and more scientists concentrated on the properties of extraplanets. The transits of extrasolar planets across the stellar disk cause dips of at most a few percent in the photometric light curves of their parent stars (see Fig. 11.1). It is a very promising method to detect and characterize extrasolar planets. The transit of exoplanet can provide information about transit time, duration, relative radii and orbital inclination which can be used to derive its mass and density. Study of transit timing variation (Diaz et al. 2008; Sozzetti et al. 2009) is very important because it

---

L. Zhang (✉) • Q. Pi
Department of Physics and NAOC-GZU-Sponsored Center for Astronomy Research,
College of Science, Guizhou University, Guiyang 550025, China
e-mail: liy_zhang@hotmail.com

© Springer-Verlag Berlin Heidelberg 2015
S. Jin et al. (eds.), *Planetary Exploration and Science: Recent Results and Advances*,
Springer Geophysics, DOI 10.1007/978-3-662-45052-9_11

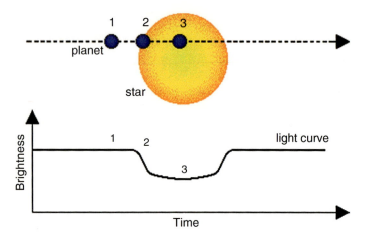

Fig. 11.1 The model of transit of extrasolar planet (From Kepler Internet)

may reveal the effect of other perturbing planets in the exoplanet systems (Steffen and Agol 2005) or moons of the transiting exoplanet (Szabo Gy et al. 2006; Simon et al. 2007; Kipping 2009a, b).

Ca $_{II}$ H&K lines are classical diagnostic and indicators of chromospheric activity and became a classical activity diagnostic. Star-planet in stellar chromospheres could produce the enhanced Ca $_{II}$ H&K flux following the periodicity of the planet orbiting its host star. If modulation of chromospheric emission has a period comparable to that of the planet's orbital, then it is likely that the planet is stimulating the activity through some tidal and magnetic mechanism (Shkolnik et al. 2008). Recently, Shkolnik et al. (2003, 2005, 2008) reported on planet-induced chromospheric activities on several stars, such as HD 179949, apparent from the night to night modulation of the Ca $_{II}$ H&K chromospheric emission phase with the hot Jupiter's orbit (Fig. 11.2). It is a convincing case of planetary-induced modulation (Shkolnik et al. 2003). The chromospheric rotational modulation will offer further physical constraints on the interaction between the extrasolar planet and its host star.

11.2 Telescope and Plans

On the one hand, we plan to monitor the extrasolar Planets using an 85-cm telescope (Zhou et al. 2009) and 60-cm telescope at Xinglong Station of the National Astronomical Observatories of China (NAOC). The photometer of the 85-cm telescope was equipped with a 1024*1024 pixel CCD and the standard Johnson-Cousin-Bessell BVRcIc filters (Zhou et al. 2009). Our follow-up photometric objects are taken from the SuperWASP project: WASP-4b, WASP-36b... (Pollacco et al. 2006; Christian et al. 2006), HATNet project: HAT-P-10b/WASP-11b, HAT-P-19b,

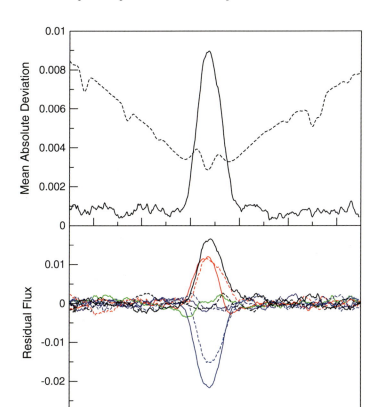

Fig. 11.2 *Top*: the mean absolute deviation of the Ca II K core (*solid line*) of HD 179949. The units are intensity as a fraction of the normalization level. Overlayed (*dashed line*) is the mean spectrum (*scaled down*) indicating that the activity on HD 179949 is confined to the K emission. *Bottom*: residuals from the normalized mean spectrum of the Ca II K core of HD 179949 (With permission from Shkolnik et al. 2003)

HAT-P-20b, HAT-P-25b ... (Bakos et al. 2006), CoRoT project (Barge et al. 2008), XO project (MvVullough et al. 2005), Kepler catalogue (Borucki et al. 2010), and other exoplanets HD 198733B (Knutson et al. 2007), and Qatar-2b (Alsubai et al. 2011). The exoplanet transit database (http://var2.astro.cz/ETD/) provided all ever observed transits of transiting exoplanets to observers and researchers. We could select most suitable observational objects from the internet (exoplanet transit database) (Fig. 11.4).

On the other hand, we will study chromospheric emission on the host star of extraplanet system in several chromospheric activity indicators – Ca II H&K lines, Na I D1 D2 lines, H alpha, H beta, and other Balmer lines, He I D3 lines, and Ca II infrared triplet lines (especially Ca II H & K emission) induced by extrasolar planets

(Montes et al. 2004; Zhang 2011). Our telescope is a 2.16-m telescope at Xinglong Station of the National Astronomical Observatories of China. A spectral resolution of the Coude Echelle Spectrograph is about 37,000 with a spectral region of 380–910 nm (Zhao and Li 2001). Our objects are bright stars (brighter than 8 mag) that have known short-period (period <4 days) giant planet (open exoplanet catalogue).

11.3 Observations

Here we present new photometric observations of several transiting planets of HAT-P-10b/WASP-11b (West et al. 2009; Bakos et al. 2009), HAT-P-19b (Hartman et al. 2011), WASP-43b (Hellier et al. 2011; Gillon et al. 2012), WASP-36b, HAT-P-25b, and HAT-P-20b. They were carried out in 2012 with the 85-cm telescope at the Xinglong Station of the National Astronomical Observatories of China. Our observations were made in R band. The exposure times are about 60 s. All observed CCD images were reduced using the Apphot sub-package of IRAF in the standard fashion (bias subtraction, cosmic ray remove, flat-field correction using twilight sky exposure, and aperture photometry). The light curves of several transiting planets (HAT-P-10b/ WASP-11b, HAT-P-19b, WASP-43b, WASP-36b, HAT-P-25b, and HAT-P-20b) are displayed in Fig. 11.3, where Del Mag represents the differential magnitude between the objects and comparison star. The errors of our observations are better than 0.01 mag in R band. The published parameters of HAT-P-10b/WASP-11b, HAT-P-19b, and WASP-43b are listed in Table 11.1.

Fig. 11.3 The pictures of 85-cm (*left*) and 60-cm (*right*) telescopes at Xinglong Station, NAOC

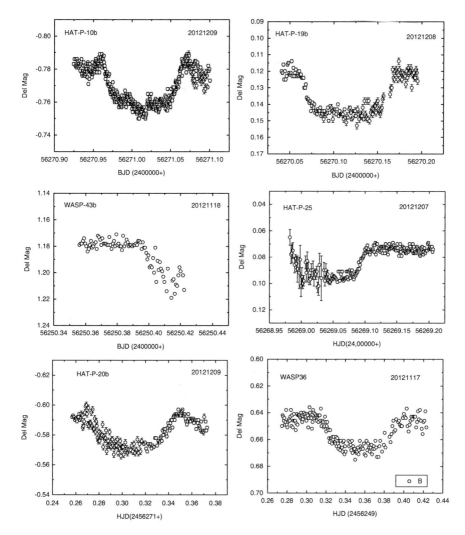

Fig. 11.4 New light curves of the observed transits (HAT-P-10b/ WASP-11b, HAT-P-19b, WASP-43b, WASP-36b, HAT-P-25b, and HAT-P-20b) using the 85-cm telescope at Xinglong Station, NAOC

11.4 Results and Analysis

11.4.1 Light Curve Analysis

Our transit curves are modeled using JKTEBOP code (Southworth 2008, 2009) and adopting quadratic limb-darkening law (Claret 2000). The JKTEBOP fits for the sum of fractional radii (ra + rb) and ratio of the radii (Southworth et al. 2005). The

Table 11.1 The published parameters of HAT-P-10b/WASP-11b, HAT-P-19b, and WASP-43b

Parameters	HAT-P-10b/WASP-11b	HAT-P-10b/WASP-11b	HAT-P-19b	WASP-43b	WASP-43b
Spectral	K dwarf	An early to mid K dwarf		K7 V	K7 V
K(=rb/rA)	0.132 ± 0.001	0.127 ± 0.017	0.1418 ± 0.002	0.160 ± 0.035	0.159 ± 0.002
Orbital inclination i(deg)	$88.6 + 0.5/-0.4$	$89.8 + 0.2/-0.8$	88.2 ± 0.4	$82.6 + 1.3/-0.9$	82.33 ± 0.20
Epoch (days) BJD	$2{,}454{,}759.68683 \pm 0.00016$	$2{,}454{,}473.05588 \pm 0.0002$	$2{,}455{,}091.53417 \pm 0.00034$	$245{,}528.86774 \pm 0.00014$	$2{,}455{,}726.54336 \pm 0.00012$
Period (days)	3.7224747 ± 0.0000065	3.722465 ± 0.000007	4.008778 ± 0.000006	0.813475 ± 0.000001	0.81347753 ± 0.0000007
Transit width		2.556 h	0.1182 ± 0.0014 days		1.2089 h $+ 0.024/-0.025$
References	Bakos et al. (2009)	West et al. (2009)	Hartman et al. (2011)	Hellier et al. (2011)	Gillon et al. (2012)

11 Photometric and Spectroscopic Observations of Exoplanet Transit Events

Table 11.2 The result about the transit and orbital parameters of HAT-P-10b/WASP-11b, HAT-P-19b, and WASP-43b

Parameters	HAT-P-10b/WASP-11b	HAT-P-19b	WASP-43b
rA + rb	0.088 ± 0.002	0.087 ± 0.003	
K(=rb/rA)	0.130 ± 0.002	0.141 ± 0.002	
Orbital inclination i(deg)	89.931 ± 8.52	89.836 ± 4.59	
Our transit minima (BJD)	2456,271.01479 ± 0.00009	2456,270.1160 ± 0.0003	2,456,250.4201 ± 0.0007
Our transit depth (mag)	0.0229 ± 0.0004	0.0251 ± 0.0006	0.026 ± 0.003
Our transit width (minute)	146.9 ± 1	162.3 ± 2	69.5 (fixed)
RMS of residuals (mmag)	3.733	3.029	
The orbital ephemeris (BJD)	2454,729.9071(2) + 3.7224810(9)	2,455,091.5354(2) + 4.0087788(1)	2,455,528.8685(1) + 0.81347438(2)

reason is that these quantities are only very weakly correlated for a wide variety of the light curve shapes, which is better than solution convergence. We only analyzed the data of HAT-P-10b/WASP-11b, HAT-P-19b, and WASP-43b. After a lot runs, we obtained the fractional radii (ra + rb), the ratio of the radii, orbital inclination, new transit minima, our transit depth, and transit width. Our new transit orbital parameters are listed in Table 11.2. The errors of the parameters on the light curves used the Monte Carlo simulation algorithm (Southworth et al. 2004, 2005). These methods could give us more reliable results (Fig. 11.6).

11.4.2 The Period Analysis

We also collected minima times from ETD (Poddany et al. 2010), AXA (Amateur Exoplanet Archive) and other published papers. The timings were transformed from HJD into BJD using the online applets2 developed by Eastman et al. (2010). Using a linear least square fit, new linear ephemeris was obtained in Table 11.2 and O-C diagrams are displayed in Fig. 11.5. The points represent the O-C data (The solid points represent our new data), and the solid lines represent polynomial fits. As can be seen from Fig. 11.5, it seems that the period variation shows polynomial curves.

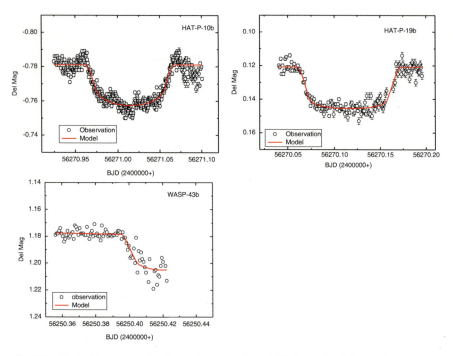

Fig. 11.5 The light curves of the observational transits and fitted models of the exoplanet systems HAT-P-10b/WASP-11b, HAT-P-19b, and WASP-43b

11.5 Conclusions

The results from the analysis of our new data can be summarized as follows:

1. Using the JKTEBOP code, we analyzed new observational light curves of HAT-P-10b/ WASP-11b, HAT-P-19b, and WASP-43b. The fractional radii (ra + rb), the ratio of the radii, the orbital inclination, the new transit minima, our transit depth, and transit width were obtained. The new orbital parameters (see Tables 11.1 and 11.2) of these systems are in agreement with the previous results (West et al. 2009; Bakos et al. 2009; Hellier et al. 2011; Gillon et al. 2012; etc).
2. The timing residuals giving a hint about a period change (see Fig. 11.5), which might be due to the presence of a second planet in the system or the sporadic asymmetries of transit curves due to star spots on stellar surface result from magnetic activity (Steffen and Agol 2005; Lee et al. 2012; etc). More photometric data are needed to confirm that.

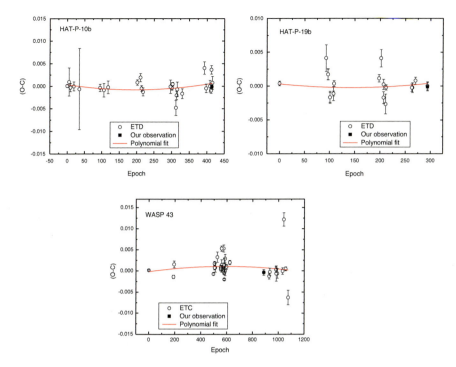

Fig. 11.6 O-C diagrams of the exoplanet systems of HAT-P-10b/ WASP-11b, HAT-P-19b, and WASP-43b

In the future, we will monitor the exoplanet transit events by photometry using an 85- and 60-cm telescope at Xinglong Station, NAOC, and magnetic interaction between an extrasolar planet and its parent star using high-resolution spectra.

Acknowledgements The authors would like to thank the observing assists of Xinglong Station for their help during our observations. This work is supported by the NSFC of No 10978010 and 11263001.

References

Alsubai KA, Parley NR, Bramich DM et al (2011) Qatar-1b: a hot Jupiter orbiting a metal-rich K dwarf star. MNRAS 417:709

Bakos GA, Noyes RW, Kovacs G et al (2006) HAT-P-1b: a large-radius, low-density exoplanet transiting one member of a stellar binary. Astrophys J 656:552

Bakos GA, Pal A, Torres G et al (2009) HAT-P-10b: a light and moderately hot Jupiter transiting A K Dwarf. Astrophys J 696:1950

Barge P, Baglin A, Auvergne M et al (2008) Transiting exoplanets from the CoRoT space mission. I. CoRoT-Exo-1b: a low-density short-period planet around a G0V star. AA 482L:17B

Borucki WJ, Koch D, Basri G et al (2010) Kepler planet-detection mission: introduction and first result. Science 327:977

Christian DJ, Polaacco DL, Skillen I et al (2006) The SuperWASP wide-field exoplanetary transit survey: candidates from fields 23 h < RA < 03 h. MNRAS 372:1117

Claret A (2000) A new non-linear limb-darkening law for LTE stellar atmosphere models. Calculations for -5.0 < = log[M/H] < = +1, 2000 K < = Teff < = 50000 K at several surface gravities. Astron Astrophys 363:1081

Diaz RF, Rojo P, Melita M et al (2008) Detection of period variations in extrasolar transiting planet OGLE-TR-111b. Astrophys J 682:L49

Eastman J, Siverd R, Gaudi BS (2010) Achieving better than 1 minute accuracy in the Heliocentric and Barycentric Julian Dates. PASP 122:935

Gillon M, Triaud AHMJ, Fortney JJ, The TRAPPIST et al (2012) The TRAPPIST survey of southern transiting planets. I. Thirty eclipses of the ultra-short period planet WASP-43 b. Astron Astrophys 542:4

Hartman JD, Bakos GA, Sato B et al (2011) HAT-P-18b and HAT-P-19b: two low-density Saturn-mass planets transiting metal-rich K stars. Astrophys J 726:52

Hellier C, Anderson DR, Collier Cameron A et al (2011) WASP-43b: the closest-orbiting hot Jupiter. Astron Astrophys 535:L7

Kipping DM (2009a) Transit timing effects due to an exomoon. MNRAS 392:181

Kipping DM (2009b) Transit timing effects due to an exomoon – II. MNRAS 396:1797

Knutson HA, Charbonneau D, Allen LE et al (2007) A map of the day-night contrast of the extrasolar planet HD 189733b. Nature 447:183

Lee JW, Youn JH, Kim SL et al (2012) The sub-Saturn mass transiting planet HAT-P-12b. Astrophys J 143:5

Montes D, Crepo-Chacon I, Galvez MC et al (2004) Cool stars: chromospheric activity, rotation, kinematic and age. LNEA 1:119

MvVullough PR, Stys JE, Valenti JA et al (2005) The XO project: searching for transiting extrasolar planet candidates. PASP 117:783

Poddany S, Brat L, Pejcha O (2010) Exoplanet transit database. Reduction and processing of the photometric data of exoplanet transits. New Astron 15:297

Pollacco DL, Skillen I, Collier CA et al (2006) The WASP project and the SuperWASP cameras. PASP 118:1407

Shkolnik E, Walker GAH, Bohlender DA (2003) Evidence for planet-induced chromospheric activity on HD 179949. Astrophys J 597:1092

Shkolnik E, Walker GAH, Bohlender DA et al (2005) Hot Jupiters and hot spots: the short- and long-term chromospheric activity on stars with giant planets. Astrophys J 622:1075

Shkolnik E, Bohlender DA, Walker GAH et al (2008) The on/off nature of star-planet interactions. Astrophys J 676:628

Simon A, Szatmary K, Szabo Gy M (2007) Determination of the size, mass, and density of "exomoons" from photometric transit timing variations. Astron Astrophys 470:727

Southworth J (2008) Homogeneous studies of transiting extrasolar planets – I. Light-curve analyses. MNRAS 386:1644

Southworth J (2009) Homogeneous studies of transiting extrasolar planets – II. Physical properties. MNRAS 394:272

Southworth J, Maxted PFL, Smalley B (2004) Eclipsing binaries in open clusers-I. V615 Per and V 618 Per in h persei. Mon Not Roy Astron Soc 349:547

Southworth J, Smalley B, Maxted PEL et al (2005) Absolute dimensions of detached eclipsing binaries-I the metallic-lined system WW Aurigae. Mon Not Roy Astron Soc 363:529

Sozzetti A (2009) A new spectroscopic and photometric analysis of the transiting planet systems TrES-3 and TrES-4. Astrophys J 691:1145

Steffen JH, Agol E (2005) An analysis of the transit times of TrES-1b. MNRAS 364:96

Szabo Gy M, Szatmary K, Zs D, Simon A (2006) Possibility of a photometric detection of "exomoons". Astron Astrophys 450:395

West RG, Collier CA, Heb L et al (2009) The sub-Jupiter mass transiting exoplanet WASP-11b. Astron Astrophys 502:395

Zhang L (2011) The chromospheric activity of late-type stars of the different rotational periods. Asrton Soc Pac Conf 451:123

Zhao G, Li HB (2001) The Coudé Echelle spectrograph for the Xinglong 2.16 telescope. Chin J Astron Astrophys 1:555

Zhou AY, Jiang XJ, Zhang YP et al (2009) MiCPhot: a prime-focus multicolor CCD photometer on the 85-cm telescope. Res Astron Astrophys 9:349

Chapter 12
Photochemistry of Terrestrial Exoplanet Atmospheres

Renyu Hu

Abstract Terrestrial exoplanets are exciting objects to study because they could be potential habitats for extraterrestrial life. Both the search and the characterization of terrestrial exoplanets are flourishing. Particularly, NASA's *Kepler* spacecraft has discovered Earth-sized planets receiving similar amount of radiative heat as Earth. Central in the studies of terrestrial exoplanets is to characterize their atmospheres and to search for potential biosignature gases (the atmospheric components that indicate biogenic surface emissions). To achieve this goal, a deep understanding of the key physical and chemical processes that control the atmospheric composition and the atmosphere-surface interaction is pivotal.

Keywords Terrestrial exoplanets • Astrobiology • Photochemistry • Radiative transfer

12.1 Terrestrial Exoplanets in Our Interstellar Neighborhood

One of the most exciting progresses in planetary exploration in the past decade is the discovery of terrestrial exoplanets. These celestial objects are planetary bodies outside the Solar System with masses within ten times Earth's mass or with radii within two times Earth's radius. Terrestrial exoplanets were first discovered by the "radial velocity" method (Rivera et al. 2005), i.e., by measuring the wobbling of a star via the Doppler shift of the stellar spectrum induced by an orbiting planet's gravitational force. The precision required for detecting an Earth-mass planet at the 1-AU orbit to a Sun-like star (referred to as a "true Earth analogue") is on the order of 0.1 m s^{-1}. The precision of the radial velocity measurements of bright stars is getting close to this requirement, and as a result, the radial velocity search has been finding terrestrial exoplanets that have sizes closer and closer to Earth (Udry et al.

R. Hu (✉)
Division of Geological and Planetary Sciences, California Institute of Technology,
Pasadena, CA 91125, USA
e-mail: ryh@gps.caltech.edu

© Springer-Verlag Berlin Heidelberg 2015
S. Jin et al. (eds.), *Planetary Exploration and Science: Recent Results and Advances*,
Springer Geophysics, DOI 10.1007/978-3-662-45052-9_12

2007; Mayor et al. 2009; Vogt et al. 2010; Rivera et al. 2010; Dawson and Fabrycky 2010; Howard et al. 2011; Bonfils et al. 2011). Recently, an Earth-mass terrestrial exoplanet was reported around Alpha Centauri B, one of the closest stellar systems from Earth (Dumusque et al. 2012).

Another method to search for terrestrial exoplanets is to observe the dimming of a star when a planet passes in front of the star as viewed from Earth (i.e., the transit method). The signal of the transit is proportional to the ratio between the size of the planet and the size of its parent star. Earth transiting the Sun as viewed from another planetary system would have a transit signal of $\sim$80 parts per million (ppm). Modern photometry technique has been able to provide this level of precision and therefore enabled the detection of terrestrial exoplanets via transits (Leger et al. 2009; Charbonneau et al. 2009; Winn et al. 2011; Demory et al. 2011; Dragomir et al. 2012; Van Grootel et al. 2014). In recent years, the exoplanet community has witnessed an explosive increase of the number of terrestrial exoplanets discovered by transits as a result of the *Kepler* mission that monitored 160,000 stars in the sky (Batalha et al. 2011; Lissauer et al. 2011; Cochran et al. 2011; Fressin et al. 2012; Gautier et al. 2012; Borucki et al. 2012; Muirhead et al. 2012; Borucki et al. 2013; Gilliland et al. 2013; Swift et al. 2013; Rowe et al. 2014). The smallest transiting planet that has been confirmed is only slightly larger than the Moon (Barclay et al. 2013a, b).

Based on the discoveries made by *Kepler*, statistically, we now know that a large number of stars in our interstellar neighborhood have terrestrial exoplanets. The occurrence rate of terrestrial exoplanets can be estimated based on the *Kepler* observations, with correction of the geometric effect (due to the fact that the transit technique is only sensitive to those planets that pass in front of their host stars periodically), the incompleteness of detection, and the false positive of signals (Howard et al. 2012; Fressin et al. 2013). It has been estimated that $\sim$30 % of stars in our interstellar neighborhood have terrestrial exoplanets that have radii within two times Earth's radius and orbital periods within 85 days (Fressin et al. 2013). It also turns out that the planets below twice the size of Earth are more populous than the planets above (Fressin et al. 2013; Petigura et al. 2013a, b). The occurrence rate of exoplanets found by the Kepler transit survey is also consistent with the finding of radial velocity surveys that are sensitive to very different observational biases, supporting the fidelity of this result (Figueira et al. 2012). When calculating the occurrence rate, terrestrial exoplanets are usually defined by their radii, because the transit technique can only measure the radii but not the masses. Measuring the masses of these small planets and confirming their rocky nature are ongoing and have been successful for a number of close-in objects (Pepe et al. 2013; Howard et al. 2013; Marcy et al. 2014).

A handful of the discovered terrestrial exoplanets are potentially habitable. A habitable planet is defined as a planet on the surface of which liquid water is stable. As the stellar radiation is the major heat source for a terrestrial exoplanet, the conventional habitable zone, the range of semimajor axes in which planets could be habitable, has been studied for main-sequence stars (Kasting et al. 1993). The conventional definition of the habitable zone relies upon the assumption that the planet has an N_2-dominated atmosphere with variable levels of CO_2 to provide appropriate levels of greenhouse effects. Such defined habitable zone around a Sun-like star is evaluated most recently at 0.99–1.70 AU (Kopparapu et al. 2013). Moreover, the range of the habitable zone can be considerably widened, if the atmosphere is H_2 dominated (with H_2-H_2 collision-induced absorption as the source of the greenhouse effect; Pierrehumbert and Gaidos 2011) or if the water content in the atmosphere is much lower than Earth (Zsom et al. 2013).

The search of habitable terrestrial exoplanets is difficult because both the radial velocity method and the transit method are more sensitive to planets that are closer to their parent stars. The amplitude of the radial velocity signal falls with the semimajor axis a as $a^{-1/2}$; the amplitude of transit signal does not depend on the semimajor axis, but the probability of transit due to the alignment between the star, the planet, and the observers is R_*/a, where R_* is the radius of the star. Several Jupiter- and Neptune-sized planets that are in the conventionally defined habitable zones of their host stars have been discovered by the radial velocity method (Udry et al. 2007; Vogt et al. 2010; Pepe et al. 2011; Bonfils et al. 2011; Tuomi et al. 2013). *Kepler* has found several sub-Neptune-sized planets in the habitable zones by the transit method (Borucki et al. 2012, 2013; Barclay et al. 2013a, b; Quintana et al. 2014). The frequency of terrestrial exoplanets in habitable zones of their host stars is estimated using *Kepler* observations to be $\sim$15 % for cool stars (Dressing and Charbonneau 2013) and $\sim$20 % for FGK stars (Traub 2012; Petigura et al. 2013b). Therefore, it is reasonable to expect at least one potentially habitable terrestrial planet in our interstellar neighborhood of a few tens of parsecs.

One sweet spot to look for habitable terrestrial exoplanets is around M dwarf stars. M dwarf stars have sizes of a fraction of that of the Sun, and M dwarf stars are the most common type of stars in the neighborhood of the Sun (Salpeter 1955). Recent surveys by *Kepler* have suggested that planets having radii within two times Earth's radius are more frequent around small M dwarfs than around FGK stars (Howard et al. 2012; Dressing and Charbonneau 2013). Moreover, because M dwarfs are considerably sub-luminous compared with the Sun, the habitable zone around an M dwarf is much closer to the star than the habitable zone around a Sun-like star (Kasting et al. 1993). As a result, habitable planets around M dwarfs would have higher transit probabilities and larger transit signals compared with their counterparts around FGK stars. In fact, the first Earth-sized exoplanet in a star's habitable zone was detected around an M dwarf (Quintana et al. 2014). However, this planet is too far away from Earth to allow atmospheric characterization. In parallel with *Kepler*, ground-based searches for terrestrial exoplanets around nearby M dwarfs have been ongoing (e.g., Nutzman and Charbonneau 2008), which have resulted in the discovery of a 2.7 Earth-radius planet orbiting an M4.5 star only 13 parsecs away (GJ 1214 b; Charbonneau et al. 2009).

The discovery of terrestrial exoplanets, especially those potentially habitable, has impacted profoundly our inquiry of the Universe. It is the first time when the human being can say for sure there are locations outside the Solar System that may have rocky environment, widely accepted as a prerequisite for life to emerge. As a milestone in the search for planets that might harbor life, terrestrial exoplanets are important objects to study for at least three reasons: (1) the detection of terrestrial exoplanets provides a large ensemble of planets and planetary systems that enable comparative planetology of Earth-like planets beyond the Solar System; (2) terrestrial exoplanets may themselves be habitable if they receive appropriate radiative heat from their parent stars; (3) characterization of atmospheres and surfaces of terrestrial exoplanets, starting from planets that are larger than Earths (referred to as "super-Earths"), and the expertise of instrumentation, observation, and data interpretation techniques gained will serve as indispensable stepping stones for eventually characterizing Earth-sized planets that are potentially habitable.

12.2 Observations of Exoplanet Atmospheres

The atmosphere on an exoplanet can be analyzed by spectroscopy. If an exoplanet could be directly imaged, the light from the planet's atmosphere, either planetary thermal emission or reflection of the stellar light, could be analyzed via spectroscopy to determine the composition of its atmosphere. Such observations are extremely challenging due to the existence of a much stronger radiation source at close angular proximity (i.e., the host star). Therefore, direct spectroscopy studies of exoplanets have only been performed for Jupiter-sized or even larger giant planets. The first high-resolution spectrum of a directly imaged exoplanet (a nascent gas giant 40 AU from its host star) has recently been reported (Konopacky et al. 2013).

A newly developed method to mitigate the weak signal of an exoplanet without spatially resolving the planet or nulling the stellar light is to make use the information of the planet's orbital motion. A correlation between the star's radial velocity and the radial velocity of a certain group of molecular lines (e.g., CO) was used to establish the existence of the molecule in several giant planets' atmosphere (Brogi et al. 2012; Rodler et al. 2012, 2013; de Kok et al. 2013; Birkby et al. 2013; Lockwood et al. 2014).

At current stage and in the near future, however, characterization of the atmospheres of terrestrial exoplanets focuses on the planets that transit. The predictable on-and-off features of a planet's radiation when the planet passes behind its host star (referred to as "occultation") can be observed by monitoring the total light from the star-planet system in and out of transits (e.g., Seager and Sasselov 1998; Seager et al. 2000). In addition, when the planet passes in front of its host star (referred to as "transit"), parts of the stellar radiation may transmit through the planet's atmosphere and carry the information of the atmospheric composition (Seager and Sasselov 2000). Soon after the first detection of an exoplanet atmosphere via transit (Charbonneau et al. 2002) and the first detection of thermal emission from an exoplanet atmosphere via occultation (Charbonneau et al. 2005; Deming et al. 2005), both methods have been successful in characterizing extrasolar giant planets (e.g., Seager and Deming 2010 and references therein). Recently, attempts to observe super-Earth atmospheres are growing (e.g., Demory et al. 2012; Knutson et al. 2014), and the super-Earth/mini Neptune GJ 1214 b is being observed in as much detail as possible (e.g., Bean et al. 2010; Croll et al. 2011; Desert et al. 2011; Berta et al. 2012; de Mooij et al. 2012; Kreidberg et al. 2014).

Besides spectral characterization, exoplanet atmospheres have also been studied via the phase curves (e.g., Seager et al. 2000; Knutson et al. 2007), and *Kepler* has made the first observation of a phase curve from a terrestrial exoplanet (Batalha et al. 2011; Fogtmann-Schulz et al. 2014).

12.3 Physical and Chemical Processes in Terrestrial Exoplanet Atmospheres

Central in the studies of terrestrial exoplanets is to characterize their atmospheres and to search for potential biosignature gases, i.e., the atmospheric components that indicate biogenic surface emissions. For this goal, a deep understanding of the key physical and chemical processes that control the atmospheric composition is crucial.

One important process for terrestrial exoplanet atmospheres is chemical and photochemical reactions. The network of chemical reactions in the atmosphere may serve as sources for certain gases and sinks for the others. Chemical reactions occur when two molecules collide, and the reaction rates are therefore proportional to the number density of both molecules. Certain reactions would require a third body in the collision to remove excess energy or angular momentum. The rates of such termolecular reactions are therefore also dependent on the total number density of the atmosphere. Near the top of the atmosphere, photon-driven reactions contribute dominantly to the source and sink, as ultraviolet (UV) photons from the parent star that could dissociate molecules usually penetrate to the pressure levels of ~ 0.1 bar (e.g., Yung and Demore 1999; Hu et al. 2012). The UV photodissociation produces reactive radicals that facilitate some reactions that are otherwise kinetically prohibited. A generic reaction network should include bimolecular reactions, termolecular reactions, photodissociation reactions, and thermodissociation reactions for the study of terrestrial exoplanet atmospheres (Hu 2013).

The other process that controls the compositions is transport. Both large-scale mean flows and small-scale turbulence and instability can transport molecules in the atmosphere and affect the composition (e.g., Brasseur and Solomon 2005; Seinfeld and Pandis 2006). One could focus on the transport in the vertical direction and explore the compositions of terrestrial exoplanet atmospheres as a function of altitude. Altitude is the most important dimension because the temperature and the pressure are strong functions of altitude. For example, the composition of Earth's atmosphere is primarily a function of altitude instead of longitude or latitude (Seinfeld and Pandis 2006). Also, the vertically resolved compositions are critical for prediction and interpretation of spectra of a terrestrial exoplanet, because the spectra probe different altitudes of the atmosphere depending on the wavelength (Seager 2010).

The major mechanisms for vertical transport in an irradiated atmosphere on a terrestrial exoplanet include convection (the same mechanism required to transport heat), small-scale instability driven by shear of horizontal flows, and molecular diffusion. The first two processes can be approximated by the so-called eddy diffusion coefficients (e.g., Seinfeld and Pandis 2006), and the last process can be approximated by the molecular diffusion coefficients. Therefore, the atmospheres on terrestrial exoplanets can be treated as gravitationally stratified, plan-parallel irradiated atmospheres, in which vertical mixing can be parameterized (Hu 2013).

Models for terrestrial exoplanet atmospheres can be developed to compute the chemical reaction kinetics and transport processes. Such models are often called "photochemistry-thermochemistry kinetic-transport model" or simply "photochemistry-thermochemistry model." The purpose of such models is to provide a tool to predict the amounts of component gases in the atmospheres of terrestrial exoplanets and in the meantime quantify the links between the observables (e.g., abundances of trace gases and their spectral signatures) and the fundamental unknowns (e.g., geological and biological processes on the planetary surface, mixing and escape of atmosphere gases, heat sources from planetary interior and exterior).

Photochemistry models have been successful in simulating the compositions of the atmospheres of Earth (Seinfeld and Pandis 2006) and the atmospheres of planets in our Solar System (Yung and Demore 1999). The photochemistry models are also critical for the study of molecular compositions of any exoplanet atmosphere, including the atmospheres of terrestrial exoplanets, because the composition of the observable part of an exoplanet atmosphere (0.1 mbar to 1 bar, depending on the wavelength) is controlled by the competition between chemical reaction kinetics and transport. Figure 12.1 schematically shows the architecture of such a model developed by Hu et al. (2012) and Hu and Seager (2014). A handful of other photochemistry models are available (e.g., Kasting et al. 1985; Yung and Demore 1999; Liang et al. 2003; Atreya et al. 2006; Zahnle et al. 2009; Line et al. 2010; Moses et al. 2011, 2013), and these models solve the same continuity equation in Fig. 12.1 and likely have similar architectures.

Eventually, the steady-state composition of a terrestrial exoplanet atmosphere is controlled by the boundary conditions. The upper boundary conditions are the fluxes of atmospheric escape or the material exchange fluxes between the neutral atmospheres and the ionospheres (not modeled) above. The lower boundary conditions depend on whether or not thermochemical equilibrium holds near the lower boundary. One could therefore consider the following two categories of atmospheres on terrestrial exoplanets: thin atmospheres and thick atmospheres. The thick atmospheres are defined as the atmospheres that are thick enough to maintain thermochemical equilibrium at high pressures, and the thin atmospheres are defined as the atmospheres at the surface of which achieving thermochemical equilibrium is kinetically prohibited. In the following we will focus on thin atmospheres because the thin atmospheres are more akin to creating a potentially habitable environment.

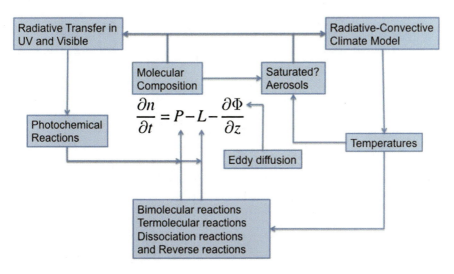

Fig. 12.1 Architecture of the photochemistry-thermochemistry model for terrestrial exoplanet atmospheres developed by Hu et al. (2012) and Hu and Seager (2014). The central equation to solve is a continuity equation that has terms for the chemical and photochemical production and loss, and terms for transport

The main components of a thin atmosphere of a terrestrial exoplanet result from its long-term geological evolution. For example, the N_2–O_2 atmosphere on Earth, the CO_2 atmosphere on Mars, and the N_2 atmosphere on Titan are results of long-term evolution (e.g., Kasting and Catling 2003; Coustenis 2005). For terrestrial exoplanets, the main components of their thin atmospheres can only be determined by observations, and the oxidation states of the thin atmospheres can range from reducing (e.g., H_2 atmospheres), to oxidized (e.g., N_2 and CO_2 atmospheres), to even oxic (O_2 atmospheres).

The photon-driven chemical reactions are especially important for thin atmospheres. UV and some visible-wavelength photons can dissociate gases, produce reactive radicals, and facilitate the conversion from emitted gases to photochemical products. These processes are key for thin atmospheres because (1) ultraviolet photons that cause photodissociation penetrate to the pressure levels of $\sim$0.1 bar, relevant to the bulk part of a thin atmospheres (Yung and Demore 1999; Hu 2013), and (2) in many cases, the photochemical processes in a thin atmosphere are irreversible. For example, the photochemical production of unsaturated hydrocarbons and haze from CH_4 occurs in the upper atmosphere of Titan, and the photochemical formation of C_2H_6 is irreversible and is therefore the dominant sink for CH_4 on Titan (Yung et al. 1984). This is in contrast to Jupiter's atmosphere, in which the photochemically formed C_2H_6 is converted back to CH_4 in deep atmosphere via pyrolysis (Strobel 1969, 1973; Gladstone et al. 1996).

The fundamental parameters that define a thin atmosphere are the surface source (i.e., emission rates), the surface sink (i.e., deposition velocities of emitted gases and their photochemical products in the atmosphere) of trace gases (Yung and Demore 1999; Seinfeld and Pandis 2006), and, in some cases, atmospheric loss to space. A photochemistry model, when applied to thin atmospheres, is to seek steady-state mixing ratios for trace gases of interest that are either emitted from the surface or produced by the chemical network in the atmosphere. The amounts of these trace gases are eventually controlled by the mass exchange between the surface and the atmosphere. It is important to study the amounts of trace gases by photochemistry models because an atmospheric spectrum may have strong features from spectroscopically active trace gases whose lifetime is controlled by the full chemical network in the atmosphere, and some of these trace gases may be hallmarks for specific atmospheric scenarios (e.g., Hu 2013).

The fact that the surface emission and deposition control the steady-state mixing ratios of trace gases in thin atmospheres on terrestrial exoplanets is pivotal to the ultimate goal of characterizing terrestrial exoplanets that might harbor life. Potential metabolic activities on a rocky planet emit a gas to the atmosphere that is otherwise not emitted or consume a gas that is otherwise not consumed. Both processes occur on Earth and regulate the composition of Earth's atmosphere. For example, Earth-based photosynthesis leads to the emission of O_2 that sustains a high O_2 mixing ratio in Earth's atmosphere, and Earth-based hydrogen-oxidizing bacteria provide an appreciable deposition velocity for H_2 from the atmosphere to the surface (e.g., Kasting and Catling 2003; Seinfeld and Pandis 2006). The photochemistry model provides the interface between the observables (atmospheric compositions) and the fundamental unknowns (surface source and sinks that may or may not be attributed to life), and the photochemistry model is therefore critical for determining the habitability of a terrestrial exoplanet and investigating whether a habitable planet is inhabited.

12.4 Exoplanet Benchmark Scenarios

Several benchmark scenarios can be set up for the atmospheres of terrestrial exoplanets (Hu et al. 2012). The goal of these benchmark scenarios is to provide baseline models to assess the stability of molecules in different kinds of atmospheres in order to calculate the lifetime of spectrally significant gases and, in particular, the lifetime of potential biosignature gases.

The most important parameter that determines the molecule lifetimes is the oxidation power of the atmosphere – the ability to reduce or oxidize a gas in the atmosphere. The main reactive species in the atmosphere provide the oxidizing or reducing power. In an oxidizing atmosphere, OH and O are created by photochemistry and are the main reactive radicals. In a reducing atmosphere, H, also created by

12 Photochemistry of Terrestrial Exoplanet Atmospheres

photochemistry, is the main reactive species. One could expect that the atmospheric composition of exoplanets will be highly varied, based on the nearly continuous range of masses and orbits of exoplanets. In this large parameter space, the primary dimension of chemical characterization for terrestrial exoplanet atmospheres is their oxidation states.

The main components of the atmosphere in large part, and the surface emission ad deposition of trace gases to a lesser extent, determine its oxidation power. In the extreme cases of the atmospheric redox state, i.e., the H_2-dominated atmospheres and the O_2-rich atmospheres, the atmospheric redox power is surely reducing or oxidizing for a wide range of surface emission or deposition fluxes. However, for an intermediate redox state, the atmosphere would be composed of redox-neutral species like N_2 and CO_2, and the redox power of the atmosphere can be mainly controlled by the emission and the deposition fluxes of trace gases (i.e., H_2, CH_4, and H_2S) from the surface. For example, the higher the emission of reducing gases is, the more reducing the atmosphere becomes.

Here we describe the benchmark scenarios proposed by Hu et al. (2012) for reducing, weakly oxidizing, and strongly oxidizing atmospheres on an Earth-size and Earth-mass habitable terrestrial exoplanet around a Sun-like star. The three scenarios are a reducing (90 % H_2–10 % N_2) atmosphere, a weakly oxidizing N2 atmosphere (>99 % N_2), and a highly oxidizing (90 % CO_2–10 % N_2) atmosphere. Hu et al. (2012) consider Earth-like volcanic gas emission rate and composition that consists of CO_2, H_2, SO_2, CH_4, and H_2S and assume that the planet surface has a substantial fraction of its surface covered by a liquid water ocean so that water is transported from the surface and buffered by the balance of evaporation/condensation. Key nonequilibrium processes in these scenarios are schematically shown in Fig. 12.2, and the molecular composition of the three benchmark scenarios is shown in Fig. 12.3.

Summarizing the benchmark scenarios, several general chemistry properties of thin atmospheres on habitable terrestrial exoplanets stand out. These properties are results of physical structures of molecules, and how they interact, and therefore are independent of detailed planetary scenarios (Hu et al. 2012; Hu 2013).

First, atomic hydrogen (H) is a more abundant reactive radical than hydroxyl radical (OH) in anoxic atmospheres. Atomic hydrogen is mainly produced by water vapor photodissociation (Hu et al. 2012; Seager et al. 2013). The production of atomic hydrogen is catalyzed by water vapor:

$$H_2O + h\nu \rightarrow H + OH,$$

$$OH + H_2 \rightarrow H_2O + H,$$

$$Net: H_2 + h\nu \rightarrow 2H.$$

It is difficult to remove hydrogen once produced in anoxic atmospheres, which is in contrast to oxygen-rich atmospheres (e.g., current Earth's atmosphere) in which H can be quickly consumed by O_2. As a result, removal of a gas by H is likely to be an important removal path for trace gases in an anoxic atmosphere. Atomic oxygen

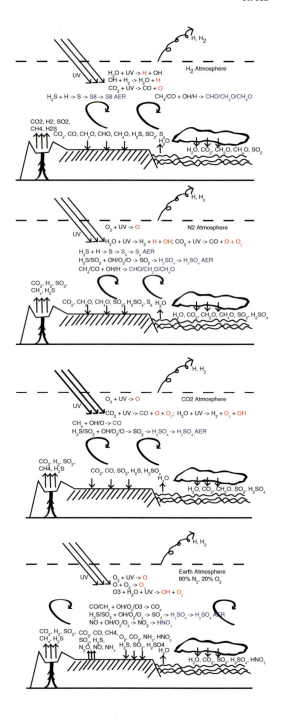

Fig. 12.2 Schematic illustrations of key nonequilibrium processes in the three scenarios of rocky exoplanet atmospheres in comparison with the current Earth. From *top* to *bottom*, the four panels correspond to the H_2, N_2, CO_2 atmospheres and the atmosphere of Earth. The *red color* highlights the reactive radicals in each atmospheric scenario, and the *blue color* highlights the major photochemical products in the atmosphere (Reproduced from Hu et al. 2012 with the permission of the AAS)

12 Photochemistry of Terrestrial Exoplanet Atmospheres

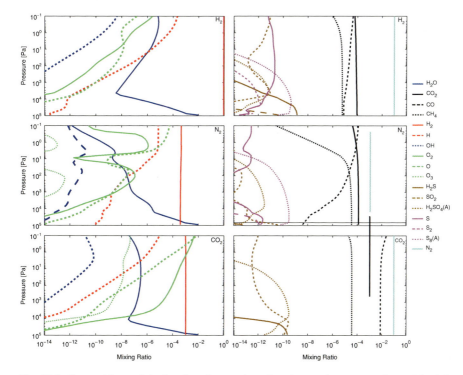

Fig. 12.3 Compositions of the benchmark scenarios of rocky exoplanet atmospheres. The *left column* shows mixing ratios of H and O species, and the *right column* shows mixing ratios of N, C, and S species. From *top* to *bottom*, the three panels correspond to the reducing (H_2-dominated), oxidized (N_2-dominated), and highly oxidized (CO_2-dominated) atmospheres. The vertical scales are expressed in pressure, which allows comparison between different scenarios that have very different mean molecular masses. *Thick lines* highlight the profiles of three reactive species, H, OH, and O (Reproduced from Hu et al. 2012 with the permission of the AAS)

is likely to be the most abundant reactive radical in CO_2-dominated atmospheres. Due to the photochemical origin of the reactive species, their abundances in the atmosphere around a quiet M dwarf star are 2 orders of magnitude lower than their abundances around a Sun-like star, because a quiet M dwarf emits much less UV radiation than a Sun-like star.

Second, dry deposition velocities of long-lived compounds, notably major volcanic carbon compounds including methane, carbon monoxide, and carbon dioxide, have significant effects on the atmospheric oxidation states. The specific choice of dry deposition velocities for emitted gases and their major photochemical by-products in the atmosphere is critical to determine the atmospheric composition and the redox power on terrestrial exoplanets (Hu et al. 2012). For example, if the dry deposition velocity of CO were greater for a CO_2-dominated atmosphere, the steady-state mixing ratio of CO in the atmosphere would be lower, and consequently, the atmosphere would have less H_2 and more O_2.

Third, volcanic carbon compounds (i.e., CH_4 and CO_2) are chemically long-lived and tend to be well mixed in terrestrial exoplanet atmospheres, whereas volcanic sulfur compounds (i.e., H_2S and SO_2) are short-lived (Fig. 12.3). CH_4 and CO_2 have chemical lifetime longer than 10,000 years in all three benchmark atmospheres ranging from reducing to oxidizing, implying that a relatively small volcanic input can result in a high steady-state mixing ratio. The chemical lifetime CO, another possible volcanic carbon compound, ranges from 0.1 to 700 years depending on the OH abundance in the atmosphere. Unlike carbon compounds, both H_2S and SO_2 are chemically short-lived in virtually all types of atmospheres on terrestrial exoplanets (Hu et al. 2013). This implies that the carbon compounds are more likely to be spectroscopically detected than the sulfur compounds. In particular, direct detection of surface sulfur emission is unlikely, as their surface emission rates need to be extremely high ($>1,000$ times Earth's volcanic sulfur emission) for these gases to build up to a detectable level. Sulfur compounds emitted from the surface will lead to photochemical formation of elemental sulfur and sulfuric acid in the atmosphere, which would condense to form aerosols if saturated.

12.5 Is O_2 a Biosignature Gas?

Oxygen and ozone are the most studied biosignature gases for terrestrial exoplanet characterization, due to their biogenic origin on Earth and their strong spectral features at visible and infrared wavelengths (e.g., Angel et al. 1986; Leger et al. 1993, 1996; Beichman et al. 1999; Snellen et al. 2013). When we consider using O_2/O_3 as biosignature gases, a natural question is whether O_2 may be produced without involving life. Indeed, photodissociation of H_2O and CO_2 produce free oxygen in the atmosphere (Fig. 12.3).

The abiotic production of oxygen in terrestrial atmospheres has been studied either for understanding prebiotic Earth's atmosphere (e.g., Walker 1977; Kasting et al. 1979; Kasting and Catling 2003) or for assessing whether abiotic oxygen can be a false positive for detecting photosynthesis on habitable exoplanets (Selsis et al. 2002; Segura et al. 2007; Hu et al. 2012; Tian et al. 2014; Wordsworth and Pierrehumbert 2014). Selsis et al. (2002) found that photochemically produced oxygen may build up in CO_2-dominated atmospheres, if there is no surface emission or deposition. The results of Selsis et al. (2002) was later challenged by Segura et al. (2007), who had additionally considered the surface emission of reducing gases including H_2 and CH_4, and found that abiotic oxygen would not build up in the atmosphere on a planet having active hydrological cycle.

Hu et al. (2012) first pointed out that the steady-state number density of O_2 and O_3 in the CO_2-dominated atmosphere is mainly controlled by the surface emission of reducing gases such as H_2 and CH_4, and without surface emission of reducing gas, photochemically produced O_2 can build up in a 1-bar CO_2-dominated atmosphere. Figure 12.4 shows the simulated CO_2-dominated atmospheres with relatively low and zero emission rates of H_2 and CH_4. The O_2 mixing ratio near the surface increases dramatically in 1-bar CO_2 atmospheres when the emission of

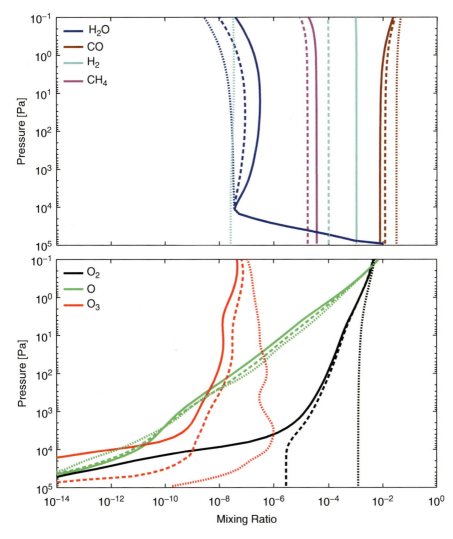

Fig. 12.4 Effects of the surface emission reducing gases on CO_2-dominated atmospheres of rocky exoplanets. The *upper panel* shows mixing ratios of H_2O, CO, H_2, and CH_4, and the *lower panel* shows mixing ratios of O_2, O, and O_3. The *solid lines* show the chemical composition of the benchmark scenario. In particular the emission rate of H_2 is 3×10^{-10} cm^{-2} s^{-1}. The *dashed lines* show the chemical composition of the same scenario, but with an H2 emission rate of 3×10^{-9} cm^{-2} s^{-1}, and the *dotted lines* show the chemical composition for zero emission of H_2 and CH_4. A dramatic increase of O_2 and O_3 mixing ratios is caused by a decrease of the surface emission of reduced gases (Reproduced from Hu et al. 2012 with the permission of the AAS)

reducing gases decreases. O_2 is virtually nonexistent at the surface for the Earth-like emission rates of H_2 and CH_4, but O_2 mixing ratio can be as high as 10^{-3} if no H_2 or CH_4 is emitted (Fig. 12.4). In particular, if no H_2 or CH_4 is emitted, the O_3 column integrated number density can reach one third of the present-day Earth's atmospheric levels, which constitutes a potential false positive.

Figure 12.4 shows that O_3 can potentially build up in the 1-bar CO_2-dominated atmosphere to a false positive level even on a planet with active hydrological cycle. Segura et al. (2007) have based their conclusion on simulations of 20 % CO_2 1-bar atmospheres with and without emission of H_2 and CH_4 and simulations of 2-bar CO_2 atmospheres with emission of H_2 and CH_4. Where Hu et al. (2012) models differ from Segura et al. (2007) is that Hu et al. (2012) model successfully simulated 90 % CO_2 1-bar atmospheres with minimal volcanic reducing gas emission. This is a parameter space that Segura et al. (2007) did not cover, but this is the parameter space for high abiotic O_2. Recently, it has further been found that around M dwarf stars that have low near-UV radiation and strong far-UV radiation, O_2 produced from CO_2 photodissociation is even easier to build up in the atmosphere (Tian et al. 2014).

Therefore one should exercise caution to use spectral features of O_2 as a probe of oxygenic photosynthesis on a terrestrial exoplanet. The risk of such false positives would affect the inference of photosynthesis via O_2 features detected in the visible wavelengths, potentially by either the Terrestrial Planet Finder – Coronagraph (e.g., Beichman et al. 2006) or the cross-correlation method applied to high-resolution spectroscopy on the 40-m class telescopes (Snellen et al. 2013). The risk of false positive is however not relevant to the detection of O_3 (a photochemical derivative of O_2) features in the mid-infrared, because the O_3 feature would be masked by strong CO_2 features and therefore not detectable for CO_2-dominated atmospheres (Selsis et al. 2002). Eventually, detecting both O_2 features and CH_4 features may mitigate the risk of false positive. A methane mixing ratio of $\sim$10 ppm would imply a surface source of reducing gases that could prevent the abiotic buildup of O_2 (Fig. 12.4).

12.6 Prospect of Terrestrial Exoplanet Characterization

Determination of the atmospheric compositions on terrestrial exoplanets is one of the most significant challenges facing astronomers. To achieve this goal, both advanced observational techniques and suitable targets are required (e.g., Deming et al. 2009). The ratio between the radiation from a terrestrial exoplanet and that from its parent star is in the orders of 10^{-10} in the visible wavelengths and 10^{-7} in the mid-infrared wavelengths. This means in the transit scenario, many observations of transits need to be stacked to lower the noise level in order to reveal the planet's signal (e.g., Seager and Deming 2010; Kreidberg et al. 2014), and in the direct imaging scenario, the stellar radiation has to be almost perfectly annihilated in order to reveal the planet (e.g., Kuchner and Traub 2002; Lawson and Dooley 2005; Trauger and Traub 2007). Even with advanced observational techniques, atmospheric characterization of terrestrial exoplanets will be confined to nearby systems (e.g., within tens of parsecs; Guyon et al. 2006; Oppenheimer and Hinkley 2009; Belu et al. 2011).

Full-sky surveys intended to discover terrestrial exoplanets around nearby main-sequence stars are being planned. The survey of terrestrial exoplanets around nearby systems is far from complete in several important areas: (1) Earth-mass planets around FGK stars have just become in the reach of the radial velocity method (Dumusque et al. 2012); (2) precise spectroscopic and photometric measurements of M dwarfs, despite their dominant numbers in our interstellar neighborhood, have been long impeded by the faintness of these stars and the concentration of their radiation in the near-infrared wavelengths that are strongly contaminated by Earth's atmosphere (Nutzman and Charbonneau 2008); (3) *Kepler*, although sensitive to Earth-sized planets, targets a small patch of sky and focuses on faint stars to maximize its scientific return (Batalha et al. 2010). One could expect rapid developments in all these areas. In particular, an all-sky space-based TESS mission (Transiting Exoplanet Survey Satellite) has recently been selected by NASA for launch in 2017 (Ricker et al. 2010). And the CHaracterising ExOPlanet Satellite (CHEOPS) and the PLAnetary Transits and Oscillations of stars (PLATO), also designed to search for terrestrial exoplanets around nearby bright stars, have been selected by European Space Agency (ESA) for launch in the next decade (Broeg et al. 2013; Rauer et al. 2013). One could expect a rapid growth in the number of terrestrial exoplanets that are suitable for follow-up observations of their atmospheres in the coming years.

The next-generation observation facility will allow thick atmospheres to be observed in great detail and even allow characterization of thin atmospheres on terrestrial exoplanets around late-type stars. Today's studies on hot Jupiter's atmospheres are flourishing with the Hubble Space Telescope and the Spitzer Space Telescope (see Seager and Deming 2010 and references therein), but much detection of atmospheric molecules remains controversial (Deming et al. 2013). In 5–10 years, larger and more sophisticated facilities will allow measurements of molecular abundances and characterization of atmospheric chemistry in thick atmospheres of gas giants, and super-Earths around M dwarf stars, to great detail (Traub et al. 2008; Kaltenegger and Traub 2009; Belu et al. 2011). These anticipated facilities include the James Webb Space Telescope (JWST) slated for launch in 2018 (Gardner et al. 2006), and the giant 20- to 40-m class ground-based telescopes that include the Extremely Large Telescope (Gilmozzi and Spyromilio 2008), the Giant Magellan Telescope (Johns et al. 2012), and the Thirty Meter Telescope (Crampton and Simard 2006).

In the more distant future, the community still holds hope that a direct-imaging space-based mission under the Terrestrial Planet Finder concept (e.g., Traub et al. 2006; Beichman et al. 2006; Levine et al. 2009) will enable Earth-like terrestrial exoplanets to be characterized. The technique of exoplanet direct imaging has been advancing rapidly and proceeding into spectroscopic observations of giant planets (Konopacky et al. 2013). A number of coronagraph instruments are mounted on state-of-the-art 10-m class telescopes, which will enable spectroscopic studies of extrasolar gas giants (e.g., Gemini Planet Imager, Chilcote et al. 2012). Notably, both coronagraph and external occultor instruments are being studied for space flights in the next decade, for directly imaging and spectrally characterizing Neptune- and Jupiter-sized exoplanets (Spergel et al. 2013; Stapelfeldt et al. 2014; Seager et al. 2014). These ground-based and space-based efforts of direct imaging and spectroscopic measurements of exoplanets will pave the way to a future space-based direct-imaging mission that will allow characterization of true Earth analogues.

References

Angel JRP, Cheng AYS, Woolf NJ (1986) Nature 322:341–343
Atreya SK, Adams EY, Niemann HB et al (2006) Planet Space Sci 54:1177
Barclay T, Rowe JF, Lissauer JJ et al (2013a) Nature 494:452–454
Barclay T, Burke C, Howell SB et al (2013b) Astrophys J 768:101
Batalha NM, Borucki WJ, Koch DG et al (2010) Astrophys J Lett 713:L109–L114
Batalha NM, Borucki WJ, Bryson ST et al (2011) Astrophys J 729:27
Bean JL, Miller-Ricci Kempton E, Homeier D (2010) Nature 468:669–672
Beichman CA, Woolf NJ, Lindensmith CA (1999) The Terrestrial Planet Finder (TPF): a NASA Origins Program to search for habitable planets
Beichman C, Lawson P, Lay O et al (2006). Status of the terrestrial planet under interferometer (TPF-I). In: Society of Photo-Optical Instrumentation Engineers (SPIE) conference series, vol 6268
Belu AR, Selsis F, Morales J-C et al (2011) Astron Astrophys 525:A83
Berta ZK, Charbonneau D, Desert J-M et al (2012) Astrophys J 747:35
Birkby JL, de Kok RJ, Brogi M et al (2013) Mon Not R Astron Soc 436:L35
Bonfils X, Gillon M, Forveille T et al (2011) Astron Astrophys 528:A111
Borucki WJ, Koch DG, Batalha N et al (2012) Astrophys J 745:120
Borucki WJ, Agol E, Fressin F et al (2013) Science 340(6132):587–590
Brasseur GP, Solomon S (2005) Aeronomy of the middle atmosphere. Springer, Dordrecht
Broeg C, Fortier A, Ehrenreich D et al (2013). CHEOPS: a transit photometry mission for ESA's small mission programme. In: European Physical Journal web of conferences, vol 47, p 3005
Brogi M, Snellen IAG, de Kok RJ et al (2012) Nature 486:502–504
Charbonneau D, Brown TM, Noyes RW, Gilliland RL (2002) Astrophys J 568:377–384
Charbonneau D, Allen LE, Megeath ST et al (2005) Astrophys J 626:523–529
Charbonneau D, Berta ZK, Irwin J et al (2009) Nature 462:891–894
Chilcote, J. K., Larkin, J. E., Maire, J et al Performance of the integral field spectrograph for the Gemini Planet Imager. In volume 8446 of Society of Photo-Optical Instrumentation Engineers (SPIE) Conference Series.
Cochran WD, Fabrycky DC, Torres G et al (2011) Astrophys J Suppl Ser 197:7
Coustenis A (2005) Space Sci Rev 116:171–184
Crampton, D. and Simard, L. (2006). Instrument concepts and scientific opportunities for TMT. In volume 6269 of Society of Photo-Optical Instrumentation Engineers (SPIE) Conference Series.
Croll B, Albert L, Jayawardhana R et al (2011) Astrophys J 736:78
Dawson RI, Fabrycky DC (2010) Astrophys J 722:937–953
de Kok RJ, Brogi M, Snellen IAG et al (2013) Astron Astrophys 554:A82
de Mooij EJW, Brogi M, de Kok RJ et al (2012) Astron Astrophys 538:A46
Deming D, Seager S, Richardson LJ, Harrington J (2005) Nature 434:740–743
Deming D, Seager S, Winn J et al (2009) Publ Astron Soc Pac 121:952–967
Deming D, Wilkins A, McCullough P et al (2013) Astrophys J 774:95
Demory B-O, Gillon M, Deming D et al (2011) Astron Astrophys 533:A114
Demory B-O, Gillon M, Seager S et al (2012) Astrophys J Lett 751:L28

12 Photochemistry of Terrestrial Exoplanet Atmospheres

Desert J-M, Bean J, Miller-Ricci Kempton E et al (2011) Astrophys J Lett 731:L40
Dragomir D, Mathews JM, Howard AW et al (2012) Astrophys J 759:L41
Dressing CD, Charbonneau D (2013) Astrophys J 767:95
Dumusque X, Pepe F, Lovis C et al (2012) Nature 491:207–211
Figueira P, Marmier M, Boue G et al (2012) Astron Astrophys 541:A139
Fogtmann-Schulz A, Hinrup B, Van Eylen V et al (2014) Astrophys J 781:67
Fressin F, Torres G, Rowe JF et al (2012) Nature 482:195–198
Fressin F, Torres G, Charbonneau D et al (2013) Astrophys J 766:81
Gardner JP, Mather JC, Clampin M et al (2006) Space Sci Rev 123:485–606
Gautier TN III, Charbonneau D, Rowe JF et al (2012) Astrophys J 749:15
Gilliland RL, Marcy GW, Rowe JF et al (2013) Astrophys J 766:40
Gilmozzi R, Spyromilio J (2008) The 42 m European ELT: status. In: Society of Photo-Optical
    Instrumentation Engineers (SPIE) conference series, vol 7012
Gladstone GR, Allen M, Yung YL (1996) Icarus 119:1–52
Guyon O, Pluzhnik EA, Kuchner MJ et al (2006) Astrophys J Suppl Ser 167:81–99
Howard AW, Johnson JA, Marcy GW et al (2011) Astrophys J 730:10
Howard AW, Marcy GW, Bryson ST et al (2012) Astrophys J Suppl Ser 201:15
Howard AW, Sanchis-Ojeda R, Marcy GW et al (2013) Nature 503:381–384
Hu R (2013) Atmospheric photochemistry, surface features, and potential biosignature gases of
    terrestrial exoplanets, Ph.D. thesis, MIT
Hu R, Seager S (2014) Astrophys J 784:63
Hu R, Seager S, Bains W (2012) Astrophys J 761:166
Hu R, Seager S, Bains W (2013) Astrophys J 769:6
Johns M, McCarthy P, Raybould K et al (2012) Giant Magellan Telescope: overview. In: Society
    of Photo-Optical Instrumentation Engineers (SPIE) conference series, vol 8444
Kaltenegger L, Traub WA (2009) Astrophys J 698:519–527
Kasting JF, Catling D (2003) Annu Rev Astron Astrophys 41:429–463
Kasting JF, Liu SC, Donahue TM (1979) J Geophys Res 84:3097–3107
Kasting JF, Holland HD, Pinto JP (1985) J Geophys Res 90:10497
Kasting JF, Whitmire DP, Reynolds RT (1993) Icarus 101:108–128
Knutson HA, Charbonneau D, Allen LE et al (2007) Nature 447:183–186
Knutson HA, Dragomir D, Kreidberg L et al (2014) ApJ 794:795
Konopacky QM, Barman TS, Macintosh BA, Marois C (2013) Science 339:1398–1401
Kopparapu RK, Ramirez R, Kasting JF et al (2013) Astrophys J 765:131
Kreidberg L, Bean JL, Desert JM et al (2014) Nature 505:69–72
Kuchner MJ, Traub WA (2002) Astrophys J 570:900–908
Lawson PR, Dooley JA (2005) Technology plan for the terrestrial planet finder interferometer.
    NASA STI/Recon Tech Rep News 6:15630
Leger A, Pirre M, Marceau FJ (1993) Astron Astrophys 277:309
Leger A, Mariotti JM, Mennesson B et al (1996) Icarus 123:249–255
Leger A, Rouan D, Schneider J et al (2009) Astron Astrophys 506:287–302
Levine M, Lisman D, Shaklan S et al (2009) Terrestrial Planet Finder Coronagraph (TPF-C) flight
    baseline concept. ArXiv e-prints
Liang M-C, Parkinson CD, Lee AY-T, Yung YL, Seager S (2003) ApJ 596:247
Line MR, Liang MC, Yung YL (2010) ApJ 717:496
Lissauer JJ, Fabrycky DC, Ford EB et al (2011) Nature 470:53–58
Lockwood AC, Johnson JA, Bender CF et al (2014) Astrophys J 783:L29
Marcy GW, Issacson H, Howard AW et al (2014) Astrophys J Suppl Ser 210:20
Mayor M, Udry S, Lovis C et al (2009) Astron Astrophys 493:639–644
Moses JI, Visscher C, Fortney JJ et al (2011) ApJ 737:15
Moses JI, Line MR, Visscher C et al (2013) ApJ 777:34
Muirhead PS, Hamren K, Schlawin E, et al (2012), ApJ 750:L37
Nutzman P, Charbonneau D (2008) Publ Astron Soc Pac 120:317–327
Oppenheimer BR, Hinkley S (2009) Annu Rev Astron Astrophys 47:253–289
Pepe F, Lovis C, Segransan D et al (2011) Astron Astrophys 534:A58

Pepe F, Cameron AC, Latham DW et al (2013) Nature 503:377–380

Petigura EA, Marcy GW, Howard AW (2013a) Astrophys J 770:69

Petigura EA, Howard AW, Marcy GW (2013b) PNAS 110:19273–19278

Pierrehumbert R, Gaidos E (2011) Astrophys J Lett 734:L13

Quintana EV, Barclay T, Raymond SN et al (2014) Science 344:277–280

Rauer H, Catala C, Aerts C et al (2013) Exp Astron (submitted)

Ricker GR, Latham DW, Vanderspek RK et al (2010) Transiting Exoplanet Survey Satellite (TESS). In: Bulletin of the American Astronomical Society, vol 42, p 450.06

Rivera EJ, Lissauer JJ, Butler RP et al (2005) Astrophys J 634:625–640

Rivera EJ, Butler RP, Vogt SS et al (2010) Astrophys J 708:1492–1499

Rodler F, Lopez-Morales M, Ribas I (2012) Astrophys J Lett 753:L25

Rodler F, Kurster M, Barnes JR (2013) Mon Not R Astron Soc 432:1980–1988

Rowe JF, Bryson ST, Marcy GW (2014) Astrophys J 784:45

Salpeter EE (1955) Astrophys J 121:161

Seager S (2010) Exoplanet atmospheres: physical processes. Princeton University Press, Princeton

Seager S, Deming D (2010) Annu Rev Astron Astrophys 48:631–672

Seager S, Sasselov DD (1998) Astrophys J Lett 502:L157

Seager S, Sasselov DD (2000) Astrophys J 537:916–921

Seager S, Whitney BA, Sasselov DD (2000) Astrophys J 540:504–520

Seager S, Bains W, Hu R (2013) Astrophys J 777:95

Seager S, Turnbull M, Sparks W et al (2014) Exo-S: Starshade probe-class exoplanet direct imaging mission concept – interim report

Segura A, Meadows VS, Kasting JF et al (2007) Astron Astrophys 472:665–679

Seinfeld JH, Pandis SN (2006) Atmospheric chemistry and physics: from air pollution to climate change, 2nd edn. Wiley, Hoboken

Selsis F, Despois D, Parisot J-P (2002) Astron Astrophys 388:985–1003

Snellen IAG, de Kok RJ, le Poole R et al (2013) Astrophys J 764:182

Spergel D, Gehrels N, Breckinridge J et al (2013) Wide-field infrared survey telescope-astrophysics focused telescope assets. WFIRST-AFTA final report. ArXiv e-prints

Stapelfeldt K, Belikov R, Bryden G et al (2014) Exoplanet direct imaging: coronagraph probe mission study "Exo-C" – Interim report

Strobel DF (1969) J Atmos Sci 26:906–911

Strobel DF (1973) J Atmos Sci 30:489–498

Swift JJ, Johnson JA, Morton TD et al (2013) Astrophys J 764:105

Tian F, France K, Linsky J et al (2014) Earth Planet Sci Lett 385:22–27

Traub WA (2012) Astrophys J 745:20

Traub WA, Levine M, Shaklan S et al (2006) TPF-C: status and recent progress. In: Society of Photo-Optical Instrumentation Engineers (SPIE) conference series, vol 6268

Traub WA, Kaltenegger L, Jucks KW (2008) Spectral characterization of Earth-like transiting exoplanets. In: Society of Photo-Optical Instrumentation Engineers (SPIE) conference series, vol 7010

Trauger JT, Traub WA (2007) Nature 446:771–773

Tuomi M, Anglada-Escude G, Gerlach E et al (2013) Astron Astrophys 549:A48

Udry S, Bonfils X, Delfosse X et al (2007) Astron Astrophys 469:L43–L47

Van Grootel V, Gillon M, Valencia D et al (2014) Astrophys J 786:2

Vogt SS, Wittenmyer RA, Butler RP et al (2010) Astrophys J 708:1366–1375

Walker JCG (1977) Evolution of the atmosphere. Wiley, New York

Winn JN, Matthews JM, Dawson RI et al (2011) Astrophys J Lett 737:L18

Wordsworth R, Pierrehumbert R (2014) Astrophys J Lett 785:L20

Yung YL, Demore WB (1999) Photochemistry of planetary atmospheres. Oxford University Press, New York

Yung YL, Allen M, Pinto JP (1984) Astrophys J Suppl Ser 55:465–506

Zahnle K, Marley MS, Freedman RS, Lodders K, Fortney J (2009) ApJ 701:L20

Zsom A, Seager S, de Wit J, Stamenkovic V (2013) Astrophys J 778:109

Chapter 13
Planet Formation in Binaries

P. Thebault and N. Haghighipour

Abstract Spurred by the discovery of more than 60 exoplanets in multiple systems, binaries have become in recent years one of the main topics in planet-formation research. Numerous studies have investigated to what extent the presence of a stellar companion can affect the planet-formation process. Such studies have implications that can reach beyond the sole context of binaries, as they allow to test certain aspects of the planet-formation scenario by submitting them to extreme environments. We review here the current understanding on this complex problem. We show in particular how each of the different stages of the planet-formation process is affected differently by binary perturbations. We focus especially on the intermediate stage of kilometre-sized planetesimal accretion, which has proven to be the most sensitive to binarity and for which the presence of some exoplanets observed in tight binaries is difficult to explain by in situ formation following the "standard" planet-formation scenario. Some tentative solutions to this apparent paradox are presented. The last part of our review presents a thorough description of the problem of planet habitability, for which the binary environment creates a complex situation because of the presence of two irradiation sources of varying distance.

Keywords Planetary systems • Binary stars

13.1 Introduction

About half of solar-type stars reside in multiple stellar systems (Raghavan et al. 2010). As a consequence, one of the most generic environments to be considered for studying planet formation should in principle be that of a binary. However, the

---

P. Thebault (✉)
LESIA, Observatoire de Paris, F-92195 Meudon Principal Cedex, France
e-mail: philippe.thebault@obspm.fr

N. Haghighipour
Institute for Astronomy and NASA Astrobiology Institute, University of Hawaii-Manoa, 2680 Woodlawn Drive, Honolulu, HI 96822, USA

© Springer-Verlag Berlin Heidelberg 2015
S. Jin et al. (eds.), *Planetary Exploration and Science: Recent Results and Advances*,
Springer Geophysics, DOI 10.1007/978-3-662-45052-9_13

"standard" scenario of planet formation by core accretion that has been developed over the past decades (e.g. Safronov 1972; Lissauer 1993; Pollack et al. 1996; Hubickyj et al. 2005) is so far restricted to the case of a single star. Of course, this bias is a direct consequence of the fact that planet-formation theories were initially designed to understand the formation of our own solar system. For the most part, however, this bias is still present today, nearly two decades after the discovery of the first exoplanets. The important updates and revisions of the standard model, such as planetary migration, planet scattering, etc., that have been developed as a consequence of exoplanet discoveries have mostly been investigated for a single-star environment. This single-star-environment tropism does also affect the alternative planet-forming scenario by gravitational instabilities (Boss 1997), which has witnessed a renewed interest after the discovery of Jovian exoplanets at large radial distances from their star (Boss 2011).

However, the gradual discovery of exoplanets in multiple star systems (Desidera and Barbieri 2007; Mugrauer and Neuhäuser 2009, and references therein), and especially in close binaries of separation $\sim$20 AU, has triggered the arrival of studies investigating how such planets could come about and, more generally, about how planet formation is affected by binarity. The latter issue is a vast and difficult one. Planet formation is indeed a complex process, believed to be the succession of several stages (e.g. Haghighipour 2011), each of which could be affected in very different ways by the perturbations of a secondary star.

Not surprisingly, the effect of binarity on each of these different stages is usually investigated in separate studies. A majority of these investigations have focused on the intermediate stage of kilometre-sized planetesimal accretion, as this stage has been shown to be extremely sensitive to stellar companion perturbations. But other key stages have also been explored, from the initial formation of protoplanetary discs to the final evolution of massive planetary embryos.

The scope of such planet-in-binary studies has been recently broadened by the discovery of several *circumbinary* exoplanets (also known as P-type orbits) by the Kepler space telescope, most of which are located relatively close to the central stellar couple. The issues related to the formation of these objects are often very different from those related to circumprimary exoplanets (also known as S-type orbits), and in-depth investigations of this issue have only just begun.

Studying how both circumprimary and circumbinary planets form is of great interest, not only to explore the history of specific planets in binaries but also for our understanding of planet formation in general. These studies can indeed be used as a test bench for planet-formation models, by confronting them to an unusual and sometimes "extreme" environment where some crucial parameters might be pushed to extreme values. We present here a review of the current state of research on planets in binary star systems. We focus our analysis on the S-type systems in which planets orbit one star of the binary.

13.2 Observational Constraints: Planets in Binaries

Exoplanet search surveys were initially strongly biased against binary systems of separation ≲200 AU (Eggenberger and Udry 2010), in great part because these searches were focusing on stellar environments as similar as possible to the solar system. In 2003, however, the first exoplanet in a close binary was detected in the γ Cephei system (Hatzes et al. 2003), and today more than 60 exoplanets are known to inhabit multiple star systems (Roell et al. 2012). Note that, in many cases, these exoplanets were detected *before* the presence of a stellar companion was later established by imaging campaigns (Mugrauer and Neuhäuser 2009). As a result, for most of these systems, the separation of the binary is indeed relatively large, often in excess of 500 AU (Roell et al. 2012). However, ~10 of these planet-bearing binaries have a separation of less than 100 AU, with 5 exoplanets in close binaries with separations of ~20 AU (Fig. 13.1): Gl86 (Queloz et al. 2000; Lagrange et al.

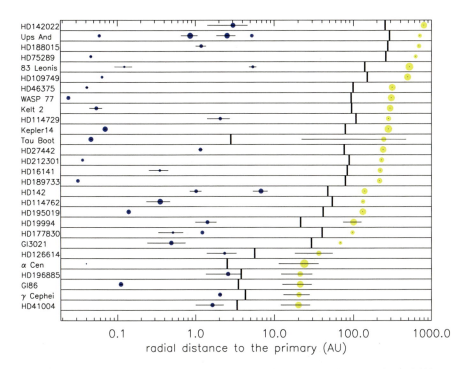

Fig. 13.1 Architecture of all circumprimary planet-bearing binaries with separation ≤1,000 AU (as of July 2013). Companion stars are displayed as *yellow circles*, whose radius is proportional to $(M_2/M_1)^{1/3}$. Planets are marked as *blue circles* whose radius is proportional to $(m_{pl}/m_{Jup})^{1/3}$. The *horizontal lines* represent the radial excursion of the planets' and stars' orbits (when they are known). For most binaries of separation ≥100 AU, the orbit is not known and the displayed value corresponds to the projected current separation. The *short horizontal lines* correspond to the outer limit of the orbital stability region around the primary, as estimated by Holman and Wiegert (1999)

2006), HD 41004 (Zucker et al. 2004), γ Cephei (Hatzes et al. 2003; Neuhäuser et al. 2007; Endl et al. 2011), HD 196885 (Correia et al. 2008; Chauvin et al. 2011) and α Centauri B (Dumusque et al. 2012).

As soon as their number became statistically significant, the characteristics of these planets in binaries have been investigated in order to derive possible specificities as compared to planets around single stars. Desidera and Barbieri (2007) and Roell et al. (2012) have shown that whilst the distribution of planets in wide ($\geq 100\,\mathrm{AU}$) systems is identical to that of planets around single stars, the characteristics of exoplanets in close binaries are significantly different. The main trend seems to be that planetary masses increase with decreasing stellar separation. According to Roell et al. (2012), the minimum planet mass scales approximately as $(10\,\mathrm{AU}/a_{\mathrm{bin}})M_{\mathrm{Jup}}$. However, these trends should be taken with caution, as the number of planets in tight binaries is still very limited. Furthermore, the probable detection of the Earth-sized planet around α Cen B in late 2012 might significantly weaken this result. As for the global occurrence of planets in binaries, Roell et al. (2012) have found that multiplicity rate amongst planet-hosting stars is $\sim 12\,\%$, approximately four times smaller than for main-field solar-type stars (Raghavan et al. 2010). But as pointed out by Duchêne (2010): "...the small sample size, adverse selection biases, and incompleteness of current multiplicity surveys are such that it is premature to reach definitive conclusions". As a consequence, future surveys should probably increase both the number of exoplanets in close binaries and the number of stellar companions in known exoplanet systems.

Besides this statistical exploration, another crucial issue that has been investigated early on is that of the long-term orbital stability of these planets in binaries. The reference work on this issue remains probably that of Holman and Wiegert (1999),[1] who derived empirical expressions for orbital stability as a function of binary semimajor axis a_B, eccentricity e_B and mass ratio μ. Later studies have shown that, reassuringly, all known exoplanets in multiple systems are on stable orbits (e.g. Dvorak et al. 2003; Haghighipour et al. 2010), although the case for HD 41004 is not fully settled yet, as it depends on the yet unconstrained eccentricity of the binary orbit (Haghighipour et al. 2010).

An important recent development in terms of observations is the discovery of several exoplanets in P-type orbits. The first confirmed such planet orbits around the cataclysmic binary DP Leonis (Qian et al. 2010), but most circumbinary planets around binaries with main-sequence stars have been detected by the Kepler space telescope (Doyle et al. 2011; Welsh et al. 2012; Orosz et al. 2012a,b; Schwamb et al. 2013; Kostov et al. 2013, 2014a,b). Here again, dynamical studies have shown that all known circumbinary planets are on long-term stable orbits.

However, even if the question of long-term stability seems to be settled for all known exoplanets in binaries, the question of their *formation* is a much more complex issue. It is true that, for many S-type planets, the stellar separation is so

---

[1]Although similar pioneering work on this issue had already been performed a decade earlier by Dvorak (1984, 1986) and Dvorak et al. (1989).

large, often exceeding 100 times the radial distance of the exoplanet to the primary, that binarity should have had a very limited effect in the planet-forming regions. The situation should, however, be radically different for the handful of planets in ~20 AU binaries, notably for γ Cephei Ab, HD 196885 Ab and HD 41004 Ab, where the planet is located close to the orbital stability limit (Fig. 13.1). It is unlikely that planet formation in these highly perturbed environments could proceed unaffected by the presence of the companion star.

13.3 Early Stages of Planet Formation

13.3.1 Protoplanetary Disc Truncation

The planet-formation process can be affected by binarity right from the start, during the formation of the initial massive gaseous protoplanetary disc. Artymowicz and Lubow (1994) and Savonije et al. (1994) have shown that such a disc can be tidally truncated by a stellar companion. This truncation distance depends on several parameters, such as the disc's viscosity, but is in most cases roughly comparable to the outer limit for dynamical stability (see previous section), i.e. typically at 1/3–1/4 of the binary's separation for non-extreme value of the orbital eccentricity.

Surveys of discs around young stellar objects (YSOs) have given observational confirmation that discs in close binaries indeed tend to be both less frequent and less massive than around single stars. For the 2 Myr old Taurus-Auriga association, Kraus et al. (2012) have shown that whilst the disc frequency in wide binaries remains comparable to that of single stars (~80 %), it significantly drops for separations $\lesssim$40 AU and is as low as ~35 % for binaries tighter than 10 AU (Fig. 13.2 left). Moreover, the millimetre-wave dust continuum imaging survey of Harris et al. (2012) for discs in the same Taurus-Auriga cluster has shown that the estimated dust mass contained in these discs strongly decreases with decreasing binary separation (Fig. 13.2 right). This agrees very well with the theoretical prediction of more compact and thus less massive discs in tight binaries.

This less frequent and more tenuous disc trend leads to two major problems when considering planet formation. This first one is that truncated discs might not contain enough mass to form planets, especially Jovian objects such as those that have been observed for the three most "extreme" systems: HD 196885, HD 41004 and γ Cephei. This issue has been numerically investigated for the specific case of γ Ceph-type systems, and Jang-Condell et al. (2008) have found that for the most reasonable assumptions regarding the disc's viscosity and accretion rate, there is just enough mass left in the truncated disc to form the observed giant planet. This encouraging result has been later confirmed by Müller and Kley (2012) using a different numerical approach. Note, however, that there might not be enough mass left in the truncated γ Ceph disc to form another Jovian planet.

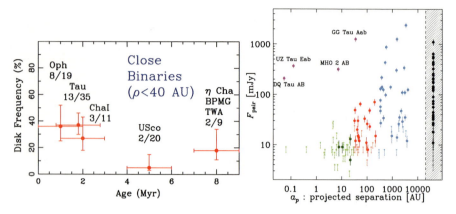

Fig. 13.2 Observational constraints on discs in binaries. *Left panel*: Frequency of disc-bearing systems in close binaries as a function of age in several young stellar clusters (From Kraus et al. 2012, courtesy of the Astrophysical Journal). *Right panel*: Measured excess flux in the millimetre as a function of binary projected separation (From Harris et al. 2012, courtesy of the Astrophysical Journal)

There is, however, a potentially bigger problem that is inherent to truncated discs, i.e. that they should be short-lived. The viscous evolution of a compact disc is indeed much faster than that of an extended system, and its mass gets drained, by accretion onto the central star, on shorter timescales. This reduces the timespan within which gaseous planet can form. As an example, Müller and Kley (2012) have shown that for γ Cephei-type systems, only for unrealistically low disc viscosities do they obtain disc lifetimes that are long enough to allow for in situ giant planet formation by core accretion. Observational confirmation of this short-lived-disc trend has been obtained by Kraus et al. (2012), who compared the disc frequency as a function of age for close ($\leq$40 AU) binaries in several nearby young associations. They found that the majority of such systems lose their disc in less than 1 Myr, even if a small fraction is able to retain discs to ages close to 10 Myr. Their preliminary conclusion is that "$\sim$2/3 of all close binary systems clear their disks extremely quickly, within 1 Myr of the end of envelope accretion. The other $\sim$1/3 of close binary systems evolve on a timescale similar to that of single stars". There seems thus to exist a disc-in-binary category for which the theoretically expected faster mass drain does not occur and which could thus be more friendly to planet formation. A good example for this category could be the young L1551 system, harbouring two resolved $\sim$10 AU wide circumstellar discs in a binary of 45 AU separation (Rodriguez et al. 1998). Note, however, that even for this population of long-lived binary discs, most estimated disc masses are much lower than for single-star cases (with, however, some important exceptions, such as L1551).

13.3.2 Grain Condensation and Growth

The next stage of planet formation, the condensation of small grains and their growth into larger pebbles and eventually kilometre-sized planetesimals, has not been extensively studied in the context of binary systems. One main reason is probably that this stage is the one that is currently the least understood even in the "normal" context of single stars (e.g. Blum and Wurm 2008), so that extrapolating it to perturbed binaries might seem premature. A noteworthy exception is the study by Nelson (2000) showing that for an equal-mass binary of separation 50 AU, temperatures in the disc might stay too high to allow grains to condense. These results were recently confirmed by the sophisticated 3-D modelling of radiative discs by Picogna and Marzari (2013), who found strong disc heating, due to spiral chocks and mass streaming between the circumprimary and circumsecondary discs, in binaries of separation 30 and 50 AU. On a related note, Zsom et al. (2011) showed that even if grains can condense, binary perturbations might impend their growth by mutual sticking because of too high impact velocities.

13.4 Planetesimal Accumulation

13.4.1 Context

The next stage of planet formation starts once "planetesimals", i.e. objects large enough (typically sub-kilometre to kilometre sized) to decouple from the gas, have formed. Within the standard planet-formation scenario, this stage leads, by mutual accretion amongst these km-sized planetesimals, to the formation of large planetary embryos. This stage is the one that has been by far the most extensively studied within the context of binary systems. The main reason is that the accretion of planetesimals and their growth to planetary embryos is potentially extremely sensitive to dynamical perturbations, since it does not take much to hamper or even halt the mutual accretion of kilometre-sized objects. Indeed, in the standard simulations of planet formation around single stars, this stage proceeds through fast runaway and oligarchic growth that require very low impact velocities between colliding bodies, typically smaller than their escape velocity, i.e. just a few m.s^{-1} for kilometre-sized objects (e.g. Lissauer 1993; Kokubo and Ida 2000). These values are less than $\sim 10^{-4}$ that of typical orbital velocities, so that even moderate dynamical perturbations can have a dramatic effect by increasing impact velocities beyond an accretion-hostile threshold.

The crucial parameter sealing the fate of the planetesimal swarm is thus their encounter velocities, and most studies of this stage have numerically investigated how this parameter evolves under the coupled effect of stellar perturbations, gas drag and physical collisions.

13.4.2 Differential Orbital Phasing: The Negative Effect of Gas Drag

In a coplanar system, when neglecting all other effects, the response of a planetesimal disc to the pure gravitational perturbations of a stellar companion is secular perturbations that lead to oscillations of their eccentricity and longitude of periastron ω (Heppenheimer 1978; Thebault et al. 2006; Giuppone et al. 2011). These secular oscillations are strongly phased in ω, so that they initially do not lead to large impact velocities Δv for planetesimal encounters. However, as these oscillations get narrower with time, at some point they are narrow enough for orbital crossing to occur, leading to a sudden increase of Δv (see Thebault et al. 2006, for a detailed discussion on this issue).

The presence of a gas disc radically alters this simple picture. As shown by Marzari and Scholl (2000), gas drag progressively damps the secular oscillations, and planetesimals converge towards an asymptotic equilibrium orbit e_g and ω_g (Paardekooper et al. 2008). However, this gas-induced evolution is *size dependent*, so that planetesimals of different sizes end up on different orbits (Thebault et al. 2004, 2006). This is clearly illustrated in Fig. 13.3, which also shows how small bodies reach their equilibrium orbits faster than larger objects (in the present example, 1 km objects are already at e_g and ω_g, whilst 10 km ones still undergo residual secular oscillations). As a consequence, Δv are small between equal-sized objects but can reach very high values for bodies of different sizes. For most "reasonable" size distributions within the planetesimal population, the differential phasing effect is expected to be the dominant one, and the dynamical environment can thus be strongly hostile to planetesimal accretion in vast regions of the circumstellar disc (Thebault et al. 2006; Thebault 2011). These accretion-hostile regions are in general much more extended than the region of orbital stability. Furthermore, even in the regions where planetesimal accretion *is* possible, it can often not proceed in the same way as around a single star, because the Δv increase is still enough to slow down and impede the runaway growth mode. A worrying result is that for the emblematic γ Cephei and HD 196885 cases, the locations at which the planets are observed are probably too perturbed to allow for this planetesimal accretion stage to proceed (Paardekooper et al. 2008; Thebault 2011). As for the arguably most famous binary system, α Centauri, simulations have shown that planetesimal accretion is very difficult in the habitable zone (Thebault et al. 2008, 2009). See for instance Fig. 13.4, showing that in the disc around α Cen B the whole region beyond $\sim$0.5 AU is globally hostile to km-sized planetesimal accretion. Note, however, that the innermost regions, where the planet α Cen Bb has been recently detected, are almost unaffected by binary perturbations.

Xie and Zhou (2009) and Xie et al. (2010a) showed that a small inclination of a few degrees between the circumprimary gas disc and the binary orbital plane can help accretion. This is because planetesimal orbital inclinations are segregated by size, thus favouring low-Δv impacts between equal-sized bodies over high-Δv impacts between differently sized objects. This reduction of high-velocity impacts

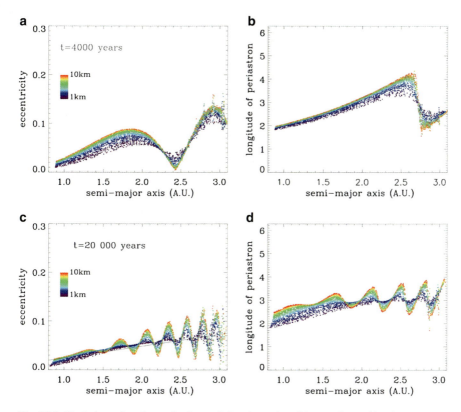

Fig. 13.3 Evolution of a planetesimal population in a close binary when taking into account gas drag. *Left panel*: eccentricity. *Right panel*: longitude of periastron (the binary's longitude of periastron is 0). The binary configuration corresponds to the HD 196885 case, i.e. $M_2/M_1 = 0.35$, $a_b = 21$ AU and $e_b = 0.42$. The gas disc is assumed to be axisymmetric and has a density profile corresponding to the theoretical Minimum Mass Solar Nebula (MMSN, see Hayashi 1981). Planetesimals have sizes $1 \leq s \leq 10$ km (Modified from Thebault 2011)

comes, however, at a price, which is that the collision rate strongly decreases, so that accretion can only proceed very slowly. Moreover, the results of Fragner et al. (2011) seem to indicate that taking into account the effect of the gas disc's gravity could offset this positive effect of orbital inclination, leading to high-Δv dynamical environments (see below).

13.4.3 More Sophisticated Models: Gas Is Still a Problem

This globally negative effect of the gas on planetesimal accretion had been identified in studies considering fiducial static and axisymmetric gas discs, and one could not rule out that this result could be an artefact due to the simplified gas-disc

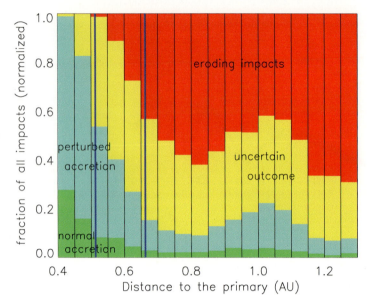

Fig. 13.4 Planetesimal disc around α Cen B. Numerical simulations with gas drag. Relative importance of different type of mutual impacts as a function of radial distance. *Red*: impacts for which $\Delta v_{s_1,s_2} \geq v_{\text{ero}-M}$, where $v_{\text{ero}-M}$ is the threshold velocity beyond which an impact between two objects of sizes s_1 and s_2 always results in net mass loss. *Yellow*: $v_{\text{esc}-m} \leq \Delta v_{s_1,s_2} \leq v_{\text{ero}-M}$, where $v_{\text{esc}-m}$ is the erosion threshold considering poorly collision-resistant material. *Green*: $\Delta v_{s_1,s_2} \leq v_{\text{esc}}$ where v_{esc} is the escape velocity of the (s_1, s_2) pair. Accretion can here proceed unimpededly, in a "runaway growth" way, as around a single star. *Light blue*: $v_{\text{esc}} \leq \Delta v_{s_1,s_2} \leq v_{\text{ero}-m}$. Collisions result in net accretion, but Δv are high enough to cancel off the fast runaway growth mode. The two thick blue lines denote the location of the inner limit of the "empirical" and "narrow" habitable zones (see Sect. 13.7). The planetesimal size distribution is assumed to be a Maxwellian centred on 5 km (Modified from Thebault et al. 2009)

prescription. In recent years, new studies have investigated this issue by considering increasingly sophisticated gas-disc models. For the most part, these studies have confirmed the accretion-hostile effect of the gas. Paardekooper et al. (2008) have considered a system with an evolving gas disc that also feels the pull of the binary. They have shown that the situation gets even worse for planetesimal accretion: gas streamlines follow paths very different from the planetesimal orbits, so that gas drag is enhanced, and so is the accretion-hostile differential phasing effect (see Fig. 13.5).

Another important mechanism that had been neglected in most early planetesimal+gas studies is that of the gas disc's gravity. Due to the difficulty of incorporating this effect, there have only been two attempts at numerically investigating this issue and only for very small populations of test planetesimals. The pioneering study by Kley and Nelson (2007) have shown that for many setups, the effect of gas-disc gravity can dominate that of gas drag in controlling planetesimal dynamics. A later study by the same team (Fragner et al. 2011) found that the net effect of disc gravity is to further increase impact speeds between planetesimals and this even between equal-sized bodies. The recent analytical exploration of Rafikov (2013) has

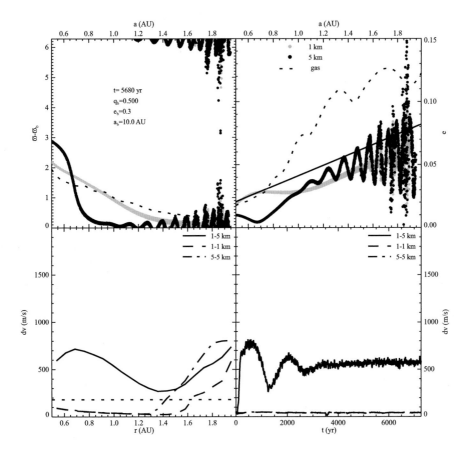

Fig. 13.5 Planetesimal evolution in an *evolving* gas disc for a tight binary of separation 10 AU. *Top left*: longitude of periastron, after 5,680 years, for 1 km planetesimals, for 5 km ones and for the gas disc. *Top right*: orbital eccentricity. *Bottom left*: encounter velocities between planetesimals as a function of their sizes. *Bottom right*: average encounter velocities at 1 AU as a function of time (From Paardekooper et al. 2008, courtesy of MNRAS)

nevertheless shown that if the gas disc is very massive and almost axisymmetric, its gravity could in fact reduce encounter velocities amongst planetesimals. However, this massive axisymmetric disc prerequisite is probably not likely to be generic, as all hydrodynamical studies of gas discs in binaries have shown that they reach eccentricities that are at least one order of magnitude higher than the few 10^{-3} needed in the Rafikov (2013) model (Kley and Nelson 2007; Marzari et al. 2009, 2012; Zsom et al. 2011; Müller and Kley 2012).

On a more positive note, Xie and Zhou (2008) showed that during the late stages when gas dissipates from the disc, the dispersal of planetesimal orbits narrows and the systems can get accretion friendly again. However, this behaviour probably occurs too late, once most planetesimals have already been either grounded to dust or spiralled onto the star because of gas friction (Thebault et al. 2008).

13.5 Late Stages

The stage for which the influence of a companion is probably best understood is the final step of planetary accretion, leading from lunar-sized embryos to fully formed planets. Several studies have shown that the regions where embryo accretion can proceed roughly correspond to those for orbital stability (Barbieri et al. 2002; Quintana et al. 2002, 2007; Haghighipour and Raymond 2007; Guedes et al. 2008; Haghighipour et al. 2010). This is a further reassuring result for all known exoplanets in binaries. As an illustration, we show in Fig. 13.6 the results obtained by Quintana et al. (2002) for the late stages of planet formation around an αCen-like binary. As can be clearly seen, even in such a tight binary, planetary objects can grow within $\sim$2 AU region from the primary star.

More recent work on these late stages has been devoted to more specific issues. As an illustration, we present here results obtained by Haghighipour and Raymond (2007) that focus on the level of water delivery and water mixing in the terrestrial regions within a binary, assuming that a giant planet could form further out in the disc. The authors considered a binary with a solar-mass primary and a Jupiter-mass giant planet at 5 AU. The mass of the secondary star was taken to be 0.5,

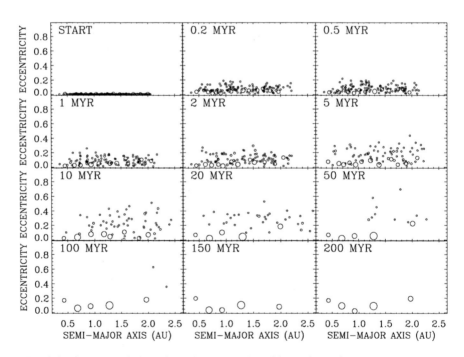

Fig. 13.6 Late stages of planet formation around the primary of an α Centauri-like binary. The initial disc consists of a mixture of 14 already formed embryos of lunar mass and 140 massive planetesimals of mass $9.33 \times 10^{-3} M_\oplus$ (Results taken from Quintana et al. (2002), courtesy of the Astrophysical Journal)

1.0 and 1.5 solar masses, and the semimajor axis and eccentricity of the binary were varied in the range of 20–40 AU and 0–0.4, respectively. For each value of the mass of the primary, Haghighipour and Raymond (2007) used the planetary orbit stability criterion, given by Rabl and Dvorak (1988) and Holman and Wiegert (1999), and identified the combination of the mass, semimajor axis and eccentricity of the binary for which the giant planet would maintain a stable orbit at 5 AU. They then considered a disc of 115 Moon-to-Mars-sized bodies with masses ranging from 0.01 to 0.1 Earth masses.

Figure 13.7 shows the results of a sample of their simulations for a binary with a mass ratio of 0.5 and for different values of its semimajor axis and eccentricity. The grey area corresponds to the boundaries of the habitable zone of the primary star (Kasting et al. 1993). As a point of comparison, the inner planets of the solar system are also shown. As shown here, many planets of the same or similar size as Earth and with substantial amount of water formed in and around 1 AU. Figure 13.8 shows

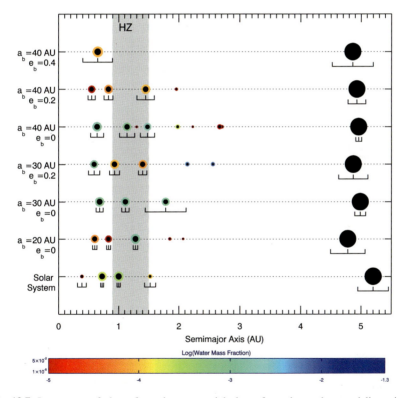

Fig. 13.7 Late stages of planet formation: terrestrial planet formation and water delivery in the presence of an already formed giant planet. Results of simulations in a binary system with a mass ratio of 0.5 and for different values of the eccentricity (e_b) and semimajor axis (a_b) of the binary. The solar system's configuration is given as a comparison (Taken from Haghighipour and Raymond 2007, courtesy of the Astrophysical Journal)

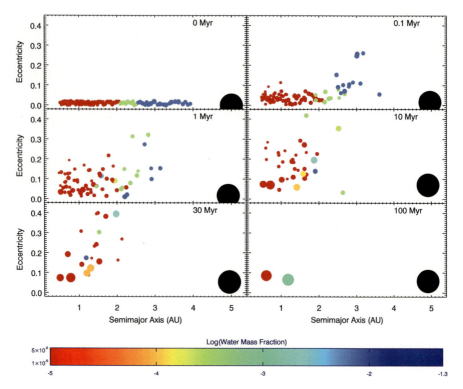

Fig. 13.8 Temporal evolution of the terrestrial planet growth in a binary system with a mass ratio of 0.5 and for different values of the eccentricity (e_b) and semimajor axis (a_b) of the binary (Taken from Haghighipour and Raymond 2007, courtesy of the Astrophysical Journal)

the snapshots of one of such simulations. During the course of the integration, the embryos in part of the disc close to the giant planet are dynamically excited. As a result of the interaction of embryos with one another, the orbital excitation of these objects is transmitted to other bodies in closer orbits causing many of them to be scattered out of the system or undergo radial mixing. In the simulation of Fig. 13.8, this process results in formation of a planet slightly larger than Earth and with a large reservoir of water.

An interesting result of the Haghighipour and Raymond (2007) numerical exploration is that *assuming the first stages of planet formation were successful,* binaries with moderate to large perihelia and with giant planets on low-eccentricity orbits are most favourable for Earth-like planet formation. Similar to the formation of terrestrial planets around single stars, where giant planets, in general, play destructive roles, a strong interaction between the secondary star and the giant planet in a binary planetary system (i.e. a small binary perihelion) increases the orbital eccentricity of this object and results in the removal of the terrestrial planet-forming materials from the system.

13.6 Planets in Formation-Hostile Regions: Solving the Paradox

As has been shown in the previous sections, with the exception of its very late stages, the formation of S-type planets in close binaries has many obstacles to overcome. This is in particular true for the planetesimal accretion stage, which seems unlikely to proceed at the location where the γ Ceph, HD 196885 and HD 41004 planets have been detected. To explain this paradox of having planets in planet-formation-hostile environments, several solutions can be considered.

A first possibility is that these planets were not formed in situ but in friendlier regions closer to the primary and later migrated outward to their present position. Classical type I or type II migrations in the gas disc would usually move the planet in the "wrong" direction, i.e. inward. There exists, however, possibilities for these migrations to revert direction (e.g. Masset and Snellgrove 2001; Pierens et al. 2012), but these scenarios have not been explored yet in the context of binaries. The outward migration process that has been explored for binaries regards a later phase, i.e. the planetesimal-driven migration of terrestrial embryos once the gas disc has dissipated. Payne et al. (2009) have shown that such a mechanism can indeed move some large embryos beyond their initial position, over timescales of typically 10^6 years. However, the maximum outward displacement does not exceed $\sim$30 % of the initial semimajor axis and is not enough to explain the formation of extreme planets such as HD 196885 or γ Ceph. Another potential way of moving planets outward is by planet-planet scattering. Marzari et al. (2005) have studied the dynamical evolution of unstable multi-planet systems in binaries and showed that for three neighbouring Jovian planets, a possible outcome is the ejection of one planet, the inward jump of a second one and the outward displacement of the third one. In this case, one would end up with two giant planets, one of which potentially far outside the region where it accreted. However, the major caveat of this scenario is that it requires the presence of a second giant planet on a close circumprimary orbit and that such a planet should in principle be more easily detected that the outer one. As such, the non-detection of any second Jovian planet in HD 196885 or γ Cephei is a problematic issue. A related, albeit theoretical, problem is that this scenario requires the truncated protoplanetary disc to be massive enough to form not just one but *two* giant planets.

Another solution is simply to bypass the critical km-sized planetesimal range. If indeed the "initial" planetesimals are large, typically $\geq$100 km, then they are big enough to sustain high-velocity collisions and can survive and grow in the perturbed environment of a binary. Of course, for this scenario to work, such large planetesimals should be formed directly from much smaller boulder- and pebble-sized bodies, so that the whole dangerous 100 m-to-10 km range is avoided. Interestingly, two recent planetesimal-forming models, the streaming-instability model of Johansen et al. (2007) and the clumping mechanism of Cuzzi et al. (2008), advocate such a km-size-straddling formation mode. However, it remains to be seen if these formation mechanisms can operate in the highly perturbed environment of a binary.

Even if the kilometre-sized planetesimal phase cannot be bypassed, recent studies have shown that there is a possibility that these bodies might after all grow even in the presence of high-Δv collisions. Xie et al. (2010b) have analytically investigated the "snowball" growth mode, in which planetesimals accrete mass preferentially by sweeping up of dust particles instead of mutual collisions with other planetesimals. If the mass density contained in dust exceeds that contained in large bodies, then this mechanism could provide a viable growth mode in binaries, because dust-on-planetesimal impacts can result in accretion for Δv much higher than planetesimal-on-planetesimal ones. These results have been strengthened by Paardekooper and Leinhardt (2010), who incorporated for the first time a collisional evolution model in their simulations, taking into account the fragments produced by high-velocity impacts. Their runs have shown that many planetesimals are able to re-accrete, by sweeping, a large fraction of the mass that has been lost to small fragments by earlier high-Δv collisions. Moreover, frequent collisions with these small fragments can prevent km-sized planetesimals from reaching their equilibrium secular+gas-drag imposed orbits, so that their *mutual* impact velocities never fully reach the high values resulting from the differential phasing effect. However, the collision outcome prescription of Paardekooper and Leinhardt (2010) is still relatively simplistic, with all bodies smaller than ~ 1 km being treated at dust coupled to the gas, and more sophisticated models have to be tested.

Another promising explanation for the presence of S-type planets in very close binaries is that these binaries were initially wider than today. As it happens, there is in fact theoretical support for this hypothesis. If indeed most stars are born in clusters, then they should experience many close encounters in their early history. Malmberg et al. (2007) have shown that, for a typical stellar cluster, some very wide binaries get broken by close encounters, but for binaries that survive, the main effect of these encounters is to shrink their orbits. Interestingly, for the cluster setup they considered, Malmberg et al. (2007) found that around half of the binaries having a present separation of ~ 20 AU have had a significant orbit-shrinking encounter (see Fig. 13.9).[2] However, for the specific case of HD 196885, we run a series of simulations, each time increasing the binary separation, and found that the present-day planet location became accretion friendly only for a binary separation of ~ 45 AU. This means that this separation should have shrunk by at least a factor 2 during the system's evolution, and it remains to be seen if such a change is realistic. Perspectives are however more optimistic for the HZ of the α Centauri system, for which only a moderately larger and/or less eccentric initial binary orbit would be enough to allow planetary accretion there (see Fig. 13.10).

---

[2]The simulations of Malmberg et al. (2007) were, however, only assuming a single, hypothetical, initial setup. More generic numerical investigations, exploring a wider range of possible initial conditions, should be carried out.

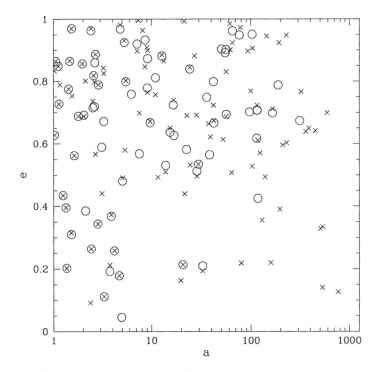

Fig. 13.9 Initial (*crosses*) and final (*circles*) binary orbital parameters obtained in the stellar cluster evolution simulations of Malmberg et al. (2007) (Figure reproduced from the Figure 4 of that paper in the MNRAS)

If none of the aforementioned solution works, then a more radical explanation should be considered, i.e. that planets in close binaries form by a different channel than the usual core-accretion scenario. An obvious alternative would be the gravitational instability scenario (e.g. Boss 1997; Mayer et al. 2010), in which planets form very quickly by direct disc fragmentation. Although this model still has several issues that need to be solved, in particular the cooling of collapsing protoplanetary clumps, it has recently been advocated as a possible explanation for the presence of giant planets at large radial distances from their stars, such as in the HR 8799 system (Boss 2011). Duchêne (2010) argues that this alternative formation mechanism could be at play in $a_b \leq 100$ AU binaries and that it could explain why planets in tight binaries are more massive than around single stars. In such close binaries, the instability mechanism could indeed be more efficient than around single stars, because circumprimary discs might be more compact, and thus more prone to fragmentation, than non-truncated ones, but also because binary perturbations could act as an additional trigger to instabilities. As a result, even close-in planets could be formed this way in binaries (Duchêne 2010). However, several studies have also shown that the instability scenario does also encounter major difficulties in the context of close binaries and that no circumprimary disc

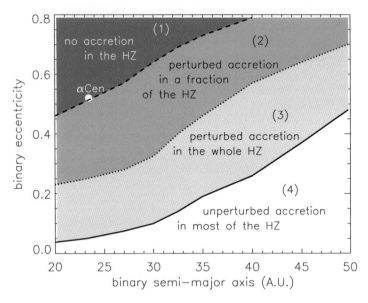

Fig. 13.10 Accretion vs. Erosion behaviour of a population of kilometre-sized planetesimals in the habitable zone of the α Centauri when varying the binary's separation and eccentricity (Taken from Thebault et al. 2009, courtesy of the MNRAS)

gets dense enough to be unstable. See, for example, Nelson (2000) or Mayer et al. (2005, 2010), whose main conclusions are that the presence of a close ($\leq$50–60 AU) companion could greatly hinder the development of instabilities. This issue is thus far from being settled yet, and further, more detailed investigations are needed to assess if disc fragmentation can be considered as a viable alternative formation scenario in close binaries.

13.7 Habitability

Although Earth-like planets are yet to be discovered in the habitable zone (HZ) of binary star systems, many planet-formation-in-binary studies have rightfully paid special attention to the HZ (see previous sections). In most of these simulations, it has been generally assumed that the HZ of a binary is equivalent to the single-star HZ of its planet-hosting star. Although in binaries with separations smaller than 50 AU, the secondary star plays an important role in the formation, long-term stability and water content of a planet in the HZ of the primary; the effect of the secondary on the range and extent of the HZ in these systems was ignored. However, the fact that this star can affect planet formation around the primary, and can also perturb the orbit of a planet in the primary's HZ in binaries with moderate eccentricities, implies that the secondary may play a non-negligible role in the habitability of the system as well.

13 Planet Formation in Binaries

Unlike around single stars where the HZ is a spherical shell with a distance determined by the host star alone, in binary star systems, the radiation from the stellar companion can influence the extent and location of the HZ of the system. Especially for planet-hosting binaries with small stellar separations and/or in binaries where the planet orbits the less luminous star, the amount of the flux received by the planet from the secondary star may become non-negligible.

In addition, effects such as the gravitational perturbation of the secondary star (see, e.g. Georgakarakos 2002; Eggl et al. 2012) can influence a planet's orbit in the binary HZ and lead to temperature fluctuations if the planetary atmosphere cannot buffer the change in the combined insolation. Since in an S-type system, the secondary orbits more slowly than the planet, during one period of the binary, the planet may experience the effects of the secondary several times. The latter, when combined with the atmospheric response of the planet, defines the HZ of the system.

Despite the fact that as a result of the orbital architecture and dynamics of the binary, at times the total radiation received by the planet exceeds the radiation that it receives from its parent star alone by a non-negligible amount, the boundaries of the actual HZ of the binary cannot be obtained by a simple extrapolation of the boundaries of the HZ of its planet-hosting star. Similar to the HZ around single stars, converting from insolation to equilibrium temperature of the planet depends strongly on the planet's atmospheric composition, cloud fraction and star's spectral type. A planet's atmosphere responds differently to stars with different spectral distribution of incident energy. Different stellar types will therefore contribute differently to the total amount of energy absorbed by the planetary atmosphere (see, e.g. Kasting et al. 1993). A complete and realistic calculation of the HZ has to take into account the spectral energy distribution (SED) of the binary stars as well as the planet's atmospheric response. In this section, we address these issues and present a coherent and self-consistent model for determining the boundaries of the HZ of S-type binary systems.

13.7.1 Calculation of the Binary Habitable Zone

Habitability and the location of the HZ depend on the stellar flux at the planet's location as well as the planet's atmospheric composition. The latter determines the albedo and the greenhouse effect in the planet's atmosphere and as such plays a strong role in determining the boundaries of the HZ. Examples of atmospheres with different chemical compositions include the original $CO_2/H_2O/N_2$ model (Kasting et al. 1993; Selsis et al. 2007; Kopparapu et al. 2013a) with a water reservoir like Earth's and model atmospheres with high H_2/He concentrations (Pierrehumbert and Gaidos 2011) or with limited water supply (Abe et al. 2011).

At present, the recent update to the Sun's HZ given by Kopparapu et al. (2013a,b) presents the best model. According to this model, the HZ is an annulus around a star where a rocky planet with a $CO_2/H_2O/N_2$ atmosphere and sufficiently large water content (such as on Earth) can host liquid water permanently on its solid surface. This definition of the HZ assumes that the abundance of CO_2 and H_2O in the atmosphere is regulated by a geophysical cycle similar to Earth's carbonate-silicate cycle. The inner and outer boundaries of the HZ in this model are associated with a H_2O- and CO_2-dominated atmosphere, respectively. Between those limits on a geologically active planet, climate stability is provided by a feedback mechanism in which atmospheric CO_2 concentration varies inversely with planetary surface temperature.

The locations of the inner and outer boundaries of a single star's as well as a binary's HZ depend also on the cloud fraction in the planet's atmosphere. That is because the overall planetary albedo is a function of the chemical composition of the clear atmosphere as well as the additional cooling or warming of the atmosphere due to clouds. Since the model by Kopparapu et al. (2013a,b) does not include clouds, it is customary to define two types of HZ: the *narrow* HZ which is considered to be the region between the limits of runaway and maximum greenhouse effect in the model by Kopparapu et al. (2013a,b) and the *empirical* HZ, as a proxy to the effect of clouds, that is derived using the fluxes received by Mars and Venus at 3.5 and 1.0 Gyr, respectively (the region between recent Mars and early Venus). At these times, the two planets do not show indications for liquid water on their surfaces (see Kasting et al. 1993). In these definitions, the locations of the HZs are determined based on the flux received by the planet (see, e.g. Kasting et al. 1993; Selsis et al. 2007; Kaltenegger and Sasselov 2011; and Kopparapu et al. 2013a).

13.7.2 Effect of Star's Spectral Energy Distribution (SED)

The locations of the boundaries of the HZ depend on the flux of the star at the orbit of the planet. In a binary star system where the planet is subject to radiation from two stars, the flux of the secondary star has to be added to that of the primary (planet-hosting star), and the total flux can then be used to calculate the boundaries of the HZ. However, because the response of a planet's atmosphere to the radiation from a star depends strongly on the star's SED, a simple summation of fluxes is not applicable. The absorbed fraction of the absolute incident flux of each star at the top of the planet's atmosphere will differ for different SEDs. Therefore, in order to add the absorbed flux of two different stars and derive the limits of the HZ for a binary system, one has to weigh the flux of each star according to the star's SED. The relevant flux received by a planet in this case is the sum of the spectrally weighted stellar flux, separately received from each star of the binary, as given by (see Kaltenegger and Haghighipour 2013 for details),

$$F_{Pl}(f, T_{Pr}, T_{Sec}) = W_{Pr}(f, T_{Pr}) \frac{L_{Pr}(T_{Pr})}{r_{Pl-Pr}^2} + W_{Sec}(f, T_{Sec}) \frac{L_{Sec}(T_{Sec})}{r_{Pl-Sec}^2}. \quad (13.1)$$

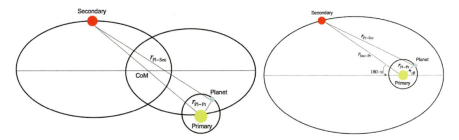

Fig. 13.11 *Left panel*: Schematic presentation of an S-type system. The two stars of the binary, primary and secondary, revolve around their centre of mass (*CoM*) whilst the planet orbits only one of the stars (*top panel*). It is, however, customary to neglect the motion of the binary around its CoM and consider the motion of the secondary around a stationary primary (*bottom panel*)

In this equation, F_{Pl} is the total flux received by the planet, L_i and T_i (i=Pr, Sec) represent the luminosity and effective temperature of the primary and secondary stars, f is the cloud fraction of the planet's atmosphere and $W_i(f, T_i)$ is the binary stars' spectral weight factor. The quantities r_{Pl-Pr} and r_{Pl-Sec} in Eq. (13.1) represent the distances between the planet and the primary and secondary stars, respectively (Fig. 13.11). In using Eq. (13.1), we normalise the weighting factor to the flux of the Sun.

From Eq. (13.1), the boundaries of the HZ of the binary can be defined as distances where the total flux received by the planet is equal to the flux that Earth receives from the Sun at the inner and outer edges of its HZ. Since in an S-type system, the planet revolves around one star of the binary, we determine the inner and outer edges of the HZ with respect to the planet-hosting star. As shown in Fig. 13.11, it is customary to consider the primary of the system to be stationary and calculate the orbital elements with respect to the stationary primary star. In the rest of this section, we will follow this convention and consider the planet-hosting star to be the primary star as well. In that case, the range of the HZ of the binary can be obtained from

$$W_{Pr}(f, T_{Pr}) \frac{L_{Pr}(T_{Pr})}{l_{x-Bin}^2} + W_{Sec}(f, T_{Sec}) \frac{L_{Sec}(T_{Sec})}{r_{Pl-Sec}^2} = \frac{L_{Sun}}{l_{x-Sun}^2}. \quad (13.2)$$

In Eq. (13.2), the quantity l_x represents the inner and outer edges of the HZ with x=(in,out). As mentioned earlier, the values of l_{in} and l_{out} are model dependent and change for different values of cloud fraction, f, and atmosphere composition.

13.7.3 Calculation of Spectral Weight Factors

To calculate the spectral weight factor $W(f, T)$ for each star of the binary and in terms of their SEDs, we calculate the stellar flux at the top of the atmosphere of an

Earth-like planet at the limits of the HZ, in terms of the stellar effective temperature. To determine the locations of the inner and outer boundaries of the HZ of a main-sequence star with an effective temperature of $2,600\,\mathrm{K} < T_{\mathrm{Star}} < 7,200\,\mathrm{K}$, we use Eq. (13.3) (see Kopparapu et al. 2013a)

$$l_{\mathrm{x-Star}} = l_{\mathrm{x-Sun}} \left[\frac{L/L_{\mathrm{Sun}}}{1 + \alpha_{\mathrm{x}}(T_i)\, l_{\mathrm{x-Sun}}^2} \right]^{1/2} \tag{13.3}$$

In this equation, $l_{\mathrm{x}} = (l_{\mathrm{in}}, l_{\mathrm{out}})$ is in AU, $T_i\,(\mathrm{K}) = T_{\mathrm{Star}}(\mathrm{K}) - 5,780$, and

$$\alpha_{\mathrm{x}}(T_i) = a_{\mathrm{x}} T_i + b_{\mathrm{x}} T_i^2 + c_{\mathrm{x}} T_i^3 + d_{\mathrm{x}} T_i^4, \tag{13.4}$$

where the values of coefficients $a_{\mathrm{x}}, b_{\mathrm{x}}, c_{\mathrm{x}}, d_{\mathrm{x}}$, and $l_{\mathrm{x-Sun}}$ are given in Table 13.1 (see Kopparapu et al. 2013b). From Eq. (13.3), the flux received by the planet from its host star at the limits of the habitable zone can be calculated using Eq. (13.5). The results are given in Table 13.1;

$$F_{\mathrm{x-Star}}(f, T_{\mathrm{Star}}) = F_{\mathrm{x-Sun}}(f, T_{\mathrm{Star}}) \left[1 + \alpha_{\mathrm{x}}(T_i)\, l_{\mathrm{x-Sun}}^2 \right]. \tag{13.5}$$

From Eq. (13.5), the spectral weight factor $W(f, T)$ can be written as

$$W_i(f, T_i) = \left[1 + \alpha_{\mathrm{x}}(T_i)\, l_{\mathrm{x-Sun}}^2 \right]^{-1}. \tag{13.6}$$

Table 13.2 and Fig. 13.12 show $W(f, T)$ as a function of the effective temperature of a main-sequence planet-hosting star for the narrow (left panel) and empirical (right panel) boundaries of the HZ. As expected, hotter stars have weighting factors smaller than 1, whereas the weighting factors of cooler stars are larger than 1.

Table 13.1 Values of the coefficients of equation (13.4). See Kopparapu et al. (2013b) for details.

	Narrow HZ		Empirical HZ	
	Runaway greenhouse	Maximum greenhouse	Recent Venus	Early Mars
$l_{\mathrm{x-Sun}}$ (AU)	0.97	1.67	0.75	1.77
Flux (Solar flux @ Earth)	1.06	0.36	1.78	0.32
a	1.2456×10^{-4}	5.9578×10^{-5}	1.4335×10^{-4}	5.4471×10^{-5}
b	1.4612×10^{-8}	1.6707×10^{-9}	3.3954×10^{-9}	1.5275×10^{-9}
c	-7.6345×10^{-12}	-3.0058×10^{-12}	-7.6364×10^{-12}	-2.1709×10^{-12}
d	-1.7511×10^{-15}	-5.1925×10^{-16}	-1.1950×10^{-15}	-3.8282×10^{-16}

Table 13.2 Samples of spectral weight factors

Star	Eff. temp	Narrow HZ Inner	Narrow HZ Outer	Empirical HZ Inner	Empirical HZ Outer
F0	7,300	0.850	0.815	0.902	0.806
F8V (HD 196885 A)	6,340	0.936	0.915	0.957	0.913
G0	5,940	0.981	0.974	0.987	0.973
G2V (α Cen A)	5,790	0.999	0.998	0.999	0.998
K1V (α Cen B)	5,214	1.065	1.100	1.046	1.103
K3	4,800	1.107	1.179	1.079	1.186
M1V (HD 196885 B)	3,700	1.177	1.383	1.154	1.419
M5	3,170	1.192	1.471	1.179	1.532
LM2	3,520	1.183	1.414	1.163	1.458

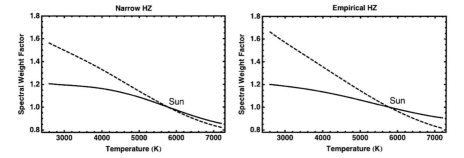

Fig. 13.12 Spectral weight factor $W(f, T)$ as a function of stellar effective temperature for the narrow (*left*) and empirical (*right*) HZs. The *solid line* corresponds to the inner and the *dashed line* corresponds to the outer boundaries of HZ. We have normalised $W(f, T)$ to its solar value, indicated on the graphs (*Sun*)

13.7.4 Effect of Binary Eccentricity

To use Eq. (13.2) to calculate the boundaries of the HZ, we assume that the orbit of the (fictitious) Earth-like planet around its host star is circular. In a close binary system, the gravitational effect of the secondary may deviate the motion of the planet from a circle and cause its orbit to become eccentric. In a binary with a given semimajor axis and mass ratio, the eccentricity has to stay within a small to moderate level to avoid strong interactions between the secondary star and the planet and to allow the planet to maintain a long-term stable orbit (with a low eccentricity) in the primary's HZ. The binary eccentricity itself is constrained by the fact that in highly eccentric systems, periodic close approaches of the two stars truncate their circumstellar discs depleting them from planet-forming material (Artymowicz and Lubow 1994) and restricting the delivery of water-carrying objects to an accreting terrestrial planet in the binary HZ (Haghighipour and Raymond 2007).

This all implies that in order for the binary to be able to form a terrestrial planet in its HZ, its eccentricity cannot have large values. In a binary with a small eccentricity, the deviation of the planet's orbit from circular is also small and appears in the form of secular changes with long periods (see, e.g. Eggl et al. 2012). Therefore, to use Eq. (13.2), one can approximate the orbit of the planet by a circle without the loss of generality.

The habitability of a planet in a binary system also requires long-term stability in the HZ. For a given semimajor axis a_{Bin}, eccentricity e_{Bin} and mass ratio μ of the binary, there is an upper limit for the semimajor axis of the planet beyond which the perturbing effect of the secondary star will make the orbit of the planet unstable. This maximum or *critical* semimajor axis (a_{Max}) is given by (Rabl and Dvorak 1988; Holman and Wiegert 1999)

$$a_{Max} = a_{Bin}\left(0.464 - 0.38\,\mu - 0.631\,e_{Bin} + 0.586\,\mu\,e_{Bin} + 0.15\,e_{Bin}^2 - 0.198\,\mu\,e_{Bin}^2\right).$$
(13.7)

In Eq. (13.7), $\mu = m_2/(m_1 + m_2)$ where m_1 and m_2 are the masses of the primary (planet hosting) and secondary stars, respectively. One can use Eq. (13.7) to determine the maximum binary eccentricity that would allow the planet to have a stable orbit in the HZ ($l_{out} \leq a_{Max}$). For any smaller value of the binary eccentricity, the entire HZ will be stable.

13.8 Examples of the Habitable Zone of Main Sequence S-Type Binaries

As mentioned earlier, we assume that the orbit of the planet around its host star is circular. Without knowing the exact orbital configuration of the planet, one can only estimate the boundaries of the binary HZ by calculating the maximum and minimum additional flux from the secondary star at its closest and furthest distances from a fictitious Earth-like planet, as a first-order approximation. Note that using the maximum flux of the secondary onto the planet for calculating the new binary HZ overestimates the shift of the HZ from the single star's HZ to the binary HZ due to the secondary because the planet's atmosphere can buffer an increase in radiation temporarily. This shift is underestimated when one uses the minimum flux received from the secondary star onto the planet. To improve on this estimation, one needs to know the orbital positions of the planet as well as the stars in the binary. That way one can determine the exact flux over time reaching the planet as well as the number of planetary orbits over which the secondary's flux can be averaged. This depends on the system's geometry (both stellar and planetary parameters) and needs to be calculated for each planet-hosting S-type system, individually.

To explore the maximum effect of the binary semimajor axis and eccentricity on the contribution of one star to the extent of the HZ around the other component, we consider three extreme cases: an M2-M2, an F0-F0 and an F8-M1 binary.

13 Planet Formation in Binaries 333

We consider the M2 and F0 stars to have effective temperatures of 3,520 (K) and 7,300 (k), respectively, and their luminosities to be 0.035 and 6.56 of that of the Sun for our general examples here. We note that in calculating the boundaries of the HZ, the orbital (in)stability of the fictitious Earth-like planet is not considered. As a result, depending on the orbital elements and mass ratio of the binary, its HZ may be unstable.

To demonstrate the effect of the secondary on the boundaries of the HZ, we calculate the binary HZ of the systems mentioned above considering the minimum value of the binary semimajor axis for which the outer edge of the primary's empirical HZ will be on the stability limit. Figure 13.14 shows the results for the case of a circular binary. The top panels in this figure correspond to an M2-M2 (left) and F0-F0 (right) binary system. As shown here, the secondary does not have a noticeable effect on the extent of the HZ around the primary. The binary HZ around each star is equivalent to its single-star HZ. The bottom panels in Fig. 13.14 correspond to an F8-M1 binary (left) where the primary is the F star and an F8-M1 binary (right) where the primary is the M star. As expected, the effect of the M star on the extension of the single-star HZ around the F star is negligible. However, at its closest distances, the F star can extend the outer limit of the single-star HZ around the M star so far out that at the binary periastron, the two HZs merge.

To explore the effect of binary eccentricity in a system with a hot and cool star, we carried out similar calculations as those in Fig. 13.13, for the F8-M1 binary,

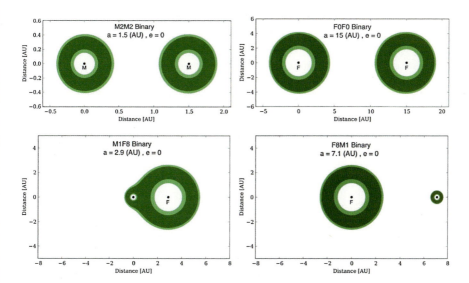

Fig. 13.13 Boundaries of the narrow (*dark green*) and empirical (*light green*) HZs in an M2-M2 (*top left*), F0-F0 (*top right*) and M1-F8 S-type binary star system (*bottom two panels*). Note that the primary is the star at (0,0). The primary star in the bottom panels is the F8 star (*left*) and the M1 star (*right*). The semimajor axis of the binary has been chosen to be the minimum value that allows the region out to the outer edge of the primary's empirical HZ to be stabile in a circular binary

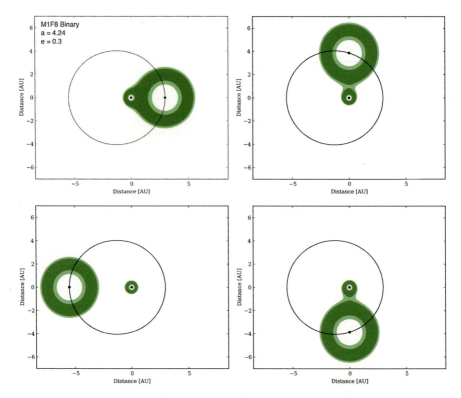

Fig. 13.14 Boundaries of the narrow (*dark green*) and empirical (*light green*) HZs in an F8-M1 binary. Note that the primary is the M1 star at (0,0). The panels show the effect of the F8 star whilst orbiting the primary starting from the *top left panel* when the secondary is at the binary periastron. The semimajor axis of the binary has been chosen to be the minimum value that allows the region out to the outer edge of the primary's empirical single-star HZ to be stable for a binary eccentricity of 0.3

assuming the binary eccentricity to be 0.3. Figures 13.14 and 13.15 show the results for four different relative positions of the two stars. In Fig. 13.14, the primary is the M star and in Fig. 13.15, the primary is the F star. As shown in these figures, when the secondary is more luminous, it will have considerable effects on the shape and location of the single-star HZ around the other star. However, a cool and less luminous secondary will not change the limits of the primary's single-star HZ.

13.8.1 Habitable Zone of α Centauri

The α Centauri system consists of the close binary α Cen AB and a farther M dwarf companion known as Proxima Centauri at approximately 15,000 AU. The semimajor axis of the binary is ∼23.5 AU and its eccentricity is ∼0.518. The

13 Planet Formation in Binaries

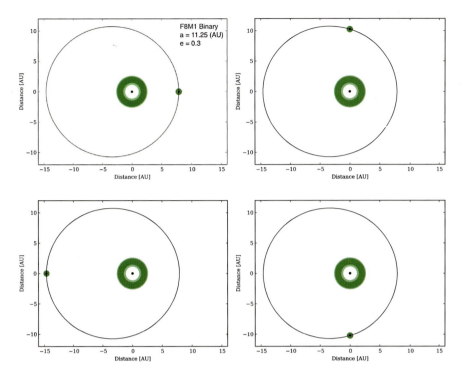

Fig. 13.15 Same as Fig. 13.14, with the F star as the primary

component A of this system is a G2V star with a mass of $1.1\,M_{Sun}$, luminosity of $1.519\,L_{Sun}$ and an effective temperature of 5,790 K. Its component B has a spectral type of K1V, and its mass, luminosity and effective temperature are equal to $0.934\,M_{Sun}$, $0.5\,L_{Sun}$ and 5,214 K, respectively.

The announcement of a probable super-Earth planet with a mass larger than 1.13 Earth masses around α Cen B (Dumusque et al. 2012) indicates that unlike the region around α Cen A where terrestrial planet formation encounters complications, planet formation is efficient around this star (Guedes et al. 2008; Thebault et al. 2009) and it could also host a terrestrial planet in its HZ. Here we assume that planet formation around both stars of this binary can proceed successfully, and they both can host Earth-like planets. We calculate the spectral weight factors of both α Cen A and B (Table 13.2) and, using the minimum and maximum added flux of the secondary star, estimate the limits of their binary HZs using Eq. (13.2).

Table 13.3 shows the estimates of the boundaries of the binary HZ around each star. The terms Max and Min in this table correspond to the planet-binary configurations of $(\theta, \nu) = (0, 0)$ and $(0, 180°)$ where the planet receives the maximum and minimum total flux from the secondary star, respectively. As shown here, each star of the α Centauri system contributes to increasing the limits of the binary HZ around the other star. Although these contributions are small, they extend

336 P. Thebault and N. Haghighipour

Table 13.3 Estimates of the boundaries of the binary HZ calculated using Eq. (13.2) for the maximum and minimum amount of flux received by the planet from the secondary star corresponding to the closest and farthest distances between the two bodies

| Host star | Estimates of narrow HZ (AU) | | | | Estimates of empirical HZ (AU) | | | |
| | With secondary | | Without secondary | | With secondary | | Without secondary | |
	Inner	Outer	Inner	Outer	Inner	Outer	Inner	Outer
α Cen A (Max)	1.197	2.068	1.195	2.056	0.925	2.194	0.924	2.179
α Cen A (Min)	1.195	2.057	1.195	2.056	0.924	2.181	0.924	2.179
α Cen B (Max)	0.712	1.259	0.708	1.238	0.544	1.340	0.542	1.315
α Cen B (Min)	0.708	1.241	0.708	1.238	0.543	1.317	0.542	1.315
HD 196885 A (Max)	1.454	2.477	1.454	2.475	1.137	2.622	1.137	2.620
HD 196885 A (Min)	1.454	2.475	1.454	2.475	1.137	2.620	1.137	2.620
HD 196885 B (Max)	0.260	0.491	0.258	0.481	0.198	0.529	0.197	0.516
HD 196885 B (Min)	0.258	0.483	0.258	0.481	0.197	0.518	0.197	0.516

the estimated limits of the binary HZ by a noticeable amount. This is primarily due to the high luminosity of α Cen A and the relatively large eccentricity of the binary which brings the two stars as close as $\sim$11.3 AU from one another (and as such making planet formation very difficult around α Cen A).

Given the eccentricity of the α Cen binary ($e_{Bin} = 0.518$), both narrow and nominal HZs for α Cen A and B are stable. The stability limit around the primary G2V star is at $\sim$2.768 AU. This limit is slightly exterior to the outer boundary of the star's narrow and empirical HZs. Although the latter suggests that the HZ of α Cen A is dynamically stable, the close proximity of this region to the stability limit may have strong consequences on the actual formation of an Earth-like planet in this region (see, e.g. Thebault et al. 2008, Eggl et al. 2012).

13.8.2 Habitable Zone of HD 196885

HD 196885 is a close main-sequence S-type binary system with a semimajor axis of 21 AU and eccentricity of 0.42 (Chauvin et al. 2011). The primary of this system (HD 169885 A) is an F8V star with a T_{Star} of 6,340 K, mass of 1.33 M_{Sun} and luminosity of 2.4 L_{Sun}. The secondary star (HD 196885 B) is an M1V dwarf with a mass of 0.45 M_{Sun}. Using the mass-luminosity relation $L \sim M^{3.5}$ where L and M are in solar units, the luminosity of this star is approximately 0.06 L_{Sun}, and we consider its effective temperature to be $T_{Star} = 3,700$ K. The primary of HD 196885 hosts a Jovian-type planet suggesting that the mass ratio and orbital elements of this binary allow planet formation to proceed successfully around its primary star. We assume that terrestrial planet formation can also successfully proceed around both stars of this binary and can result in the formation of Earth-sized objects.

To estimate the boundaries of the binary HZ of this eccentric system, we ignore its known giant planet and use Eq. (13.2) considering a fictitious Earth-like planet in the HZ. We calculate the spectral weight factor $W(f, T)$ for both stars of this system (Table 13.2) and estimate the locations of the inner and outer boundaries of the binary's HZ (Table 13.3).

As expected (because of the large periastron distance of the binary and the secondary star being a cool M dwarf), even the maximum flux from the secondary star does not have a noticeable contribution to the location of the HZ around the F8V primary of HD 196886. However, being a luminous F star, the primary shows a small effect on the location of the HZ around the M1V secondary star (Table 13.3).

References

Abe Y, Abe-Ouchi A, Sleep NH, Zahnle KJ (2011) Habitable zone limits for dry planets. AsBio 11:443

Artymowicz P, Lubow SH (1994) Dynamics of binary-disk interaction. 1: resonances and disk gap sizes. ApJ 421:621

Barbieri M, Marzari F, Scholl H (2002) Formation of terrestrial planets in close binary systems: the case of alpha Centauri A. A&A 396:219

Blum J, Wurm G (2008) The growth mechanisms of macroscopic bodies in protoplanetary disks. ARA&A 46:21

Boss AP (1997) Giant planet formation by gravitational instability. Science 276:1836

Boss AP (2011) Formation of giant planets by disk instability on wide orbits around protostars with varied masses. ApJ 731:74

Chauvin G, Beust H, Lagrange A-M, Eggenberger A (2011) Planetary systems in close binary stars: the case of HD 196885. Combined astrometric and radial velocity study. A&A 528:8

Correia AC et al (2008) The ELODIE survey for northern extra-solar planets. IV. HD 196885, a close binary star with a 3.7-year planet. A&A:479:271

Cuzzi JN, Hogan RC, Shariff K (2008) Toward planetesimals: dense chondrule clumps in the protoplanetary nebula. ApJ 687:1432

Desidera S, Barbieri M (2007) Properties of planets in binary systems. The role of binary separation. A&A 462:345–353

Doyle L, Carter J et al (2011) Kepler-16: a transiting circumbinary planet. Science 333:1602

Duchêne G (2010) Planet formation in binary systems: a separation-dependent mechanism? ApJ 709:L114

Dumusque X, Pepe F et al (2012) An Earth-mass planet orbiting alpha Centauri B. Nature 491:207

Dvorak R (1984) Numerical experiments on planetary orbits in double stars. CeMec 34:369

Dvorak R (1986) Critical orbits in the elliptic restricted three-body problem. A&A 167:379

Dvorak R, Froeschle C, Froeschle Ch (1989) Stability of outer planetary orbits (P-types) in binaries. A&A 226:335

Dvorak R, Pilat-Lohinger E, Funk B, Freistetter F (2003) Planets in habitable zones: a study of the binary Gamma Cephei. A&A 398:L1

Eggl S, Pilat-Lohinger E, Georgakarakos N, Gyergyovits M, Funk B (2012) An analytic method to determine habitable zones for S-type planetary orbits in binary star systems. ApJ 752:74

Eggenberger A, Udry S (2010) Probing the impact of stellar duplicity on planet occurrence with spectroscopic and imaging observations. In: Haghighipour N (ed) Planets in binary star systems. Springer, New York, p 19

Endl M, Cochran W, Hatzes A, Wittenmeyer RA (2011) News from the γ Cephei planetary system. In: Proceedings of the international conference on planetary systems beyond the main sequence. AIP conference proceedings, vol 1331, p 88

Fragner M, Nelson R, Kley W (2011) On the dynamics and collisional growth of planetesimals in misaligned binary systems. A&A 528:40

Georgakarakos N (2002) Eccentricity generation in hierarchical triple systems with coplanar and initially circular orbits. MNRAS 337:559

Giuppone CA, Leiva AM, Correa-Otto J, Beaugé C (2011) Secular dynamics of planetesimals in tight binary systems: application to γ Cephei. A&A 530:103

Guedes JM, Rivera EJ, Davis E, Laughlin G, Quintana EV, Fischer DA (2008) Formation and detectability of terrestrial planets around alpha centauri B. ApJ 679:1582

Haghighipour N (2011) Super-Earths: a new class of planetary bodies. Contemp Phys 52:403–38

Haghighipour N, Raymond SN (2007) Habitable planet formation in binary planetary systems. ApJ 666:436–446

Haghighipour N, Dvorak R, Pilat-Lohinger E (2010) Planetary dynamics and habitable planet formation in binary star systems. ASSL 366:285

Harris RJ, Andrews SM, Wilner DJ, Kraus AL (2012) A resolved census of millimeter emission from taurus multiple star systems. ApJ 751:115

Hatzes AP, Cochran WD, Endl M, McArthur B, Paulson DB, Walker GAH, Campbell B, Yang S (2003) A planetary companion to Gamma Cephei A. ApJ 599:1383

Hayashi C (1981) Structure of the solar nebula, growth and decay of magnetic fields and effects of magnetic and turbulent viscosities on the nebula. PthPS 70:35

Heppenheimer TA (1978) On the formation of planets in binary star systems. A&A 65:421

Holman MJ, Wiegert PA (1999) Long-term stability of planets in binary systems. AJ 117:621

Hubickyj O, Bodenheimer P, Lissauer JJ (2005) Accretion of the gaseous envelope of Jupiter around a 5–10 Earth-mass core. Icarus 179:415–431

Jang-Condell H, Mugrauer M, Schmidt T (2008) Disk truncation and planet formation in Gamma Cephei. ApJ 683:L191

Johansen A, Oishi JS, Mac Low M-M, Klahr H, Henning T, Youdin A (2007) Rapid planetesimal formation in turbulent circumstellar disks. Nature 448:1022

Kaltenegger L, Sasselov D (2011) Exploring the habitable zone for Kepler planetary candidates. ApJL 736:L25

Kaltenegger L, Haghighipour N (2013) Calculating the habitable zone of binary star systems. I. S-type binaries ApJ 777, article id. 165

Kasting JF, Whitmire DP, Reynolds RT (1993) Habitable zones around main sequence stars. Icarus 101:108

Kley W, Nelson RP (2007) Early evolution of planets in binaries: planet-disk interaction. In: Haghighipour N (ed) Planets in binary star systems. Springer, Dordrecht

Kokubo E, Ida S (2000) Formation of protoplanets from planetesimals in the solar nebula. Icarus 297:1067

Kopparapu RK, Ramirez R, Kasting JF et al (2013a) Habitable zones around main-sequence stars: new estimates. ApJ 765:131

Kopparapu RK, Ramirez R, Kasting JF et al (2013b) Erratum: "habitable zones around main-sequence stars: new estimates". ApJ 770:82

Kostov VB et al (2013) A gas giant circumbinary planet transiting the F star primary of the eclipsing binary star KIC 4862625 and the independent discovery and characterization of the two transiting planets in the Kepler-47 system. ApJ 770:52

Kostov VB et al (2014a) Kepler-413b: a slightly misaligned, Neptune-size transiting circumbinary planet. ApJ 784:14

Kostov VB et al (2014b) Erratum: "Kepler-413b: a slightly misaligned, Neptune-size transiting circumbinary planet". ApJ 787:93

Kraus AL, Ireland MJ, Hillenbrand LA, Martinache F (2012) The role of multiplicity in disk evolution and planet formation. ApJ 745:12

Lagrange A-M, Beust H, Udry S, Chauvin G, Mayor M (2006) New constrains on Gliese 86 B. VLT near infrared coronographic imaging survey of planetary hosts. A&A 459:955

Lissauer JJ (1993) Planet formation. ARA&A 31:129

Malmberg D, Davies MB, Chambers JE (2007) Close encounters in young stellar clusters: implications for planetary systems in the solar neighbourhood. MNRAS 378:1207

Marzari F, Scholl H (2000) Planetesimal accretion in binary star systems. ApJ 543:328

Marzari F, Weidenschilling SJ, Barbieri M, Granata V (2005) Jumping Jupiters in binary star systems. ApJ 618:502

Marzari F, Scholl H, Thebault P, Baruteau C (2009) On the eccentricity of self-gravitating circumstellar disks in eccentric binary systems. A&A 508:1493

Marzari F, Baruteau C, Scholl H, Thebault P (2012) Eccentricity of radiative disks in close binary-star systems. A&A 539:98

Masset F, Snellgrove M (2001) Reversing type II migration: resonance trapping of a lighter giant protoplanet. MNRAS 320:55

Mayer L, Wadsley J, Quinn T, Stadel J (2005) Gravitational instability in binary protoplanetary discs: new constraints on giant planet formation. MNRAS 363:641

Mayer L, Boss A, Nelson AF (2010) Gravitational instability in binary protoplanetary disks. In: Haghighipour N (ed) Planets in binary star systems. Astrophysics and space science library, vol 366. Springer, Dordrecht. ISBN 978-90-481-8686-0

Mugrauer M, Neuhäuser R (2009) The multiplicity of exoplanet host stars. New low-mass stellar companions of the exoplanet host stars HD 125612 and HD 212301. A&A 494:373

Müller TWA, Kley W (2012) Circumstellar disks in binary star systems. Models for γ Cephei and α Centauri. A&A 539:18

Nelson A (2000) Planet formation is unlikely in equal-mass binary systems with a = 50 AU. ApJ 537:65

Neuhäuser R, Mugrauer M, Fukagawa M, Torres G, Schmidt T (2007) Direct detection of exoplanet host star companion Gamma Cep B and revised masses for both stars and the sub-stellar object. A&A 462:777

Orosz JA, Welsh WF et al (2012a) Kepler-47: a transiting circumbinary multiplanet system. Science 337:1511

Orosz JA, Welsh WF et al (2012b) The Neptune-sized Circumbinary Planet Kepler-38b. ApJ 758:87

Paardekooper S-J, Thebault P, Mellema G (2008) Planetesimal and gas dynamics in binaries. MNRAS 386:973

Paardekooper S-J, Leinhardt ZM (2010) Planetesimal collisions in binary systems. MNRAS 403:L64

Payne MJ, Wyatt MC, Thebault P (2009) Outward migration of terrestrial embryos in binary systems. MNRAS 400:1936

Picogna G, Marzari F (2013) Three-dimensional modeling of radiative disks in binaries. A&A 556:148

Pierens A, Baruteau C, Hersant F (2012) Protoplanetary migration in non-isothermal discs with turbulence driven by stochastic forcing. MNRAS 427:1562

Pierrehumbert R, Gaidos E (2011) Hydrogen greenhouse planets beyond the habitable zone. ApJL 734:L13

Pollack JB, Hubickyj O, Bodenheimer P, Lissauer JJ, Podolak M, Greenzweig Y (1996) Formation of the giant planets by concurrent accretion of solids and gas. Icarus 124:62–85

Queloz D, Mayor M, Weber L, Blécha A, Burnet M, Confino B, Naef D, Pepe F, Santos N, Udry S (2000) The CORALIE survey for southern extra-solar planets. I. A planet orbiting the star Gliese 86. A&A 354:99

Qian S-B, Lia W-P, Zhu L-Y, Dai Z-B (2010) Detection of a giant extrasolar planet orbiting the eclipsing polar DP Leo. ApJL 708, 66

Quintana EV, Lissauer JJ, Chambers JE, Duncan MJ (2002) Terrestrial planet formation in the alpha Centauri system. ApJ 576:982

Quintana EV, Adams FC, Lissauer JJ, Chambers JE (2007) Terrestrial planet formation around individual stars within binary star systems. ApJ:660, 807

Rabl G, Dvorak R (1988) Satellite-type planetary orbits in double stars – a numerical approach. A&A 191:385–391

Rafikov R (2013) Planet formation in small separation binaries: not so secularly excited by the companion. ApJ 768:112

Raghavan D, McAlister HA, Henry TJ, Latham DW, Marcy GW, Mason BD, Gies DR, White RJ, ten Brummelaar TA (2010) A survey of stellar families: multiplicity of solar-type stars. ApJS 190:1

Roell T, Neuhäuser R, Seifahrt A, Mugrauer M (2012) Extrasolar planets in stellar multiple systems. A&A 542:92

Rodríguez LF, D'Alessio P, Wilner DJ, Ho PTP, Torrelles JM, Curiel S, Gómez Y, Lizano S, Pedlar A, Cantó J, Raga AC (1998) Compact protoplanetary disks around the stars of a young binary system. Nature 395:355

Safronov VS (1972) Evolution of the protoplanetary cloud and formation of the earth and planets. (NASA-TTF-677)

Savonije GJ, Papaloizou JCB, Lin DNC (1994) On tidally induced shocks in accretion discs in close binary systems. MNRAS 268:13

Schwamb ME, Orosz JA, Carter JA et al (2013) Planet hunters: a transiting circumbinary planet in a quadruple star system. ApJ 768:127

Selsis F, Kasting JF, Levrard B et al (2007) Habitable planets around the star Gliese 581? A&A 476:1373

Thebault P (2011) Against all odds? Forming the planet of the HD 196885 binary. CeMDA 111:29

Thebault P, Marzari F, Scholl H, Turrini D, Barbieri M (2004) Planetary formation in the Gamma Cephei system. A&A 427:1097

Thebault P, Marzari F, Scholl H (2006) Relative velocities among accreting planetesimals in binary systems: the circumprimary case. Icarus 183:193

Thebault P, Marzari F, Scholl H (2008) Planet formation in alpha Centauri A revisited: not so accretion friendly after all. MNRAS 388:1528

Thebault P, Marzari F, Scholl H (2009) Planet formation in the habitable zone of alpha Centauri B. MNRAS 393:L21

Welsh WF, Orosz JA et al (2012) Transiting circumbinary planets Kepler-34 b and Kepler-35 b. Nature 481:475

Xie J-W, Zhou J-L (2008) Planetesimal accretion in binary systems: the effects of gas dissipation. ApJ 686:570

Xie J-W, Zhou J-L (2009) Planetesimal accretion in binary systems: role of the companion's orbital inclination. ApJ 698:2066

Xie J-W, Zhou J-L, Ge J (2010a) Planetesimal accretion in binary systems: could planets form around alpha Centauri B? ApJ 708:1566

Xie J-W, Payne MJ, Thebault P, Zhou J-L, Ge J (2010b) From dust to planetesimal: the snowball phase? ApJ 724:1153

Zsom A, Sandor Z, Dullemond C (2011) The first stages of planet formation in binary systems: how far can dust coagulation proceed? A&A 527:10

Zucker S, Mazeh T, Santos NC, Udry S, Mayor M (2004) Multi-order TODCOR: application to observations taken with the CORALIE echelle spectrograph. II. A planet in the system HD 41004. A&A 426:695